Between Human and Machine

Johns Hopkins Studies in the History of Technology

Merritt Roe Smith, Series Editor

# between

# human

# and

# machine

Feedback, Control, and Computing

before Cybernetics

David A. Mindell

The Johns Hopkins University Press, Baltimore and London

© 2002 David A. Mindell

All rights reserved. Published 2002

Printed in the United States of America on acid-free paper

9 8 7 6 5 4 3 2 1

The Johns Hopkins University Press

2715 North Charles Street

Baltimore, Maryland 21218-4363

www.press.jhu.edu

Library of Congress Cataloging-in-Publication Data

Mindell, David A.

  Between human and machine : feedback, control, and computing before cybernetics / David A. Mindell.

      p.  cm. — (Johns Hopkins studies in the history of technology)

Includes bibliographic references and index.

ISBN 0-8018-6895-5 (alk. paper)

  1. Computers—History.   2. Electronic data processing  -History.   I. Title.   II. Series.

  QA76.17 .M46 2002

   004'.09—dc21

2001004203

A catalog record for this book is available from the British Library.

To my parents

Marvin and Phyllis Mindell

My first teachers, my first students, and my first colleagues

If you want the truth—I know I presume—you must look into the technology of these matters. Even into the hearts of certain molecules—it is they after all which dictate temperatures, pressures, rates of flow, costs, profits, the shapes of towers . . .
You must ask two questions. First what is the real nature of synthesis? And then: What is the real nature of control?
—Ghost of Walter Rathenau to the Nazi elite, Thomas Pynchon, *Gravity's Rainbow*


People "track" during every conscious moment. . . . alignment processes, in which the alignment error serves as datum for its own annihilation, are forever being carried out in the familiar operations of living. . . . The needs and nature of the interpretive and computing equipment cannot finally be separated from those of tracking controls.
—George Philbrick, 1945

# Contents

# Preface and Acknowledgments

As I put the finishing touches on this book in September 2001, our world was changed by a horrific new human-machine combination. A familiar civilian technology, the commercial airliner, has been grotesquely transformed into a murderous weapon. Security officials proclaimed in shock that their systems worked but that they had never envisioned aircraft as suicide missiles. Writing in the *New York Times,* Thomas Friedman reminded us that it was not our systems that failed us so much as our imaginations: we simply could not imagine such an atrocity and hence could not defend against it.*

Indeed the terrorists here were terribly imaginative, concocting an aggregate of human and machine that we will never forget. Some of their tools were similar to those described in this book: simulation, training, control systems. Although these technologies are not always destructive, the situation today resonates with that when the U.S. Navy faced kamikaze suicide attacks in the Pacific during World War II, described in chapter 10. The computers and control systems designed to fend off air attacks embodied a set of assumptions about the enemy; the systems failed when those assumptions were undermined by a new human-machine combination, developed in a different cultural setting with different values. This book, a history of control systems, examines people, events, and technologies of an earlier era, both in peace and in war, but its central topic still permeates today's world: the relationship between people and machines.

This book sees the history of control systems and computing through the lenses of human-machine interaction and machine representation. As the child of a writer and an engineer, I came to this topic via several routes. One began in elementary school, when my father and I put together a science fair project on "simple computers"—Napier's bones, slide rules, and the like. That was my first exposure to continuous computing, as well as to binary arithmetic. In high school I made a minor hobby out

---

*Thomas Friedman, "Terrorism Game Theory," *New York Times,* 25 September 2001.

of building analog-to-digital converter circuits. As an engineering student at Yale I gravitated toward control systems and robotics because they inhabit both the continuous, analog world and the discrete, quantized digital domain. As a student of literature I wrote my senior thesis on Thomas Pynchon's *Gravity's Rainbow,* a book that shaped my perspective on technology. It lurks in the background here.

In the late 1980s I began working at the Woods Hole Oceanographic Institution, developing control systems for robots and exploring the deepest parts of the ocean. Initially it seemed that human presence in the deep ocean, in deep-diving submersibles, would be replaced by robotic vehicles operated remotely from the surface. Over time it became clear that the two technologies were complementary. The whole process of automation entailed more than simply replacing a human with a machine. The trade-offs between remote and manned had not been anticipated by designers; they had to do with new work practices, scientists' roles, and institutional commitments as much as with the simple replacement of humans by machines. My mentor at Woods Hole, Dr. Dana Yoerger, introduced me to the sociotechnical problems of control systems during late-night lab sessions and operations at sea. Dana had been a student of Professor Tom Sheridan, of the Massachusetts Institute of Technology, who elaborated the idea of *telerobotics,* according to which machines do not replace human operators but rather enhance their powers and allow them to work in remote or dangerous environments.

In 1991 I came to MIT as a graduate student to work with Leo Marx and Merritt Roe Smith in the history of technology. In a first-year course taught by Sherry Turkle, I began studying Norbert Wiener's book *Cybernetics,* where he mentions the problem that led him to think about human-machine interfaces: how to shoot down an attacking aircraft by predicting its future position and firing a shell to arrive at that point at some time in the future. Yet nowhere could I find any published discussions of the technologies designed to do this or of the people who had asked Wiener to look at the problem. When I began digging, I came across all kinds of curious human-machine interfaces—as "cybernetic" as anything in Wiener—that were absent from the historical literature.

In a course with Thomas P. Hughes on large technological systems I was intrigued to learn that MIT's first digital computer had emerged from the Servomechanisms Laboratory. Other than a simple origin story, little was known about the institutional context and engineering culture that generated the laboratory and its famous computer. When I interviewed Jay Forrester about his efforts to model the world using techniques derived from control theory, he made no mention of cybernetics. The "systems sciences" of the Cold War all worked from similar principles, but each had its own unique style and approaches, reflecting institutional, cultural, and personal legacies. Though my interest was piqued by Cold War technologies, I found almost nothing on the cultures, practices, and technologies of control engineering before World War II or on how they contributed to and shaped the more familiar postwar history. Stuart Bennett's 1993 book on the history of control engineering further excited my curiosity, and the more I looked, the more the connections grew. The records of the National Defense Research Committee, only recently declassified, revealed a rich set of interrelated projects and technologies.

Through these routes, then, I came to write a dissertation on the history of control systems, "Datum for its Own Annihilation: Feedback, Control, and Computing, 1916–1945," in 1996 and then expanded and rewrote it as this book. Portions of chapter 4 appeared in "Opening Black's Box: Rethinking Feedback's Myth of Origin," *Technology and Culture* 41, no. 3 (July 2000). Portions of chapters 9 and 10 appear in "Automation's Finest Hour: Radar and System Integration In World War II," in *Systems, Experts, and Computers: The Systems Approach in Management and Engineering, World War II and After*, ed. Agatha C. Hughes and Thomas Parke Hughes (Cambridge: MIT Press, 2000). And portions of chapter 11 appear in David A. Mindell, Jérôme Segal, and Slava Gerovitch, "From Communications Engineering to Communications Science: Cybernetics and Information Theory in the United States, France, and the Soviet Union," in *Science and Ideology*, ed. Mark Walker (2002).

Several institutions helped along the way. The Department of the History of Science and Technology at the Royal Institute of Technology in Stockholm, Sweden, the Lemelson Center for Innovation at the Smithsonian Institution in Washington, D.C., and the Dibner Institute for the History of Science in Cambridge, Massachusetts, were my home for short periods. The Program in Science, Technology, and Society at MIT has been my home for the past decade, and MIT overall has provided a wonderfully supportive environment for someone with hybrid interests. The Frances and David Dibner chair in the history of engineering and manufacturing, which I have been privileged to hold since 1996, provided valuable support and sparked my interest in manufacturing. I have also been aided by a National Science Foundation graduate fellowship, a Rockefeller Archive Center travel grant, a Dibner graduate fellowship, an Old Dominion fellowship from the dean of humanities and social sciences at MIT, an MIT Provost's junior faculty leave, and NSF Grant #6759400. Several archivists were particularly helpful in uncovering material: Helen Samuels and Elizabeth Andrews at the MIT Archives, Sheldon Hochheiser at the AT&T Archives, Michael Nash and Barbara Hall at the Hagley Museum and Library, and Marjory Ciarlante at the National Archives.

A number of friends and colleagues deeply read parts of the manuscript, provided material, or have been generous (and challenging!) interlocutors during this long project: Atsushi Akera, Stuart Bennett, Chris Bissel, Ed Eigen, Mats Fridlund, Ivan Getting, Rebecca Herzig, Arne Kaiser, Ron Kline, Svante Linquist, Jennifer Mnookin, Michael Mahoney, Axel Roch, Jean-Jacques Slotine, Judy Spitzer, Jon Sumida, Emily Thompson, Sherry Turkle, Tim Wolters, Chris Wright, and Dana Yoerger. May Maffei provided invaluable research assistance for the final phase of the book, becoming a scholar of the Chicago style manual in a remarkably short time. Joanne Allen did an outstanding and thorough job of copyediting. Henry Tom, of the Johns Hopkins University Press, kindly took on the project at a critical time and provided valuable advice on framing and finishing the book. Dr. Annette Bitsch, of the Humbolt University in Berlin, proved a most careful and attentive reader; she has considered the ideas in this book more deeply than anyone else.

Finally, I owe a special debt to my academic mentors. This study is in conversation with all of them: Tom Hughes, Leo Marx, Tom Sheridan, and Merritt Roe Smith. Each has influenced me in his own way, both when I was a student and more recently as a colleague; those who know their work will recognize its influence here. I must also mention James Snead, my undergraduate mentor in literature at Yale, who first suggested that I might be a humanist as well as an engineer and directed me to MIT. His tragic early death pains me often, but his voice echoes in this work.

My father nurtured my love of machines (historical, contemporary, and imagined). What I hope is the clarity of writing in this work owes more to my mother, Dr. Phyllis Mindell, than to anyone else.

Between Human and Machine

# Introduction

## A History of Control Systems

In 1934, at the height of the machine age, Lewis Mumford laid out his vision for technology and human development. In his landmark study, *Technics and Civilization*, Mumford proposed a theory of phases to characterize historical periods according to their core technologies. The "eotechnic" phase of water, wood, and handcrafts served as cultural preparation for the industrial era, which arrived as the "paleotechnic" world of steam, iron, and factories. The final phase, what Mumford called the "neotechnic," was still aborning as he wrote. The neotechnic era was characterized by "mathematical accuracy, physical economy, chemical purity, surgical cleanliness,"[1] as well as electric light and novel materials like celluloid and Bakelite. The world of neotechnics evoked the streamlined America of the 1930s as idealized in corporate headquarters, film sets, and world's fairs.

I begin this history of control systems with *Technics and Civilization* because historians of technology consider it a foundational text and because it overlooks human-machine interaction. For all his progressive hopes and his insight into the human dimensions of technology, Mumford defined the neotechnic by its materials and sources of energy, not by its inhabitants. In the clean spaces of the neotechnic world, no people operated the machines. According to Mumford, when the neotechnic is fully realized, "automatism" in production progresses to the point where "in the really neotechnic industries and processes, the worker has been almost eliminated."[2] In this vision people disappear from even the shiniest vehicles: for Mumford automobiles and airplanes were about gasoline and speed, not driving or piloting. However clean and electrical, machinery for Mumford remained inert and mechanistic, not actively involved with human beings.

Yet even as Mumford wrote, people were entering into new, intimate couplings with machines, with dramatic effects. Just a few years before Mumford's book, Charles Lindbergh flew across the Atlantic, blurring the boundary between pilot and machine. Lindbergh called his own account of the flight *We* to emphasize the collaborative nature of the feat—the hero was both the operator and the machine, their assemblage and their synthesis. "We shared our experiences together," he wrote, "each feeling beauty, lift, and death as keenly, each dependent on the other's loyalty. *We* have made this flight across the ocean, not *I* or *it*."[3] A few years later, Wiley Post flew around the world, his sole companion a Sperry automatic pilot that kept him on course while he slept. *Technics and Civilization* mentions neither Lindbergh nor Post, nor other technologies of machine control.

Even after the era of heroic flyers, World War II continued to blur the boundaries between mechanical and organic. Radar operators manipulated blips on screens as they fought automated attackers, and aircraft made human bodies into new and terrible weapons. During the Cold War, computerized command and control systems generated a vision (à la *Dr. Strangelove*) wherein the end of the world would be directed, in real time, by men in a data-processing center. By 1969, barely a generation after Mumford's book, Neil Armstrong's landing on the moon presented a cybernetic image to succeed Lindbergh's: his vision structured by instruments, his hand on a stick controlling a powerful engine, his actions aided and mediated by a digital computer and a room of ground controllers a world away. As American icons, Lindbergh and Armstrong allude to a long tradition of images of the helmsman. From sea captains and riverboat pilots to aviators and computer operators, these figures stood for a masculine ideal of control over two worlds, the natural and the technological.[4]

Ironically, while overlooking the helmsman, Mumford observed other phenomena remarkably relevant to the unfolding technological world. *Technics and Civilization* anticipated that automated factories would replace workers, that the importance of machines would no longer be measured by their bulk, and that instant communications would expand human capacities and quicken the pace of life. Mumford notably defined technology as both constitutive and symbolic of a culture and its values. He saw not only force and speed in technology but also form, mentality, and human consciousness. "Men had become mechanical," he declared, "before they perfected complicated machines to express their new bent and interest." Mumford identified the significance of a machine not so much in its material potency as in its cultural and psychological underpinnings, leading him to famously reject the steam engine in favor of the clock as the "outstanding fact and the typical symbol" of the modern industrial age.[5]

Most important, the "technics" of *Technics and Civilization* refers to the ability to represent the world with symbols and to manipulate those symbols with great facility. "The specific triumph of the technical imagination," Mumford wrote, "rested on the ability to dissociate lifting power from the arm and create a crane: to dissociate work from the action of men and animals and create the water-mill: to dissociate light from the combustion of wood and oil and create the electric lamp."[6] Words like *divorce, dissociate,* and *division,* referring to the separation of the world of abstractions (e.g., words, symbols, and codes) from the world of the concrete, pepper his text. "Men became powerful to the extent that they neglected the real world of wheat and wool, food and clothes and centered their attention on the purely quantitative representation of it in tokens and symbols."[7] Writing was the key technology in this evolution, and for Mumford (as for many others) the printing press served as the prime example, effecting "the divorce between print and firsthand experience" by allowing texts to travel great distances from their origins.[8]

It remains curious, then, that while emphasizing what he called techniques of abstraction, Mumford appeared indifferent to technologies of control and human-machine interaction. That such a perceptive analyst did not associate the two suggests that he wrote at a moment when they were just beginning to crystallize and converge. Indeed, 1934 saw the publication of several theories of feedback, such as Harold Hazen's theory of servomechanisms and Harold Black's negative feedback telephone amplifiers, that enabled engineers to link control and communications. In following decades, these ideas became inextricable in engineering theory and, increasingly, in the popular imagination; Mumford himself acknowledged the connection in his later work.[9] The association continues today, although transformed by another generation of culture and technology. In the era of cyberspace, global networks, and what William Gibson called "jacking in," the line between technology and human identity continues to shift and erode.[10] How did this association between human-machine interaction, on one hand, and technologies of representation, on the other, come about?

### Cybernetic Synthesis

The question seems to have a simple answer. In 1948 Norbert Wiener published *Cybernetics: Or Control and Communication in the Animal and the Machine*. There he argued that "the problems of control engineering and of communications engineering were inseparable,"[11] that they were united by the fundamental notion of the message, and that feedback loops, both within machines and between machines and people, must be understood in such terms. Wiener also argued that human behavior and dynamic mechanisms operated according to similar principles, and he posited the analogy between the digital computer (then in its infancy) and the human nervous system. He famously called for a new science of feedback, human behavior, and information, for which he coined the term *cybernetics,* from the Greek word *kubernētēs,* for "steersman."

*Cybernetics* elaborated the marriage of control and communication, a vision of the human relationship with machines, for engineers and systems theorists of varying stripes. As a movement, cybernetics sought to extend its notions to social systems, science, and even human cultures.[12] Its ambitions sparked popular imagination, and the legacies of cybernetics survive today in our language (*cyborgs, cyberspace,* etc.) and also in our relationship to machinery. In 1960, for instance, the psychologist J. C. R. Licklider articulated a theory of "man-computer symbiosis," in which machines would aid people in the real-time work of thinking. Licklider's vision inspired a research program and a community of scientists that eventually led to the Internet, among other innovations. Licklider and his disciples recognized that we live in constant interaction and exchange with machines, and that the boundaries between human and mechanical continue to blur and evolve as the online world takes on an order and a reality of its own. As Donna Haraway pointed out, we are all cyborgs, shifting combinations of organism and machine.[13]

Norbert Wiener, then, seems the obvious link between Mumford's neotechnic machine world and the cybernetic decades after World War II. Wiener himself would have us believe that he effected the genesis obviously and completely in the course of his wartime research on antiaircraft prediction. "I think that I can claim credit," he wrote in his memoir, "for transferring the whole theory of the servomechanism bodily to communication engineering" (although he never explained what he meant by "bodily").[14] Indeed, scientists, engineers, and the interested public associate Wiener and cybernetics with the image of a human being dynamically coupled to a machine and the notion of the message as the fundamental unit of a system, be it natural or human-made. Accordingly, historians of science and technology have explored the genesis of Wiener's project, its relationship to other work, and its long-term effects.[15] While expanding our understanding of cybernetics itself, these studies still center on Wiener—the academic, the intellectual, the mathematician. Wiener's origin story repeats a classic foundation myth of science and technology: the scientific (mathematical) genius generates great ideas in a grand tradition—Wiener refers to Leibniz, Pascal, Maxwell, and Gibbs as "ancestors" of his new discipline—leaving it to others (usually engineers) to expand, implement, and "reduce" them to practice.

But Wiener's simple genealogy is inadequate and incomplete, for it effaces its antecedents. Figures like Lindbergh and, indeed, Wiener's own invocation of the steersman draw our attention to a long history of human-machine relationships, brought into greater intimacy during the late nineteenth and early twentieth centuries. Edison's phonograph, for example, originally intended as a dictation machine, sought to replace writing with mechanically inscribed speech. The French physiologist and photographer Etienne-Jules Marey studied physiology with paper traces and photographic sequences, defining a modernist image of the body as a mechanism, recordable and calculable with the techniques of natural science. Marey, and other scientists like him, saw in automatic instruments a "mechanical objectivity" that recorded the world exactly as it was, without human intervention. The management consultant Frederick Winslow Taylor sought to rationalize human work by redesigning both machines and bodily movement for a better match. Henry Ford's engineers reoriented their factory around a moving line, mechanizing workers' actions as well as material flows. World War I brought people and machines into still other disquieting combinations as machine guns, tanks, aircraft, and even prosthetic limbs blurred the lines between organism and machine.[16]

In the early decades of the twentieth century engineers drew on these and related phenomena as they began to think deeply about control, communications, and human-machine interaction. The research programs of World War II, in which Wiener participated, built on prior engineering cultures and human-machine relationships, brought them into new combinations, and laid foundations for the postwar systems sciences.

### Revisiting the Wienerian Account

Control and communications did indeed converge, but more broadly and gradually than Wiener's account suggests. The story includes feedback control, human-machine interfaces, and computers, which I group under the rubric "control systems" to incorporate the complexity and diversity that follow from the idea of "system." While including aspects of control theory and control engineering, the term *control systems* also describes particular technologies, for systems are things as well as ideas.[17]

During the early twentieth century, before Wiener's formulation, American technology was already suffused with what would later be called cybernetic ideas. Several interwar engineering cultures—military gunfire control, aircraft and ship controls, communications engineering, and a nascent control theory—exemplify the convergence of communications and control that predated cybernetics. Wiener was indeed a critical player; he crystallized, articulated, and popularized this convergence, and he worked out some of its underlying mathematics, but he did not originate it. Before *Cybernetics,* servo engineers had already turned to telephone techniques to characterize feedback systems. Radar engineers had already adapted communications theory to deal with noise in tracking. Psychologists had already studied human operators as necessary but problematic components of automatic control loops. The engineering cultures elucidated here constitute an indispensable prerequisite to Wiener's would-be innovation of 1948, his proudly announced beacon of a cybernetic universal science.

It is tempting to call the story that follows a prehistory of cybernetics because it traces the roots of cybernetic thinking to decades before they are usually recognized. This description is only accurate to a rough approximation, however, because the term *prehistory* implies a certain teleology that is abhorrent to the historian of technology. Cybernetics was neither an implicit nor a necessary endpoint of the developments I describe here. Other formulations, in fact, had currency in the postwar world and may even have outshone cybernetics in their effects. What we might call the *systems sciences*—cybernetics, systems engineering, control theory, systems dynamics, information theory, and operations research, among others—all drew on the approaches and technologies developed in the decades before 1948. No unified scheme spawned these similar sciences; rather, engineers worked them out in response to different problems and environments.

This history is significant, then, beyond a mere corrective to Wiener's origin story.

It contributes to a fuller understanding of American technology between the world wars. It clarifies the nature of wartime research and its continuities with the prewar world. It connects the remarkable innovations in control, computing, and human-machine interaction that shaped today's world to longer histories of engineering and manufacturing in America. It also recasts the relationship between analog and digital as a theme in the history of computing, from a linear progression to a parallel evolution. Following Mumford's interest in "technics," this history traces the emergence of electrical signals as abstract representations of the world, the technologies developed to gather, manipulate, and transmit them, and the social and organizational conditions that made those technologies possible.

## A History of Control Systems

The story takes place between 1916 and 1948 and thus includes the two world wars. Between the wars engineers produced control systems of increasing performance and delicacy and raised a number of theoretical questions. These questions were just crystallizing when World War II began, and the war spawned a set of research projects that brought together diverse threads of control. From this work emerged new theories and practices of feedback, control, and computing. As always, the continuous nature of these processes makes the choice of beginning and ending somewhat arbitrary. This narrative ends in 1945, when the Office of Scientific Research and Development (OSRD) closed down, and with the subsequent publication of *Cybernetics,* Claude Shannon's "Mathematical Theory of Communications," and the Massachusetts Institute of Technology (MIT) Radiation Laboratory series of textbooks on radar, electronics, and servomechanisms. These and other publications helped spread the results of the war's massive research and development projects and laid foundations for a new era of communications, control, and computing.

The first half of the book follows four discrete technological traditions, or engineering cultures, of control systems in the United States between the world wars, and the second part chronicles their convergence and transformation during World War II. The four traditions correspond to four institutions, with their own technical problems, organizational imperatives, engineering practices, and user environments: the U.S. Navy Bureau of Ordnance and its fire control contractors, the Sperry Gyroscope Company, the Bell Telephone Laboratories, and Vannevar Bush's laboratory at MIT.

During World War II the U.S. government's research program in control systems brought these threads together. In 1940, at Vannevar Bush's request, President Franklin Roosevelt established the National Defense Research Committee (NDRC). It included a division devoted to control systems, headed by Warren Weaver of the Rockefeller Foundation. Weaver led the NDRC to develop a broad array of automatic controls, systems, and theory, including gun directors, target predictors, radar-controlled devices, and Wiener's theoretical studies. To run the NDRC program, Weaver brought management techniques from the Rockefeller Foundation, as well as representatives from the prewar engineering cultures at Sperry Gyroscope, MIT, Bell Labs, and the navy. The NDRC was not alone, as other government agencies also worked with industry and academia and came up with their own solutions.

The second half of the book traces communications and control in several different wartime engineering cultures: Gordon Brown and the Servomechanisms Laboratory at MIT, Bell Labs' electrical computers project, tracking radars at MIT's Radiation Lab, and Norbert Wiener, George Stibitz, and others thinking about human operators, digital representation, and control systems as generalized information processors. Through the NDRC and its related control systems projects, feedback theory, electrical power, and telephone engineering contributed to conceptions of control, computing, and communications that coalesced as the war ended.

## Six Theses

I divide the historical argument into six supporting theses, listed here in roughly the order they appear in the text, although most also run through the entire story. First, in the decades before World War II engineers in a variety of settings developed ideas and technologies of feedback, control, communications, and computing. In each of these settings, according to varying institutional goals and local engineering cultures, engineers had differing conceptions of what constituted a system, the role of the human operators, and how machines represented the world. In each case the human operator was not a universal person but an ideal type that engineers created (consciously or unconsciously) as they designed machinery—sometimes an unskilled drone, a creative officer, or a belligerent enemy. Some engineers called their machines "computers," pushing back our understanding of the use of this term for calculating machines to more than a decade earlier than scholars have recognized. These notions remained largely separate before the war, though each attained a certain technical and organizational maturity.

Here I draw from thinkers like Donald MacKenzie, who argued for the importance of local cultures, what he called "gyro culture" in the history of inertial guidance, in shaping engineers' approaches and solutions. Thomas P. Hughes has also written about this period, both on the emergence of ideas of systems in electric-power engineering and on the early industry of feedback devices, in his biography of Elmer Sperry.[18] Whereas Hughes compared national contexts in *Networks of Power,* I am comparing local, institutional settings to show how they shaped engineers' ideas. Where Hughes concentrated on Elmer Sperry the man, I concentrate on the company he created and how its approach to human-machine interaction responded to military and commercial demands, with both success and failure.

Second, the technological watershed of World War II had significant continuity with its predecessors, both in technical matters and in the management of research. Wartime research programs drew on and reconfigured an established base of expertise and technologies in order to forge new solutions to wartime problems. Contributing programs included military-industrial collaborations, "big science" instrument projects, and private patronage of research. Some scholars frame the war as a discontinuous break from what came before, creating a "World War II regime" that determined the postwar world.[19] By contrast, I emphasize continuities and contribute to an emerging body of scholarship that shows the conservatism of Bush's "revolution," the struggles for legitimacy and authority it encountered, and its significant component of practical, systems-oriented research.[20] Where Paul Edwards identifies a "cyborg discourse" that hatched in the fires of World War II and developed in the Cold War, I aim to uncover the longer-term themes that contributed to that discourse and examine their technical and cultural underpinnings.[21] This is not to say that the NDRC did not transform the conduct of American research. Rather, the continuities with the prewar world help us accurately gauge the nature of that transformation.

During World War II, antiaircraft fire control posed a critical problem, hence the third thesis: while working on the antiaircraft problem, engineers merged communications and control in the realms of servomechanisms, integrated systems, human-machine interfaces, and analog and digital computing. The basic solution entailed tracking a target, predicting its future position, and calculating the "lead" so that the guns could aim at the aircraft's future position. To solve this problem, the NDRC devised a research program that merged the prewar threads of control systems.

Despite these common ideas, no single, unified set of principles emerged from the wartime program, and this constitutes my fourth thesis. New and diverse fields like control theory, systems engineering, cybernetics, and digital computing, among others, carried feedback, control, and computing into the postwar world (each also drew on threads from beyond this book). Here I am contributing to recent work on the postwar systems sciences, particularly in the fields of systems engineering, cybernetics, and digital computing.[22] A broad picture of wartime research in control systems broadens our understanding of postwar technology and its underlying epistemologies.

The fifth thesis is that histories of computing that focus solely on computers as conceptualized by mathematicians are incomplete. It is wrong to suggest that computers arose first as logic machines and then took on cybernetic or connected characteristics. Ideas of communications, systems, and human interaction were present from the early days of digital computing. Numerous American computing pioneers, including Norbert Wiener, George Stibitz, John Atanasoff, J. Presper Eckert, John Mauchly, Claude Shannon, and Jay Forrester, among others, participated in the NDRC's research program on control systems. This is more than coincidence, for these men did not build electronic digital computers simply as calculators. Nor were they generally concerned with the questions of computability and logic that occupied mathematicians like Alan Turing and John von Neumann. Rather, they drew on longstanding traditions of control engineering, especially the technologies of fire control. My point is not to rewrite the history of computing—mathematicians of course played critical roles, as did the business machine industry—but rather to establish how the era of cyberspace and the Internet, with its emphasis on the computer as a communications device and as a vehicle for human interaction, connects to a longer history of control systems that generated computers as networked communications devices.

The sixth thesis is that analog and digital arose together, as distinct but related approaches to representing the world in machines. In general, historians of computing have neglected analog computing, viewing it primarily as an obsolete predecessor to digital. One historian, for example, recently argued that we have "succeeded pretty well" in understanding analog computing and that it is time to move on to the postwar, digital period.[23] Others write about it only as an obstacle to be overcome and a source of resistance to the new, inevitable, digital techniques. On the contrary, we have not yet begun to understand the history and significance of analog computing, especially the relationship between analog and digital machines.[24] I say this not out of any antiquarian interest but because digital did not replace analog in a simple progression of superior technology. The transition to digital computers was neither instant, obvious, nor complete.

Here again I invoke Mumford's call to examine the technologies of abstraction. Doing so draws our attention not only to the symbolic but also to the material history of computing. The kinds of symbols a computer manipulates affect how it is built, by whom and with what methods, questions raised in the varying cultures of control in the twenties and thirties, especially as they also faced the problem of mechanical versus electronic computers. These questions concerned industry as much as mathematics and design: mechanical versus electrical manufacturers, precision machining versus point-to-point wiring, the skills (and errors) of machinists versus those of clerical workers and human "computers."

## Engineering Cultures

Values and technologies of control have permeated Western culture and philosophy since the Enlightenment. The term *contre-rolle* originally referred to a register for accounts, and control has a history that runs through economics, science, and politics before making its way into engineering.[25] The historian James Beniger labeled as "the control revolution" the transformation of business and technology that occurred from the late nineteenth to the early twentieth century.[26] The two preeminent historians of feedback control, Otto Mayr and Stuart Bennett, have both written extensively and insightfully about the history of feedback control. James Watt's centrifugal flyball governor for steam engines, for example, became the first feedback mechanism to be widely employed by technologists and to enter popular consciousness. The device, derived from the rotating pendulums used to regulate windmills, appeared in 1788, and the spinning balls became familiar icons of mechanical motion.

Mayr chronicles the history of these and similar feedback mechanisms before 1800 and begins to connect it to intellectual and political currents in Britain and France. Bennett's two volumes address the periods 1800–1930 and 1930–55, respectively, with a technical, engineering-oriented account depicting the origins of modern control engineering from an insider's perspective.[27] Where these two historians identify a series of technologies and thinkers that led up to modern control engineering, I examine the particular problems engineers were trying to solve when they innovated. These problems and the engineering cultures that generated them are keys to understanding the broader significance of control engineering within the history of technology.

This story focuses on engineers—Hannibal Ford, Elmer Sperry, Harold Black, Harold Hazen, Gordon Brown, Ivan Getting, George Stibitz, and a host of others—

on the machines and knowledge they produced, and on their problem-solving activities. It includes the issues raised by manufacturing and those faced by users in the field. Source materials include not only correspondence and publications but also technical data, engineering drawings, and the machines themselves. I also include information from training documents, operator feedback, and field reports, and in a few cases I have been fortunate to observe and operate the machines myself.

The reader will, I hope, take away from this book not only an understanding of the history of control systems but also a sense for the technologies themselves and an intuition for feedback. From the thermostats in our homes and the cruise controls in our cars to the subtle dynamics of the large systems on which we depend, feedback loops and control systems remain integral to today's technological life.

Despite my interest in machinery, this is not what historians call an internal account—a genealogy of hardware and ideas, separate from the world. Such histories tend to impose a logic and coherence to events that they lacked at the time. Political and military conflicts structure this book, and they had major consequences for the engineers and technologies in question. Friction, personalities, false starts, and failures appear not as distractions but as integral to technological change. At every point this story examines local institutions, engineers' backgrounds, and their connections to their social and political worlds, the engineering cultures in which they were embedded.

Engineering cultures link engineers to their surroundings and, indeed, constitute their surroundings. I use the phrase in the plural to avoid the impression that there was any single engineering culture to which all engineers subscribed. Rather, this notion of engineering cultures is local, characteristic of a particular company, laboratory, or institution and embodied in techniques, tools, knowledge, and, above all, a group of people skilled in applying feedback mechanisms.[28] Such cultures may also cut across organizations and encompass what Edward Constant called "communities of practitioners," such as the officers, engineers, and businessmen who worked on naval fire control between the world wars or the group of engineers at Bell Labs who studied transmission in the telephone network.[29]

An engineering culture interacts with technologies it produces in numerous ways, but we can describe some of the salient mechanisms. Different engineering cultures have different organizational structures, different individual career trajectories for engineers, different languages for describing systems, and different relations to the broader world of technology. Their influence on engineering practice is perhaps best summarized by the engineers' adage, "You do what you know." George Stibitz worked on switching for AT&T; when designing computers he sensibly made them out of telephone relays. Gordon Brown's landmark paper "Transient Behavior and Design of Servomechanisms" (1940) drew on MIT's longstanding emphasis on transient analysis, developed during earlier work in electric power systems. Charles Stark Draper and his lab made gyroscopic instruments for aircraft; when they made fire control computers they built them out of gyroscopes. Equally important, each organization has a set of institutional goals: universities produce knowledge and students, navies make weapons and fight wars, and industrial firms create machines that earn profits. By asking what kinds of problems engineers were trying to solve in each case, we can see how each engineering culture developed its own concepts of feedback, stability, control, and the role of the human operator. A balanced picture of control systems entails examining several worlds simultaneously.

The four traditions discussed here by no means proceeded in isolation. The borders between them were porous and shifting; individuals, information, and even hardware constantly moved between them. Sperry Gyroscope hired MIT professors as consultants, MIT taught a special course in control engineering for naval officers, the navy's Bureau of Ordnance directed computer development at Sperry, and Bell Labs had close intellectual exchange with MIT. Other factors inhibited these flows, including military secrecy, industrial concerns about patents and proprietary solutions, and plain narrow-mindedness. The technology developed as an ongoing conversation, collaboration, and competition between organizations, though through imperfect channels.

Despite the importance of these four cultures, their significance does not reside in their being first, singular, or comprehensive. That is, I do not attempt to cover all aspects of feedback, control systems, or computing. Other individuals and technical communities also played important roles. Industrial process control, for example, was arguably more common, if less sophisticated, in industry between the world wars than the forms of control I trace here.[30] It does not form a significant part of this story because it was not included in the wartime formulations of control, a decision discussed in chapter 8. Similarly, engineers building feedback devices had little interaction with biologists and their ideas of homeostasis, so I have little to say about biological regulation other than to mention Norbert Wiener's interest in it. Moreover, during the world wars secrecy confined military control engineering within national borders, although the United States and Britain shared technology in wartime. Engineers in Germany and Russia made significant contributions, especially through the German rocket program, and in both countries they established committees on automation and remote control during the 1930s. Other individuals in the United States and elsewhere played important roles in the history of control, but if their work did not appear in the engineering cultures described here, then they do not appear in this account. The Russian mathematician A. N. Kolomogorov, for example, anticipated much of Wiener's work in time series phenomena, but American theorists did not recognize it until after the war, so Kolomogorov is not part of this story.[31] This book tells what Peter Galison calls "mesoscopic history," stories that have greater significance than simply case studies but are not intended to be universally illustrative of control in all times and places.[32] Their importance ultimately lies in their connections, both among themselves and to prior and subsequent histories, through practice, pedagogy, physical artifacts, and epistemology.

An emphasis on institutions and cultures notwithstanding, this book is not an argument for the social construction of technology. The idea that technology and society are constructed simultaneously has been repeatedly demonstrated and elaborated by competent scholars and does not require reiteration here.[33] This book treats it as a starting point, not as a conclusion. What people are doing, who they are, and how they relate to each other has everything to do with what kind of technologies they build, an idea implicit in every facet of this narrative. Yet I use the word *construction* to refer only to building things.

The book's arguments are ultimately historical, both as an original history (though not a history of origins) and as a synthesis of new ways to think about the evolution of control, communications, computing, and information. This history of control systems brings new people, machines, and ideas into major currents of twentieth-century history and proposes new ways to conceptualize that history by considering how machines represent the world.

## Control Systems and Representation

Engineers building control systems consistently addressed problems of representation. How does a physical quantity or symbol stand in for (signify) something else? Does it perfectly represent that something, or does it introduce distortion? Does it decay over time and require renewal? How, and by what conventions, is the quantity or symbol to be interpreted? These questions not only allow us to reconnect historical computers to the worlds they inhabited but also force a turn away from an intellectual history toward a technical and material one. Engineers often solved problems of representation by putting people into control loops, and the methods they used had industrial implications. By exploring the detailed, practical processes by which the world is represented by physical quantities and symbols, and vice versa, a history of control systems reattaches computers to their environment, to their human operators, and to the skills that constructed them.

From Turing's machine to artificial intelligence, computers have been nothing if not symbol-manipulating machines. In this tradition any computer that fits Turing's definition of a "universal" machine is equivalent to any other, and they can all be understood in a purely symbolic realm, without reference to hardware or architecture. This realization, of course, provided a foundation for the computer and cognitive sciences and has been a major reason for their success. Yet it says nothing about what the symbols inside the machine refer to or how they travel into or out of the machine to interact with the world. Put another way, the ideal Turing machine can calculate anything, but it does not *do* anything. Turing and his successors considered the formal systems of manipulating symbols according to rules and references to other symbols, but they did not consider what or how those symbols might signify.[34] They thus effected the "separation" and "divorce" that Mumford noted between "print and first-hand experience," between the symbolic and the concrete.

Yet Mumford's interest in processes of abstraction continues a long history. The study of signs, symbols, and representation has long been a part of Western thought. Contemporary with Mumford, for example, Walter Benjamin pondered how mechanical reproduction could dissociate a work of art from its grounding in tradition.[35] Recently, historians of science have taken up questions of instruments and simulations, exploring how machines can represent, communicate, and even create truths about the natural world.[36] For them, a measurement stands for some aspect of the physical world. A "good" instrument accurately represents some physical quantity, whereas an "unreliable" one introduces uncertainty and distortion. A key point is that measurements, like all symbols, must be interpreted within some community and according to its standards and conventions (which in turn define and create the community). The people in that community must be convinced that an air pump, a cloud chamber, or an x-ray produce accurate, reliable knowledge about the world; there is always room for noise, interruption, and misunderstanding.

In this story we shall encounter devices that draw pictures of battles as they occur, mechanisms that create mechanical models of attacking aircraft, and circuits that transmit voices, images, and words through wires. In each case engineers devised machines to represent the world, and in some cases they explicitly developed common languages of signals for that purpose. How engineers generated, transmitted, interpreted, maintained, stored, and made such signals effective are questions that run throughout this book.

Unlike scholars in other fields, historians of technology by and large have not taken up questions of representation. The omissions are curious because they consider *Technics and Civilization* a seminal text, and Mumford explicitly called attention to "technics" as the set of abstractions that surrounded a new technology and to the role of machines in manipulating those representations "in tokens and symbols." Granted, historians of technology have examined public representations like world's fairs, literature, advertisements, and films, but rarely have they opened the proverbial black box of machinery to study the representations inside. By contrast, here I examine *representation in machines* as much as *representations of machines*.

Two thinkers help us to weave this theme into a historical narrative. The sociologist of science Bruno Latour speaks of scientists as "mobilizing the world" with equations, maps, writing, and other representations. For Latour, scientists control their world by drawing it (literally) into "centers of calculation," spaces where symbols can be manipulated separate from their referents in the outside world. Such accomplishments rest, in part, on making symbols that are accurate, easily combinable, and stable over time, a technical problem that itself entails a great deal of human labor (e.g., calibrating instruments, eliminating contamination, controlling variables).[37] Historians of technology generally associate Latour with his contributions to social construction and to "actor-network theory." They have not yet responded to his exhortation to examine the labors of abstraction, by which he means the details of how symbols are created, maintained, and transmitted in technological networks. Also overlooked is Latour's acknowledged role as an interpreter of the mathematician and philosopher Michel Serres, who emphasized the role of interruption and noise in the evolution of communication systems.[38]

To situate these abstractions in specific technical operations I turn not to a historian but to an ethnographer and cognitive scientist. Edwin Hutchins details how people and machines come together to process information. Examining a group of sailors aboard a navy ship as they navigate into port, Hutchins finds them generating and exchanging representations of the world through charts, instruments, radars, and similar tools. He follows information as it is converted, transformed, and communicated through a "cascade of representations" that involves both machines and people. In such operations questions of analog and digital do not apply simply to black-box computers. Rather, the sailors themselves generate, transmit, and convert representations of both types. Reading a number off a compass, for example, entails an analog-to-digital conversion, and plotting coordinates on a chart converts from digital to analog. Each mode has strengths and weaknesses, and which is used at any moment depends on a host of factors, from convenience and simplicity to deeply held cultural and historical assumptions.[39] Hutchins rethinks expertise, intelligence, and learning as embedded neither solely in the machines themselves nor in the heads of the people but in the sum of the people in interaction with the machines, a true sociotechnical system, bound together by symbolic exchange.

Conveniently, Hutchins studies navigation aboard a U.S. Navy ship, so the machinery and tasks he describes resemble the control systems examined here. While his ethnographic approach highlights the cultural assumptions embedded in navigational practices, Hutchins is not concerned with their historical dimensions. How, over time, did these computational systems evolve? What conditions and constraints lead to a particular combination of machines, people, and representations? How is control traded between them? A history of control systems as evolving hybrids of humans and machines addresses these questions.

THE next four chapters each cover an engineering culture of control systems in the United States from about 1916 to 1945. Each chapter also highlights particular engineers. The U.S. Navy Bureau of Ordnance and its contractors (like Hannibal Ford and the Ford Instrument Company) developed fire control systems for naval guns that evolved from simple sighting devices to distributed networks of perception, integration, and articulation aboard ships (chapter 2). Sperry Gyroscope, though it started out in naval fire control, found commercial success with simpler control systems that tightly coupled human operators to their machines, from automatic pilots to gun turrets in aircraft (chapter 3). Harold Black, Harry Nyquist, Hendrik Bode, and their associates at Bell Labs developed feedback amplifiers to transmit voice signals throughout the national telephone network. Their work took place within an organization devoted to generating, manipulating, and transmitting electrical signals (chapter 4). At MIT, Vannevar Bush, Harold Hazen, and their colleagues developed simulation machines, analog computers, and servomechanism theory, and they moved from industrial problems to fundamental studies in engineering science during the 1930s (chapter 5).

Chapter 6 covers a brief interlude before World War II when several of these threads began to converge. In 1939 four young gunnery lieutenants from the Bureau of Ordnance came to MIT to study servomechanisms and then carried new theories of feedback from the university setting into naval operations. Building the Palomar telescope, facilitated by Warren Weaver of the Rockefeller Foundation, also brought Vannevar Bush together with experts in naval fire control.

When Bush founded the NDRC in 1940, he immediately drew on Weaver to bring the prewar traditions of control systems to bear on what he called "the antiaircraft problem" (chapter 7). Weaver quickly surveyed the field and formulated a research program. Chapters 8–11 chronicle how this research program brought together ideas of communications and control and the solutions it produced. In chapter 8 we see Gordon Brown and MIT's new Servo Lab applying techniques developed for telephone amplifiers to hydraulic and mechanical control systems and struggling with industrial and governmental relationships. In chapter 9, engineers at Bell Labs such as Donald Parkinson and Clarence Lovell create an analog computer to track airborne targets and combine it with an innovative automatic tracking radar built at MIT's Radiation Lab by Ivan Getting and his group. In chapter 10, Getting pushes the combination further, designing radar and control systems together from the first and crafting a new role, that of system integrator, for his own organization. Chapter 10 shows the NDRC working with Norbert Wiener, George Stibitz, and others interested in questions of digital computing and in how the representations employed by their machines related to organizational contexts and industrial skills. Chapter 12 concludes by reviewing the six major theses, tracing their implications in the latter half of the twentieth century, and suggesting how this history enables us to see the emergence of computers, networks, and information in a new light.

# Naval Control Systems

## The Bureau of Ordnance and the Ford Instrument Company

### Grids on a Swirling Sea

On a July day in 1916 a group of young officers gathered around a strange new machine. They were aboard the USS *Texas,* one of the newest battleships in the fleet, embodying the age of big guns and steel armor. In the battleship era, naval strategy and tactics revolved around ships like the *Texas* and their gunnery crews. These confident young men, lieutenants well trained in ballistics and mathematics, belonged to the "gun club," the elite cadre of ordnance officers who controlled the battleships' main battery guns. These weapons could throw 14-inch, 1,500-pound projectiles more than 10 miles. Such long ranges pushed fleets further and further from their enemies, and gunnery officers increasingly became the people who actually fought naval battles—not on the bridge with eyes on the horizon but in armored rooms poring over charts and instruments. From these remote locations, aiming and firing the big guns carried the prestige of the marksman.[1]

The trouble was that hitting something that far away was no simple task. The likely target was another battleship, moving faster than 20 knots, itself shrouded in smoke and mist, firing back at you as your own platform sped through the water, pitching and rolling on a swirling sea. In the minute or so that it took a shell to fly to the target, the entire geometry of the battle could change. Two ships steaming toward each other at 25 knots would close range by nearly a mile during that time. In the more likely situation when they were on parallel or oblique courses, the officers would have

Fig. 2.1. Mark 1 Ford Rangekeeper. Note the absence of automatic inputs; all variables except ship's course were entered with the hand cranks. Courtesy of Hagley Museum and Library.

to calculate the enemy ship's course and speed, extrapolate into the future, and then aim where it was expected to be. This calculation required several consecutive measurements of the target's range and bearing, plotted in relation to one's own ship's course and speed. All in real time, in the heat of battle.

Hence the young men's interest in this new machine. A sealed box a couple of feet square, it sat on top of a stand and presented an angled face with dials and digits. A set of cranks on the perimeter surrounded dials and counters on the face (Fig. 2.1). The whole thing had a precise, mathematical flavor, as did the young inventor and entrepreneur who accompanied the machine, Hannibal Ford. He was showing off the prototype of his new company's first product, called a *rangekeeper*. Ford intended it to be the calculating center for the extensive shipboard system of instruments, data transmitters, and human operators that tracked targets and aimed guns, known as a *fire control system*.

Hannibal Ford walked the lieutenants through a typical problem. Say we have two ships approaching each other at 30 knots at a range of 18,000 yards on opposite but parallel courses separated by 10,000 yards. How will the range between them change over time? Ford showed them how to set up the problem using the machine's cranks. The dials then indicated not just one answer but a series of answers over time. If one entered an initial range, the machine calculated the range into the future. After two minutes the range would be 15,000 yards, after eight minutes they would pass at 10,000 yards, and then it would begin to open again. The lieutenants were much impressed, they wrote in their report, because "this problem could not be handled by any of the present methods."

This was no academic issue, for international events had just turned on long-range gunnery. Just two months before the *Texas* trials, in May 1916, British fire control systems, which the Americans thought to be the best in the world, had failed their acid test. At Jutland, the largest naval battle of World War I (and the only full-scale encounter between battleship fleets in history), British and German fleets had engaged at ranges from 14,000 to 18,000 yards. The British, with all their equipment, had achieved fewer than 3 percent hits on their enemy, an embarrassing and potentially scandalous performance. Though the poor results had numerous causes, the battle had exposed the weakness of the British system, which relied heavily on gunnery officers to do the plotting. The single British ship with a mechanized calculating system had turned in the best shooting performance. "Never has the potential power of naval force stood in so sharp a contrast with its actual efficiency in war," the system's frustrated but vindicated inventor, Arthur Hungerford Pollen, wrote of Jutland.[2]

The U.S. Navy reacted with new technology. The summer after Jutland, following Hannibal Ford's visit, the *Texas* took his rangekeeper to sea and conducted trials. After putting it through its paces, the gunnery lieutenants filed a favorable report with the navy's Bureau of Ordnance (BuOrd), which had requested the evaluation.[3] Unlike Pollen, Ford cleverly designed and presented his machine not as a replacement for the skilled officers but rather as an aid, an instrument that would both require and enhance the prestige of their mathematical skill. "The rangekeeper works out problems with mathematical accuracy," the young men noted, even as they underscored the importance of the human operator: "Its value therefore depends considerably upon the expertness of the operator and his skill in utilizing the data supported." They recommended that six to eight rangekeepers be supplied to the fleet as soon as possible, and that the machines be employed continuously alongside existing manual methods of range plotting. The Commander in Chief of the Atlantic Fleet supported the evaluation board's recommendations to the Chief of Naval Operations. The following month, Ford's company quoted a price of $100,000 to the navy for delivery of nine Ford Rangekeepers within eight to ten months.[4]

The Ford Instrument Company was on its way. Within a few years this private, secret, startup company became, in the words of an envious British observer, "the secret Fire Control Design Section of the U.S. Navy."[5] The first engineering culture of control systems that we shall examine, then, surrounds fire control systems in the U.S. Navy from the time of Jutland up to World War II.

This was a period of intense technological ferment, as militaries worldwide contended with the advent of radio, submarines, and internal combustion engines, among other inventions, as well as their tactical, doctrinal, and organizational implications. During this period aircraft and aircraft carriers supplanted the beloved battleships, and antiaircraft guns correspondingly replaced heavy guns ("main batteries") at the forefront of gunnery innovation. Yet curiously, control technologies remain virtually unexamined by historians of interwar militaries, partly as a result of the subtlety and secrecy of the machinery. A major recent work on technology in the interwar navy, for example, doesn't even mention fire control. Likewise, aviation historians (like aviators themselves) have never been fond of the lowly machines designed to bring airplanes back to earth, so antiaircraft systems have a low profile in histories of military technology as well.[6] Yet as Jutland demonstrated, these technologies could have a decisive effect.

In a secret sphere, Ford Instrument, Sperry Gyroscope, the Arma Engineering Company, and the General Electric Company, in close cooperation with BuOrd, designed and manufactured these technologies of control. The story revolves around the Ford Rangekeeper, which gave the U.S. Navy its first competitive advantage in fire control and formed the heart of its fire control systems until World War II. It also includes Hannibal Ford himself and the company he founded, one embedded in webs of contracting, competition, and collaboration with other firms and with the Navy's BuOrd. What began as a "rangekeeper" for main batteries evolved into the center of distributed, reconfigurable systems, eventually acquiring the name "computer."

### The Elements of Aiming: Perception, Integration, and Articulation

Warfare at sea takes place on a battlefield with no landmarks, no terrain, no features. More akin to a magnetic field than to a farmer's field, it is characterized not by stable geography but by imaginary lines of force imposed upon an otherwise smooth and turbulent space. Military technologists, striving to measure, rationalize, and order the chaotic space of war, value uniformity, measurement, and control.[7] Bringing these spaces under quantitative control required instruments to establish references for heading, horizon, and lines of fire. It also entailed representing the world relative to these quantities.

Fire control systems sought to solve the basic task of aiming a gun by breaking it into three acts: perception (looking through the sight), integration (leading the target, estimating the trajectory), and articulation (pulling the trigger). These have pertinence for the larger history of control systems as well, so it is worth examining them in detail.

*Perception* refers to how a control system apprehends and absorbs the world. I connect control systems to a broad history of quantification of the senses in which "vision, as well as the other senses, is now describable in terms of abstract and exchangeable magnitudes." Instruments of perception do not enable transparent, value-free observation but rather structure the world for the observer, focus the observer's attention, and shape his or her own act of perception as well.[8] Instruments of perception such as telescopes and rangefinders gathered data about the environment and the enemy and converted continuous physical quantities into numbers. Transmitters sent them by voice or remote indicators to a central location. As Hutchins pointed out in his study of naval navigation, these operations, when they involve human beings, often entail a series of analog-to-digital and digital-to-analog conversions.[9]

In the plotting room, machines and officers created an analog of the space of battle, either plotted on a chart or represented inside a mechanism. The Ford Rangekeeper integrated these data in two senses: combining disparate elements and also integrating mathematically. Combining means bringing elements together and tying them into a unified whole (as in "system integration"). But the Ford Rangekeeper also integrated mathematically (integrating velocity gives position), performing progressive accumulation and averaging. *Integration* implies memory (i.e., storing data for averaging), involves some type of processing, and is equivalent to low-pass filtering, namely, smoothing a signal to eliminate rapid, high-frequency fluctuations. In these two ways integration performs a kind of translation, from perception to articulation, frequently from a visual representation to a textual or mechanical one.[10]

From the rangekeeper or plotting room, numerical answers, verbal commands, and powerful servos articulated outputs to drive the guns. *Articulation* refers to the output of the human or system, effecting a concrete action, and it also has two related meanings. It derives from the Latin *articulatus,* which means to divide into joints, so articulate speech is that which is clearly segmented. The word also has a mechanical usage, referring to segmented construction. An articulated crane, for example, is divided into distinct segments; the hipbone articulates the thighbone. Thus, the controller can articulate either speech or mechanical motion; both entail movement and action.

Perception, articulation, and integration form a historical schema to organize the history of control systems; all who built control systems worked with these elements in some form, though not usually with this nomenclature. Throughout, people were the glue that held the systems together. By converting, interpreting, and communicating the data, they worked to maintain the veracity of the abstractions and their faithfulness to the outside world. For long-range gunnery this process had reached a state of maturity and technical stability by World War II. While successful by their own terms, these technologies also enshrined a set of assumptions that broke down as World War II began, and people and machines entered into combinations not envisioned by this engineering culture.

## The Other Ford: Hannibal Ford

Hannibal Choate Ford (no relation to Henry Ford), whom Vannevar Bush called "about as ingenious an individual as I ever heard of," was born in Dryden, New York, in 1877 and later moved to the nearby town of Cortland. Ford's father published the local newspaper, and young Hannibal got his introduction to machinery in the press room and on the precision lathe of the local jeweler. He attended Cornell University, where he studied electrical and mechanical engineering. In 1903, at a gathering of the American Institute for Electrical Engineers in Niagara Falls, Ford met an ambitious inventor from Cortland, Elmer Ambrose Sperry. Seventeen years older than Ford, Sperry, who had been a friend of Ford's older brother, had already started several companies to make electrical equipment, dynamos, even electric automobiles. Sperry soon became one of the premier American inventor-engineers of the early twentieth century. He had a special interest in control systems.

Ford got on well with Sperry, and the two kept in touch during the following years while Ford held a number of different engineering jobs. He patented speed control devices for the New York City subways, designed typewriter mechanisms, and worked at Westinghouse as a toolmaker. Hannibal Ford possessed a special mechanical talent augmented by an enthusiasm for machinery and fine shop skills.[11] Unsatisfied in his early engineering jobs, Ford corresponded with Elmer Sperry on matters of mechanical design.

In 1909 Sperry hired Ford as a design engineer for a new venture, building a gyrocompass. This device, a mechanically precise rotor spinning at high speed, sensed the rotation of the earth and aligned itself to the earth's rotational axis without regard to magnetic fields. By pointing to true north instead of to magnetic north it eliminated the problems that plagued magnetic compasses, a huge advantage for stable navigation, especially on steel warships and submarines.[12] Sperry soon founded a new company in New York City, the Sperry Gyroscope Company, to develop and manufacture this device. Hannibal Ford became the company's first employee and its chief engineer. Sperry and the company he founded became commercial leaders in manufacturing and marketing control systems for a variety of machines and vehicles.[13]

With Ford's help, by 1911 Elmer Sperry had completed a gyrocompass and tested it aboard one of the U.S. Navy's first dreadnought battleships, the *Delaware*. Hannibal Ford supervised the installation at the Boston Navy Yard. The gyrocompass performed well on the *Delaware,* and the navy immediately ordered six units from Sperry Gyroscope for dreadnoughts and submarines. In addition to the technical demonstration, the *Delaware* trial defined the personal core of Sperry Gyroscope and its close ties to the navy. Ford met two enthusiastic young navy men who had responsibility for the trials. Ensign Reginald E. "Foxy" Gillmor, originally from Menominee, Wisconsin, had graduated from the Naval Academy in 1907. Electrician Petty Officer Thomas Morgan, "a square hewn-country boy from North Carolina," had joined the navy right out of high school.[14] In 1912 both Gillmor and Morgan left the navy and joined Sperry Gyroscope; both would eventually become president of the company.

### The Sperry Gyrocompass

Inside the Sperry gyrocompass a feedback device called a "follow-up" or "phantom" element automatically tracked the movement of the spinning gyro and amplified its signal. The follow-up servo could drive any number of "repeater compasses," located anywhere on a ship, each of which replicated the reading of the central gyro.[15] The phantom made the gyrocompass an active instrument, effectively separating the data from the instrument itself, freeing it for transmission throughout the system.

The gyrocompass, with its ability to track true north as opposed to the unreliable magnetic north, had obvious advantages for piloting and navigation at sea. With the navy's help, Sperry developed the gyrocompass into a practical system and a viable commercial product, the Sperry Gyrocompass for Ships. The product brought Sperry Gyroscope valuable marine engineering experience and a pattern of contact and technical exchange with naval officers. The gyrocompass achieved success in the market, and by 1920 the company had installed it on more than 700 ships worldwide, both commercial and military.[16]

The navy had an additional, unique interest in the device: the gyrocompass's stable heading and connected repeaters could form the reference point for naval gunfire. Repeater compasses, because they transmitted a solid heading reference to various points on the ship, could do even more, especially within the novel systems of director fire.

### Director Fire

At the turn of the century the main guns on warships were aimed by *pointer fire;* that is, a captain at each gun sighted and fired his gun independently. In 1898 Captain (later Admiral) Percy Scott, of the Royal Navy, improved this system through an innovation in controls: by modifying the gunner's handwheel and attaching an improved telescope, Scott connected the gunners more closely to their guns. The controls allowed gunners to track a target as the ship rolled, a technique that became known as *continuous aim firing*. Scott combined his innovations in gun control with novel devices and procedures for training and became an ardent advocate of "scientific" gunnery.[17]

Continuous aim applied to short and moderate ranges. For long-range gunnery Scott actually negated his own invention, introducing *director fire*, which split apart the perception and articulation of continuous aim and put the guns under the control of a central location, or director.[18] The director, an officer aloft or on an upper deck in a gun director (an enclosed space), sighted the target with a precision telescope, calculated settings for each gun, and transmitted orders to the gunners, seeking to make the entire ship's broadside converge on a single point. By virtue of his elevated position, not only could the director see further than the gunners but he was above the smoke generated by the guns. Aiming and firing the guns all at once also simplified the task of spotting the shell splashes and converging on the proper range.[19]

In addition to centralizing control, director fire introduced new instruments of perception. Telescopes in the director tower (called "directorscopes") measured the bearing of the target to a fraction of a degree. Optical rangefinders, giant binocular-like devices with greatly exaggerated distances between the lenses, determined the target's range. The director "spotted" shell splashes as they fell near the target and called corrections to the gunners to bring the next shots closer, thereby closing a feedback loop.[20] Actually, the method had been proposed before Scott, but it could not become practical until the director could signal, with an electrical trigger, all the guns to fire simultaneously, termed *single-key* or *master-key* firing. To enable this innovation in gun control, Scott designed communications devices to transmit accurate data from the directors to the gunners.[21]

Director fire effectively displaced aiming from sailors operating guns to officers with instruments. Gunners were now only articulators, with no responsibility for perception. By contrast, the argument went, gunnery officers in the director had to exercise skill and judgment while interpreting the shell splashes and correcting the fire. Despite this centralization of authority, Percy Scott had to mount a vigorous campaign to get his ideas accepted because many officers still considered manual control heroic and more accurate. Improved gunnery scores at battle practices bolstered Scott's program, however, and by 1913 it was being installed in all new British battleships and cruisers.

Actually, orders did not go directly from the director to the guns. In the plotting

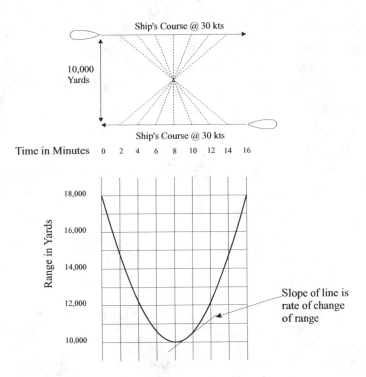

Fig. 2.2. Change in the rate of change of range for two ships approaching each other on parallel courses 10,000 yards apart at 30 knots. Redrawn from Hannibal Ford's original proposal for a range-keeper, "Ford Range and Deflection Predictor," 15 May 1915, RG 74, E-30, box 696, subject file 30199.

room, an armored room deep belowdecks, gunnery officers plotted the data, calculated firing solutions, and sent orders to the guns. They had to "lead," or predict the position of the target, which required knowing not only its range and bearing but also its course and speed. They then looked up the ballistics of the gun in numerical *firing tables* to determine what elevation would send the shell to the proper range. Officers in the plotting room originally did this work by hand in classic naval fashion, manually plotting successive ranges and bearings on a chart and measuring rates and courses with a compass or protractor.

Around 1905 the British began aiding the process with mechanical devices. Simple calculators allowed officers to estimate the target's range and the rate at which it was changing. These quantities were then set on a *range clock,* which indicated the range as it changed. This setup, however, had a critical flaw: the range to the target rarely changed at a constant rate, even if the target remained on a constant heading (Fig. 2.2).[22] The range clock ran at a fixed speed, so it had to be constantly adjusted by hand to read the proper values, which introduced significant error.

Arthur Hungerford Pollen, an English entrepreneur, understood the problem of continuously changing range rate.[23] He invented a system to predict the future position of the target based on a continuously updated derivation of the rate of change of target range. Where director firing displaced the aiming skill of gunnery crews, Pollen's invention threatened the autonomy of officers by mechanizing the subtle cycle of prediction and correction of range. Pollen found the Admiralty hostile to his ideas.[24] In 1913, after extensive trials and much debate, the Admiralty officially adopted a less automated system designed by a gunnery officer named Frederic Dreyer. The cost of the "Dreyer Table" was one-tenth that of Pollen's instrument, but more important, it preserved a greater role for the skilled operator. It was also produced by an insider, not an independent entrepreneur.[25] A bitter Pollen blamed the Dreyer Table for the poor long-range shooting of the British navy at Jutland and hence the great losses it suffered.

### Fire Control in the U.S. Navy before World War I

Before World War I the U.S. Navy lagged behind the British in fire control. The Americans did not generally recognize director fire as a significant improvement over individually controlled pointer fire. In the 1890s Bradley Fiske, of the U.S. Navy, had proposed, invented, and patented both director firing and fire control systems. Unlike Scott, however, Fiske had not managed to convince officialdom, so his ideas languished. The situation began to change in 1905, when a fire control board within the U.S. Navy formally advocated director firing, calling it "a system of fire control and a system of information." The prescient terminology "system of information" referred to how data were transmitted between the director and the gunners. Voice tubes proved inadequate, useless amid the noise of firing. The board emphasized "*rapid* and *continuous*" communications, that is, a system of visual indicators to transmit data.[26]

With a boost from another fire control board in 1910, fire control in the U.S. Navy began to take shape. The Naval Academy began teaching fire control in 1911.[27] With several years of neutrality before entering World War I, the U.S. Navy had time to catch up in technology. Between 1914 and 1917 it modernized its fleet, brought its fire control closer to British standards, and sowed seeds for its own control technologies. While the British still had superior equipment, they also had a more rigid hierarchy and a more traditional institutional culture. The Americans' relative backwardness made them more receptive to new solutions.

Within the navy, BuOrd became the driving force behind fire control in the United States. BuOrd had "cognizance," or responsibility, for the navy's guns, turrets, gun mounts, and fire control, among other things.[28] The navy bureaus were responsible for equipping and supplying the operating forces, but their powerful hold on resources earned the staff officers who manned the bureaus resentment from the line officers who ran the fleet. But unlike most of the other bureaus, BuOrd employed line officers, men from the fleet with gunnery credentials. This policy helped bring field experience to bear on weapons development and procurement, and it lessened the tensions between line and staff that plagued the bureau system. It also had a disadvantage: the officers of BuOrd, who rotated between sea duty and staff jobs, did not have engineering or management expertise in the design and construction of weapons as did staff officers in other bureaus. The officers of BuOrd, then, while knowledgeable about operations, had to rely on outside sources of technical skill to implement their ideas. They turned to small companies that specialized in engineering and manufacturing.

This problem became particularly critical during World War I, a time of both rapid expansion and technological change. During the war the U.S. Navy grew from 50,000 people to more than 500,000, and BuOrd expanded by a similar factor of ten. The fire control section went from one officer and one clerk to seven officers and eight support staff. Commander F. C. Martin headed the section until July 1917, when he was replaced by Commander Wilbur R. Van Auken, who remained until nearly the end of the war, when Commander William R. Furlong took over.[29] In later years, gun club officers acquired a reputation of technological conservatism because of their adherence to the battleship, but in the realm of fire control they innovated during the interwar period, both in technology and organization.

## Sperry Enters the Field

When the navy began to implement director firing, Elmer Sperry saw an opportunity. The 1905 fire control board had suggested replacing voice communications for gunnery data with visual indicators, and Sperry realized that a solution lay in his gyrocompass repeaters. The same devices that transmitted compass readings from the master gyro to its repeaters could also transmit fire control information from the director to the plotting room and the turrets. Building on this idea, in 1914 Sperry introduced a set of data transmitters for fire control based on the gyrocompass repeaters. The new instruments sent data for target bearing and turret train to and from the plotting room.[30]

To develop the repeater, Sperry Gyroscope relied on its technical skills and its relationship with the navy. It also had something the navy did not have: access to the technical details of British fire control. In 1913 Reginald Gillmor and Tom Morgan went to London, where they founded Sperry Gyroscope's British subsidiary, the Sperry Gyroscope Company Ltd. From England, Gillmor corresponded regularly with Elmer Sperry. The inventor frequently asked Gillmor's opinions, and he responded with detailed reports that Sperry then distributed to his naval contacts in the United States. Gillmor conducted a technical survey of European fire control systems and sent it to Sperry. Gillmor also sent Sperry details of Pollen's work and recommended that Sperry Gyroscope license and manufacture Pollen's system (it never did, and Pollen believed that Sperry Gyroscope stole his invention). In 1916 Gillmor reported about Jutland and the role of the Sperry gyrocompass in the engagement.[31] Gillmor proved so knowledgeable in these matters and had developed so many contacts within the Admiralty that in April 1917 he returned to the U.S. Navy as flag secretary on the staff of Admiral Sims, who was then serving as the American naval liaison in London. A twenty-four-page memo from Gillmor dated 1 August 1916 compared the tactical and technical issues of British fire control with those of American fire control with a clarity unsurpassed in any BuOrd documents of the time. Sperry passed the memo on to the bureau.[32] Through what Elmer Sperry called "channels which insured their freedom from censorship" Gillmor transferred fire control technology from the Royal Navy to Sperry Gyroscope, and only then to the U.S. Navy.[33]

## The Sperry Fire Control System

Gillmor's communications, combined with Sperry's own American navy contacts, gave Sperry Gyroscope an initial lead among American companies adapting the technology to the U.S. Navy.[34] In 1916 the company introduced the Sperry Fire Control System, which included gyrocompass repeaters adapted for fire control information, along with a new device: a central plotting machine designed and patented by Hannibal Ford (Fig. 2.3).[35] Ford's invention, called the Battle Tracer, plotted on paper the observed course of both the "own ship" (the ship on which it was installed) and the target ship, allowing a gunnery officer to make the prediction on paper and read the bearing to train his guns. Although less sophisticated than the Pollen system in its

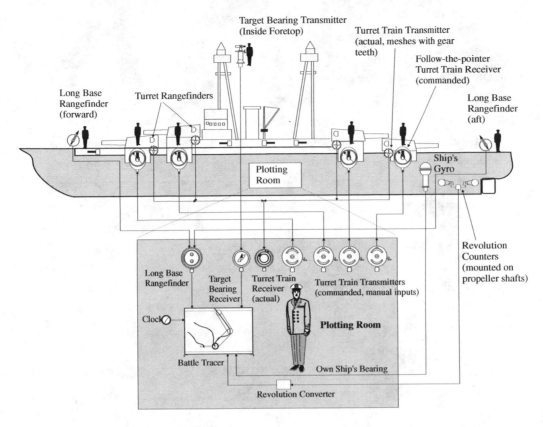

Fig. 2.3. Layout of Sperry Battle Tracer and Sperry Gun Fire Control System, ca. 1916. Drawing by the author based on Sperry technical literature.

firing solution, the Battle Tracer used the improved Sperry electrical transmissions. Its mathematics were not complicated, but the inputs came from varying sources and in varying forms. Hence the Sperry Battle Tracer and the Sperry Fire Control System integrated a diverse array of data into a single graphical representation of the field of battle. The U.S. Navy liked the Sperry system because it was lighter, less bulky, and simpler than Pollen's and also because it was of domestic origin. Stung by the loss of German optics imports, the navy sought to develop domestic sources for all fire control components.[36]

The Battle Tracer itself was an instrument of writing; it not only calculated and integrated on paper but also recorded the battle as a plot. The recording furnished "practically a bird's eye view of the engagement and maneuvers involved," and it monitored human performance, revealing carelessness or incompetence.[37] As the central part of a system for director fire, it was "designed to concentrate the control of all gunfire at one point, causing the entire battery of the ship to operate as a single unit."[38] The gunnery officer in the plotting room read the firing solution off the Battle Tracer and manually entered the data into a series of transmitters.

These values then appeared on similar receivers in the turrets, where the gunners saw the "desired" position in the form of an arrow and the "actual" turret position as a cartoon of the turret on the dial. Their only job, then, was to "follow the pointer" and bring the two dials into coincidence. This operation, also known as *pointer matching*, transformed the task of aiming the gun to one of matching visual indicators. It formed a primary task for human operators in control systems of varying types up through the end of World War II. (Pointer matching also became the basis for radar operators' method of entering radar data into control systems, known as *pip matching*.)[39] Because their job was to reduce the error between two quantities to zero, Sperry would call such operators "manual servomechanisms."

In early 1915 Sperry Gyroscope installed a Battle Tracer aboard the battleships *Utah*, *New York*, and *Arkansas* for testing, at no cost to the government. The ships employed the tracer during maneuvers in 1915–16. Since the device primarily tracked the course of the own ship, with only limited ability to track a target, the fleet found it more useful for navigation than for fire control. In general, they saw it as a "dead reckoning instrument . . . of the greatest assistance as an aid to navigation."[40] The commander of the *New York* noted that the Battle Tracer was unlikely to remain in operation during a battle due to the exposed nature of the optical instruments and that "the apparatus requires an attendant to keep it from running off the table." Still, the *New York* did use the device for fire control exercises and was able to obtain from it the course and speed of a target ship.[41] The general consensus among the ships employing the Battle Tracer was that it had potential but needed more engineering. Officially, BuOrd was less interested in the Battle Tracer itself than in the transmission instruments and the "follow-the-pointer in train" system. By the end of the war the navy had bought only 20 Battle Tracers for its battleships but several hundred of Sperry's data transmission devices (though more of these devices than Battle Tracers were required per ship) (Fig. 2.4).[42]

Sperry's system dealt only with target bearing and guns' train (i.e., rotation), not with target range or gun elevation. To overcome that limitation, in 1915 the commander of the *New York* suggested developing "a range keeper which would utilize the course and speed of the enemy," and the following year he again recommended devices to automatically track the range of the target. Gillmor made a similar suggestion

Fig. 2.4. Sperry Gun Fire Control System, 1916, a collection of apparatus, not a diagram of information flow. Courtesy of Hagley Museum and Library.

from England in his 1916 memo to Elmer Sperry.[43] BuOrd not only agreed with these suggestions, it had already started a program to build this new "range keeper." But now Sperry had competition that would grow to dwarf Sperry's efforts in fire control. Hannibal Ford had started his own firm.

### The Ford Instrument Company

Hannibal Ford left Sperry Gyroscope in 1914, immediately after designing the Battle Tracer. The following year, with the investors Jules Breuchaud and J. B. Goldsboro and $50,000 in capital, he organized the Ford Marine Appliance Corporation. Later accounts describe Ford's aim for the new company as the exclusive design of fire control instruments. Documents from the company itself, however, suggest that it originally intended to manufacture the Carrie Gyro Compass, a British device, to compete with Sperry Gyroscope in its core business (although a gyrocompass could be considered a fire control instrument).[44] The navy had become nervous about Sperry Gyroscope's American monopoly on gyrocompasses and wanted Ford to compete in that arena. The navy's desires, however, and its confidence in Ford's skills, soon changed these plans.

By 1915 the navy was installing Sperry Gyroscope's follow-the-pointer-in-train instruments throughout the fleet and employing the British range clock to track the changing range of a target.[45] Recall that this clock had the fatal flaw of requiring a manual adjustment whenever the rate of change of range was changing. To solve this problem, at the urging of the fleet officers who had tested the Battle Tracer, the navy requested proposals from both Ford Marine Appliance and Sperry Gyroscope "to develop a more efficient method of maintaining the range in action than at the present time."[46] In May 1915 both companies submitted proposals, Sperry for a "Range Clock," and Ford for a "Range and Deflection Predictor." Sperry and Ford did not become antagonists; rather, they engaged in a healthy competition, licensing each other's patents and making systems that worked together. Elmer Sperry loudly voiced his opinions when he thought his patents were violated, but no documents contain a bad word about Hannibal Ford, betraying only a mild jealousy of his success.[47]

BuOrd chose to purchase one of the Sperry Gyroscope and one of the Ford Marine Appliance machines "to encourage competition in working out future development along this line."[48] Ford Marine Appliance quickly abandoned its gyrocompass plans and regrouped in response to the navy's interest in range predictors. In late 1915 a new firm, the Ford Instrument Company, with $250,000 capital, absorbed Ford Marine Appliance. In its official announcement the new company stated that it was dedicating itself to "the inventions of Mr. Ford, in the line of scientific instruments and automatic machines involving mathematical and technical problems, intricate mechanism, epicyclic gearing, electrical devices, etc."—technical euphemisms for fire control. Signifying its expertise in precision mechanisms, the new company built its logo around a differential gear (fig. 2.5). Breuchaud remained as president, and Goldsboro as treasurer, with additional management, "men of experience in large financial and engineering operations." Hannibal Ford became vice president and general manager, reflecting his interest in the daily engineering and production of the company rather than in managing the business. The company established its factory and headquarters at 80 Lafayette Street in New York City.[49]

**SCIENTIFIC APPARATUS
AUTOMATIC MACHINES
VARIABLE SPEED DRIVE
ELECTRICAL DEVICES**

Fig. 2.5. The Ford Instrument Company logo, based on a differential gear.

In May 1916 reports trickling in from the Battle of Jutland (largely through Sperry's Gillmor) underscored the need for a device that could continuously track the changing rate of change of range. That same month, both Sperry Gyroscope and Ford Instrument demonstrated their new range predictors to the navy. According to the naval inspector, the Ford device, called the "Rangekeeper," was "entirely successful," whereas Sperry Gyroscope's machine required more work.[50] Tests showed the Ford Rangekeeper to be "far superior to present methods" of tracking target range.[51]

Ford's rangekeeper embodied a more sophisticated understanding of the problem than did his previous project, the Battle Tracer. Indeed the origin of his new insight, which closely matched Pollen's, is unclear; it was probably a combination of his own experience, Gillmor's memos, and the specifications in the navy's proposal request, which does not survive in BuOrd or Ford Instrument records. Nonetheless, in July 1916 Ford personally delivered his prototype rangekeeper to the battleship *Texas*, where, as described earlier, it was favorably evaluated by a board of gunnery lieutenants. A significant order followed, and Ford Instrument began manufacturing rangekeepers.

Sperry Gyroscope enjoyed no such luck. Its device was not ready for testing until August 1916, when the board's report on the Ford Instrument machine had already been issued. Sperry Gyroscope did not demonstrate the device aboard ship until that December, by which time contracts and production design for Ford Instrument's machine were well under way.[52] No further mention of the device exists in BuOrd records or the Sperry Gyroscope archives.

### The Ford Rangekeeper

Ford's device itself incorporated much of the function of British fire control instruments, but with a mathematical precision, mechanical elegance, and creative controls unmatched at the time. In the interest of reliability, the rangekeeper's calculations were almost entirely mechanical; the only electrical input came from the ship's gyrocompass. Ford made a point of the rangekeeper's only semi-automated calculation. He astutely perceived the technical core of the gunnery problem and automated the rangekeeper only to the degree necessary. (Perhaps he chose not to build a highly automated device in order to stay within his company's infant capability.) Unlike the Pollen system, which alienated officers by appearing to automate them out, the Ford Rangekeeper struck a mix of automation and operator control that gave it credibility with the service.

Still, the Ford Rangekeeper's success in the fleet did not derive from a radical technical superiority. Comparisons by both American and British observers found the two machines about equally accurate.[53] They found the smaller and lighter Ford machine much harder to use, having few of the automated features of the Pollen, and recommended that "no one should be allowed to touch the Ford excepting in the presence of an officer." Whereas the more automated Pollen machine broadened the pool of potential users, the Ford restricted it to officers. Nonetheless, besides the artifact itself, a key feature of the Ford Rangekeeper made it attractive to BuOrd: it was domestic technology, produced by a supplier under total control of the navy. Like Pollen, Ford was an outsider, but the U.S. Navy, unlike its British counterpart, was able to craft an arrangement to safely harness the independent entrepreneur to naval concerns.

### Feedback and Integration in the Ford Rangekeeper

Appendix A lays out the detailed functions of the Ford Rangekeeper. In short, the *own ship* does the firing (and contains the rangekeeper), shooting its guns at the *target ship*. If a shell fired from the own ship takes a minute to reach the target, then the target's course must be estimated for a minute into the future when the gun is fired. Similarly, if the target ship's course and speed are known, it can be continually tracked even when it is not under visual observation, for example if it becomes obscured by fog or smoke. Thus the prediction problem comes down to accurately estimating the course and speed of the target.

The basic problem is to use a series of range and bearing measurements to estimate the course and speed of the target. Once known, the target course and speed can be extrapolated into the future to predict the target's future position and hence its range and bearing over time. These values can then be compared with actual range and bearing measurements to continually update the course and speed estimate. Two transformations are required: after taking in data as polar coordinates (range and bearing), the machine first transforms them into Cartesian coordinates ($x$ and $y$) to do the prediction and then transforms the answer back into polar coordinates. Put another way, the rangekeeper makes a model of the target's motion and then produces an answer from that model. The data taken in are the own ship's course, from a repeater compass, and the own ship's speed, which is entered manually from an instrument connected to the ship's propellers. In the director tower, officers use optical instruments to measure the target's range and bearing and transmit them down to the plotting room, where they are manually entered into the rangekeeper. In addition to these data, a gunnery officer enters initial estimates of target course, speed, and range. The basic operation of the machine, then, consists in "tuning" these guesses based on observed data until they converge on a stable solution. The rangekeeper then calculates the *present range,* where the target is now, and the *advance range,* where the target will be at some time in the future.

Consider, for example, when the rangekeeper has the correct course and speed for the target. By integration, it calculates the expected range and bearing of the target as they change with time, while observers in the director measure the actual range and bearing. If the calculated values do not match the observed values, then the estimated course and speed of the enemy are in error and require adjustment.[54] This cycle continues until the rangekeeper's calculations match the observed data. Ideally, this process (which today we would call a feedback loop) converges on a solution: the accurate course and speed of the enemy, which may be used to predict its future position and thus to set the guns.

This cycle, of course, assumes that the target's course and speed remain constant long enough for a solution. The assumption will hold only for short periods of time during a naval engagement. The Ford Rangekeeper, however, really comes into its own when the target changes course. Say that the system has settled on a proper course and speed, with the dials that indicate the calculated course and speed matching the observed data (Fig. 2.6). When the target changes course, those dials will diverge, immediately signaling to the operator that a change has occurred. The old course and speed then serve as the initial estimates for the new solution, and the system converges on a new solution fairly quickly. Thus the rangekeeper and the operator can "track" the target as it changes direction, continuously solving for its parameters.

This cycle of measurement and correction was not radically different from the British systems, a fact not lost on the gunnery officers of the time.[55] But the Ford Rangekeeper added a crucial new ingredient: mathematically accurate integration, accomplished by a new mechanical component, the Ford integrator. In the simplest type of integrator, the wheel-and-disk type, a disk rotates at a constant speed, providing the time variable for integration. The wheel contacts the disk at a right angle and rotates at a speed proportional to its distance from the center. If the wheel is close to the center, it rotates relatively slowly; as it moves toward the outside it moves ever faster. The function to be integrated, then, determines the distance of the wheel from the

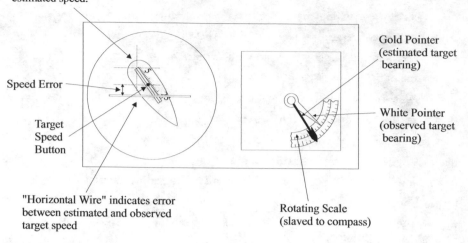

Figure of target ship represents estimated course, button indicates estimated speed.

Gold Pointer (estimated target bearing)

Speed Error

White Pointer (observed target bearing)

Target Speed Button

"Horizontal Wire" indicates error between estimated and observed target speed

Rotating Scale (slaved to compass)

Fig. 2.6. Subset of dials on Ford Rangekeeper

center, and the rotation of the wheel reflects the integral of that function. This setup works like a continuously variable transmission and is often referred to as a *variable speed drive* (Fig. 2.7).

The wheel-and-disk integrator, however, has a problem: when driving a mechanical load, the wheel slips as it rolls over the disk, compromising the accuracy of the device. Yet for the device to be useful in a calculating machine, it needs to drive other mechanisms. The Pollen machine used a ball-and-cylinder integrator, invented by James Thompson, Lord Kelvin's brother. In this device only gravity presses the ball to the disk, and the weight is not sufficient to prevent slipping.[56]

Hannibal Ford improved the Thompson integrator to eliminate this shortcoming. He added another ball, which further reduced slippage, and allowed tight springs on the cylinder bracket to hold the two balls firmly in place. This device could perform highly accurate mechanical integration and supply sufficient force to drive additional mechanisms connected to its output.[57] In fact, the two-ball integrator functioned like an amplifier. It added energy to the system (from the spinning wheel) and hence prevented the next stage of calculation from loading the previous one (and hence compromising its accuracy). Like Sperry's phantom and repeater compasses (and like Sperry Gyroscope's manual servomechanisms, Harold Black's amplifiers, and Harold Hazen's servos, which we shall examine in subsequent chapters), the Ford integrator

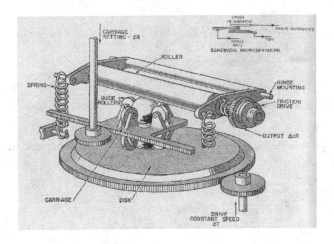

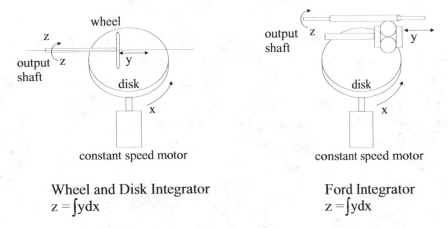

Wheel and Disk Integrator
$$z = \int y\,dx$$

Ford Integrator
$$z = \int y\,dx$$

Fig. 2.7. Wheel-and-disk integrator *(bottom left)* and Ford integrator *(top and bottom right)*. Each is equivalent to a continuously variable transmission. The rotating disk represents time, and the distance of the wheel or balls from the center of the disk represents the variable to be integrated. The wheel or balls then transmit the integral of that function to the roller, which outputs it as a shaft rotation. Reprinted from U.S. Navy, Bureau of Naval Personnel, *Naval Ordnance and Gunnery*.

renewed the data inside the machine as it calculated. This amplification decoupled the representations of the world from the mechanisms that manipulated them, thus enabling more complex mechanical calculations without loading and loss of accuracy. The integrator remained a central component in mechanical computers through World War II; some early digital computers were even referred to as "electronic integrators."

### The Secret Fire Control Design Section of the U.S. Navy

After the initial rangekeeper tests on the *Texas* in the summer of 1916 Ford Instrument geared up for production, while the prototype remained on the ship for further testing. The navy ordered more than 25 rangekeepers at a cost of $8,000 each, intending to install 4–6 units in each of its battleships (as primaries and backups). Soon an early production machine replaced the *Texas* prototype, which was transferred to the Naval Academy for use in training. (It is probably the machine now in the collection of the Naval Academy museum.) In 1917 installation began on first-line battleships, including the *New York,* the *Wyoming,* the *North Dakota,* the *Pennsylvania,* and the *Arizona.*

Ford Instrument spent the winter of 1916–17 putting the rangekeeper into production, gearing up as a manufacturer of mechanical calculators.[58] The following year the company more than quadrupled in size, from fewer than 50 workers to more than 200. In the summer of 1917 the company introduced a low-cost rangekeeper, the Mark 2, for $800. This device, nicknamed the "Baby Ford," included only the initial stages of the calculation and had no integration, no prediction, and no feedback or correction cycle. Production of the Mark 2 began in August 1917, and because of its simplicity and low cost the navy ordered Baby Fords for all its battleships, cruisers, destroyers, and gunboats. Over time a dizzying array of "mark" numbers for basic models and "mod" numbers for variations emerged, even though the function of rangekeepers changed very little. By April 1918 the navy had ordered 67 Mark 1 rangekeepers and more than 650 Mark 2 Baby Fords. Less than two years after building its original prototype, Ford Instrument was producing ten different types of equipment, including the two models of rangekeepers, several data transmitters, time-of-flight clocks, instruments, and transmitters.[59] Numerous other devices were in design or development.

The rangekeepers were high-precision mechanical instruments requiring skilled operators. Producing them required packing watchlike precision into large cabinets designed to operate under hostile conditions; this was no easy task, especially for a new company (Fig. 2.8). Ford Instrument engaged in flexible, specialty production, employing practices more akin to those of the jewelry and printing companies in surrounding New York than to the mass-production model adopted by the larger, more visible Ford Motor Company. Usually only a few machines would be made exactly alike. Assembly lines with unskilled labor would not work for these variable, low-volume products. It took seven hours simply to test the Mark 1 rangekeeper. While such specialty producers typically operated within a web of close industrial relationships, Ford Instrument had to build its machines in secrecy.[60]

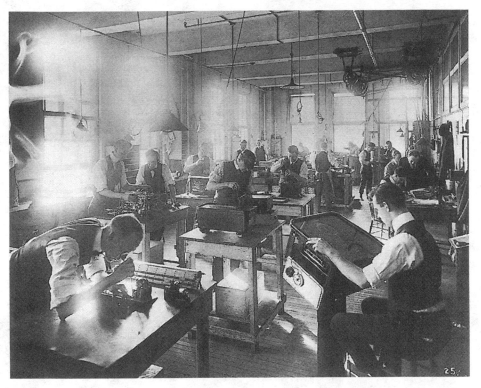

Fig. 2.8. Rangekeeper production at Ford Instrument Company, 1922. A Mark 1 rangekeeper can be seen in the right foreground, and the naval inspector's office is in the background on the right. Courtesy of Hagley Museum and Library.

Because of the navy's obsession with secrecy, manufacturing the rangekeepers also required a novel relationship between Ford Instrument and BuOrd, one that both fostered and limited the growth of the technology. Ford Instrument enjoyed a privileged position in fire control, but at the expense of a wider business. It was not allowed to sell products abroad nor even to describe them to outsiders. In exchange for this exclusivity, the company gained both profits and a favored status with its customer, as well as a corporate identity: "the secret Fire Control Design Section of the U.S. Navy."[61]

Indeed Ford Instrument enjoyed daily, immediate contact with gunnery officers. They provided not only feedback on the company's products but also a source of technical information. As was standard practice for navy contractors, the company allocated space in its factory for the naval inspector of ordnance, in this case Benjamin B. McCormick, who was also the naval inspector at Sperry Gyroscope. Before long the companies had enough business that separate inspectors were assigned, but McCormick surely passed on some knowledge from one firm to the other, if only by making comparisons. Indeed Ford Instrument credited McCormick with helping it refine production methods.[62] In addition, the navy sent gunnery officers and machinists to Ford Instrument to learn about the principles, operation, and repair of the Ford Rangekeepers. Before long Ford Instrument set up formal courses of instruction on its products.[63] Eventually the navy took over instruction of the courses, but they were still held at the Ford Instrument facility.

For new development work, the navy formulated requirements and Ford Instrument responded with a design. If BuOrd accepted the design, the company would manufacture the machine exclusively for them. To make this special institutional coupling work, BuOrd had to carefully define and control the nature of technical knowledge. Where did fire control technology originate and reside? In the tactical and strategic requirements specified by line officers? In the ability of engineers to turn those requirements into practical mechanisms? Or in the manufacturing knowledge required to produce them?

At first the navy and its contractors gave the same answer: knowledge was imported from outside the country. In 1918 those origins were swept under the rug when Ford Instrument expressed its concern to the navy that its rangekeepers might violate patents for Pollen's system, as both Sperry and Pollen believed they did. Assistant Secretary Franklin Roosevelt responded by guaranteeing "to hold and save you harmless against any and all suits" brought for infringing patents on British fire control.[64] The navy virtually stole technology from abroad and protected the companies that produced it from legal action.

At other times the location of technique proved more contentious. Despite the modest effort of the navy's Research and Development Board during World War I, which had little or no effect on fire control, BuOrd did not see research and development as a specific activity. When the navy wanted a new machine, it simply specified one and ordered it from a company. BuOrd then did what it pleased with the technology, including awarding production to other firms. By contrast, the contractors saw navy specifications as broad outlines for work that required engineering talent to implement in a working device. Companies thus claimed for their own the ideas they developed for navy specifications and challenged the navy's right to let production contracts to other manufacturers. BuOrd's contractors had numerous disputes over intellectual property, especially when one company went through a long development process only to see BuOrd award production contracts to other vendors. In the late 1930s BuOrd instituted a policy whereby the government could license inventions without paying royalties, but only within the sphere of its clique of contractors.[65]

Secrecy hampered Ford Instrument's ability to profit from its inventions beyond the confines of BuOrd. Early on, the company requested permission to make foreign sales. In response, BuOrd chief Ralph Earle ordered "that you do not disclose to any one even the fact that you are making rangekeepers, and that you do not dispose of similar instruments to other governments."[66] In a similar vein, handling patents for secret technology raised difficult contradictions. By law patents require publication, so the bureau initially insisted that no fire control devices could be patented, because of the need for secrecy. Hannibal Ford, in fact, never patented the original rangekeeper (which suggests that his idea may have had foreign influences), but only discrete parts of its computing mechanisms, usually under nondescript names like "mechanical movement," "calculating instrument," or "control system." To maintain secrecy, patent approvals were often delayed until the mechanisms were effectively declassified, sometimes more than a decade.[67] Through its fire control contractors and under sometimes conflicting conditions, BuOrd created and bought technical expertise from private companies for its unique mission.

Fire control was not the only arena where BuOrd created these special ties. Ten years later BuOrd handled bombsight development and manufacture in much the same way as fire control. The primary suppliers were Sperry Gyroscope and another captive contractor formerly at Sperry. Like Hannibal Ford, Carl Norden had been an early employee at Sperry Gyroscope, and he had left about the same time as Ford to consult for the navy. In 1927 Norden and a partner set up a company, Carl Norden, Inc., to manufacture bombsights (at 80 Lafayette Street in New York, the same building where Ford Instrument had started). Bombsights, like rangekeepers, were delicate and precise mechanical calculators that had to work under demanding conditions. They became the oxymoronic "famous secret weapons" of World War II. The navy gave production contracts to Norden without competitive bidding, and Sperry complained that Norden had stolen the technology. Like Ford Instrument, Norden's company was virtually a private extension of the navy. "Our business policy was to be controlled by the Bureau [BuOrd]," Norden's partner, Ted Barth, wrote, "and we were to function as a sort of subdivision of the Bureau as far as the bombsight problem was concerned."[68] BuOrd's relationships with Ford Instrument and Carl Norden, Inc. (and, as we shall see, with Arma Engineering) represented a concerted effort to found and foster captive contractors to make new technology in what amounted to private arsenals.[69] Sperry Gyroscope, though it started out as BuOrd's main fire control contractor, could not fill this constricting, if privileged, role.

### Sperry's Fate in Fire Control

For several years Ford's machines worked within Sperry's larger systems. By 1920 the U.S. Navy had installed fire control systems from Sperry Gyroscope on 19 dreadnoughts, 11 second-line battleships, and 9 armored cruisers. The equipment included hundreds of data transmitters, receivers, and indicators, most of them supplying data to Ford Rangekeepers.[70] Yet despite Sperry Gyroscope's strong initial position, Ford and his company eclipsed Sperry's dominance. The rise of Ford Instrument as BuOrd's favored fire control manufacturer stemmed not only from Hannibal Ford's talents and his company's production abilities but also from technical troubles with Sperry instruments, Sperry's own organizational problems, and its unwillingness to be confined to serving BuOrd.

In the original 1916 competition Sperry Gyroscope's range clock lost out to Ford's machine, but Elmer Sperry recognized the importance of the rangekeeper and still wished to compete. "The navy are obsessed by the [Ford] range clocks," he wrote to Gillmor in England, and he considered licensing Pollen's machine. Despite Pollen's urging, Elmer Sperry distrusted him, and Sperry Gyroscope never licensed Pollen's Argo Clock, believing it to be inferior to his own machine.[71]

Yet Sperry's confidence in his own products was misplaced, for they were beset with technical problems. Through 1917 and into 1918 myriad correspondence from the fleet to BuOrd, as well as directly to the company, displayed consistent frustration and disappointment with Sperry Gyroscope's fire control equipment and the company's handling of the problems. Much of the equipment was delayed, and much was returned for repair due to errors in construction and installation.[72]

BuOrd's experience with the Battle Tracer typified its relationship with Sperry Gyroscope. The fleet tested the device and deemed it potentially useful and worthy of further work, but Sperry Gyroscope did not respond. In a scathing (and typical) opinion, one commander suggested that naval officers contributed more to the technology than the company's engineers did:

The designers of the apparatus turned out by the Sperry Gyroscope Company are trial and error men. So far as is known not a single instrument or appliance that has so far been turned out by this company has been thoroughly satisfactory in its original form. The gyrocompass itself, manufactured by this company, was at first unsuccessful and has been subject to repeated modifications leading to its improvement as suggested by naval officers as a result of their experience with this instrument. The same is true of the target bearing transmitters, the turret target indicators, the optical range transmitters, the multiple turret indicators, the plotting indicators, and finally the Battle Tracer.[73]

Naval officers did not expect technical perfection straight off, and they expressed a desire to work with the company to improve its instruments. Officers reporting problems with Sperry equipment frequently stated that they believed it could be made to work and that its potential usefulness was worth the effort. Their attitude, however, was predicated on a perception of the company's willingness to cooperate and on confidence that the technical problems were tractable. Not only did Sperry Gyroscope's reputation for cooperation erode but experience at sea increasingly showed that Sperry equipment had an insoluble flaw: the problem of synchronization.

The Sperry system transmitted data in a direct-current (DC), "step-by-step" mode, derived from the original compass repeaters. When the receiver was in a different position from the master transmitter, a switch started a motor and moved the receiver dial to bring it into line. This method transmitted relative, not absolute, position, requiring that the indicators and their transmitters be synchronized by pressing a button before they were used.[74] The trouble was, these on-off elements could not withstand firing of a nearby battery of 16-inch guns; the shock of a salvo usually knocked the system out of synchronization. Gunners had to continually reset the system to get accurate readings, an operation that was annoying at best and impossible under battle conditions. Similarly, the navy was gradually incorporating the ability to swap components between multiple subsystems. With the Sperry equipment, switching a data transmitter to a new set of receivers would require resynchronization.

In response to criticism, Sperry Gyroscope initially stonewalled, then submitted a number of stopgap solutions. A "Turret synchronizing system," for instance, provided a central button to synchronize the entire system. Elmer Sperry himself patented three "synchronous data transmitters" in 1919 and 1920 alone, and seven in the decade before his death in 1930.[75] Yet the complaints continued. "Every day at General Quarters the synchronizing of the instruments is checked," the commander of the *Pennsylvania* reported, "which takes considerable time." The navy asked the National Bureau of Standards (whom it frequently consulted on fire control matters) to examine the situation. "We can make synchronizing easier," bureau engineers reported, "but we cannot make the Sperry system self-synchronizing."[76] BuOrd continued to order Sperry instruments, but the problems did not cease; by the end of 1918 confidence in the Sperry apparatus had evaporated.

Indeed the company was in a managerial crisis, and its workers were threatening to strike. At Elmer Sperry's request, Reginald Gillmor left the navy and his post in England to take control of the company in New York.[77] Gillmor soon took the bold, perhaps desperate step of exposing his company's internal problems to the navy. Gillmor admitted that the company's "efficiency and its policies have not been such as to create confidence," attributing the situation to corrupt management, which he restructured. In an attempt to win back BuOrd's favor, perhaps with a tinge of envy at Ford Instrument's new position, Sperry Gyroscope too offered "to consider itself an auxiliary of the Government service."[78]

Gillmor's actions improved the management of Sperry Gyroscope, but the company's relationship with BuOrd had already suffered lasting damage, especially as the bureau had other reasons to feel uncomfortable about the contractor. Fire control made up only part of the company's business. With its gyrocompass and other control devices, Sperry Gyroscope had a dynamic and growing commercial operation, much of it as a monopoly. BuOrd did not object to commercial sales per se, but to the corporate culture required to support them, especially internationally. Sales organizations, marketing literature, and the general level of publicity required to promote industrial products fostered the very exchange of information that BuOrd sought to restrict. BuOrd did approve some sales to allied foreign navies, but it was less comfortable with more threatening rivals. Elmer Sperry, who admired Mussolini, was beginning a personal fascination with the Japanese that resulted in sharing American technology and supplying equipment to the Imperial Navy.[79] Japanese naval officers toured the Sperry plant and inspected the Battle Tracer. Though BuOrd had not purchased the rights to it and indeed had lost interest in the device, it still worried that the demonstration would reveal to the Japanese other parts of the fire control systems that operated in the American fleet. The Russian navy also adopted the Sperry gyrocompass, deeming the device so successful that the Sperry representative in Moscow was decorated by the czar.[80] Because of these and similar connections, BuOrd saw the company as an untrustworthy source of fire control technology.

Sperry Gyroscope's technical problems, its internal difficulties, and its troubled relationship with BuOrd finally caught up with the company. In 1920 it lost a key contract for fire control systems for the new battleships *Colorado* and *Maryland*. The company dropped out of the fire control business. It continued, however, to supply the navy with gyrocompasses and gyropilots, as well as naval searchlights and some smaller naval instruments (under the cognizance of other bureaus). Ford Instrument, by contrast, was highly responsive to the demands of its sole customer and its sole source of income. Ford Instrument had no salesmen, no foreign connections, no offshore plants; in fact, Ford Instrument had no public image at all.

### Fire Control in the 1920s: General Electric and Arma

Sperry's departure from fire control was part of a broad demobilization. With the end of World War I, military work of all types became scarce. The 1922 Washington Naval Treaty (and its 1930 London successor) set strict limits on the number and size of capital ships. Numerous ships on the drawing boards were canceled, and new building during the following decades was sporadic at best. Only three new battleships—the *Maryland*, the *Colorado*, and the *West Virginia*—were laid down between the end of World War I and 1937. Still, within that limited sphere, fire control played an important role and underwent new development. Treaty limitations simplistically measured battleship power by tonnage, but improved fire control could increase the striking power of a given vessel with little added weight. The treaty also did not limit the number of cruisers, and fire control could improve the effectiveness of those ships as well.[81] With its comparatively low cost, mass, and volume, fire control thus had high leverage for sea power (as did aircraft and torpedoes). The navy, with growing understanding of how to control development within its captive contractors, made the most of its limited resources to promote exclusive American leadership. As one path toward that leadership, the navy brought in a premier American technology company.

Though BuOrd had husbanded Ford Instrument in order to foster competition, the bureau's disenchantment with Sperry Gyroscope left it again with a single contractor. In 1918 the navy sought new industrial talent to help solve the synchronization problem and create a new generation of control systems. The new contractor, General Electric, had vast technical resources and an established reputation with the navy. G.E. was widely recognized as the leader in industrial research and was certainly on the cutting edge of electrical technology. Nonetheless, BuOrd's choice represented an established route, displaying the conservatism for which the bureau system was renowned. Rather than depend solely on small companies, with their attendant friction, instability, and unreliability, the navy sought familiar expertise. Beginning in 1920, then, G.E. brought to the problem a full research organization, solid electrical skills, and experience in manufacturing expensive, specialty goods, elements lacking at both Sperry Gyroscope and Ford Instrument. As a consequence, naval control systems became both more stable and more electrical.

The choice of G.E. also reflected the background of the new postwar head of the Fire Control Section, Lieutenant William R. Furlong. A 1905 graduate of the Naval Academy, Furlong had experience in the budding technology of electronics, having earned a master's degree in electrical and radio engineering at Columbia University (he would go on to head BuOrd from 1937 to 1941 and then to oversee the salvage efforts at Pearl Harbor). Having served as a gunnery officer during World War I, he brought an electrical perspective and a keen inventive eye to his work in the fleet. In early 1918 Furlong sailed with the British Grand Fleet to evaluate fire control. He then returned to the United States to head the fire control section of BuOrd. Full of new ideas, Furlong immediately began to staff his department with electrical engineers.[82]

### General Electric's Synchronous System

G.E. had no prior experience in fire control. Yet it had built a system for commanding the motion of the doors of the locks in the Panama Canal based on a technique for transmitting information in electric power stations. Much as a fire control system represented the field of battle in the plotting room, the canal had a central control room with a miniature model of the canal's locks, doors, and water levels that both represented and commanded the canal itself.[83] Upon learning of this system in mid-1918 BuOrd procured a motor from G.E. for testing, along with samples of a position indicator the company had designed. The devices were fully synchronous and seemed suitable for fire control. Still, G.E. had no engineers with experience in the technology; they would need navy direction. As BuOrd chief Van Auken recalled, "No citizen or private manufacturer at that time had sufficient knowledge of director firing, or the complete needs of fire control, as to initiate a system." But G.E. did have credibility with the navy. It supplied a host of components to naval shipyards, including the numerous electric motors required to run modern naval vessels. The *New Mexico*, launched in 1915, had been hailed as the "all electric ship," with everything from steering gear to ammunition hoists to kitchen appliances run by G.E. electric motors. Also, the period 1916–20 saw the brief heyday of G.E.'s advanced but short-lived turboelectric power drives for ship propulsion.[84]

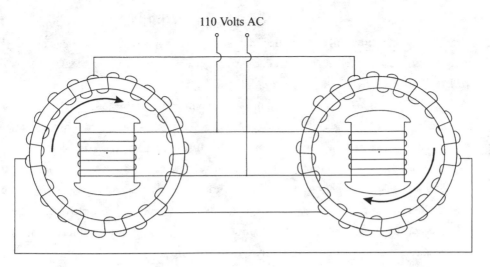

110 Volts AC

When this rotor rotates.....                    ...this rotor moves to the same place.

Fig. 2.9. The AC selsyn principle. Two dials, attached to the rotors, are synchronized at a distance with a three-wire connection and AC power. Reprinted from Hewlett, "Selsyn System of Position Indication," 210.

Furlong came across the sample synchronous device G.E. had sent to BuOrd. While in Europe, he learned that the Germans had used synchronous systems at Jutland. When he returned, Furlong immediately began development of a synchronous fire control system for the new battleships *West Virginia, Colorado,* and *Maryland.* For three years Furlong literally tutored G.E. engineer Edward Hewlett in fire control. Hewlett, the company's premier switch and switchboard designer from Schenectady, had designed the control system for the Panama Canal. Technical exchange between the navy and its contractors reduced to the interaction of these two men. They met several times a month; Furlong sketched ideas, and Hewlett designed them into a system. Under Furlong's tutelage, between 1918 and 1920 Hewlett developed a new fire control system built around the self-synchronous motor, or "Selsyn." Hewlett and his assistants designed and modeled range transmitters, bearing indicators, compass relays, directors, control towers, and a broad variety of devices and systems (Fig. 2.9).[85] Together, Furlong and Hewlett matched military experience to G.E.'s industrial expertise.

The G.E. self-synchronous system used alternating current (AC) rather than DC electrical signals to transmit rotational position over a distance through three wires. G.E. called this technology "Selsyn," but BuOrd copyrighted the name "synchro" to mean the same thing regardless of manufacturer.[86] The basic idea was not unique to G.E.; other companies, including Ford Instrument, had suggested similar devices. Selsyns, synchros, and their numerous derivatives became basic building blocks of control systems and remained so for decades. During World War II several companies, including Ford Instrument, Arma, and G.E., manufactured these devices in large numbers for a wide variety of control applications (for example, rotating radar sets).

BuOrd now favored G.E. with systems work. In 1920 the bureau contracted with G.E. instead of Sperry Gyroscope for fire control for the battleships *Colorado* and *Maryland* (violating Sperry's patents in the process).[87] By the end of that year G.E. had no less than six types of gun directors in design and a number of other fire control projects under way, all using synchronous transmission (although otherwise similar to earlier systems). Ford Instrument, as a subcontractor, provided rangekeepers and other instruments. During the 1920s G.E. converted earlier fire control systems to fully synchronous operation and also built directors for the smaller secondary, or "broadside," batteries on *West Virginia, Colorado,* and *Maryland.*[88]

Like the Ford integrator, G.E.'s synchronous systems succeeded because they transmitted data without corruption, enabling ever more complex calculation and routing. G.E. brought two other innovations from electric power into fire control that had similar effects: power drives and switching. Power drives coupled the output of the rangekeepers directly to the motion of the guns. Switching allowed the entire system to be programmed. In 1922 Ernst F. Alexanderson, the G.E. engineer who had pioneered high-frequency alternators and other fields of power electronics, became involved in Hewlett's fire control work. Alexanderson replaced human follow-the-pointer operators with electronics, making circuits that allowed vacuum tube amplifiers to drive motors. These allowed instruments, even delicate calculators, to move the turrets directly, without human intervention, and to stop and hold precise positions. In the course of this work G.E. came across a fundamental problem in control: moving large machinery with small signals easily leads to oscillation (called "hunting," as the feedback loop hunts for an equilibrium point) and instability. Much of G.E.'s research consequently related to "antihunt" devices for stabilizing these servo loops.[89]

In 1930 Alexanderson and G.E. introduced "thyratron" control for high power motors. The thyratron, whose name was derived from the Greek word for "door," a reference to opening the door to high power, was an electrical tube, but unlike a traditional vacuum tube, it contained inert gasses. It could not amplify small analog signals as a standard tube could, but the thyratron could switch high currents on and off with small electrical inputs. Later in the thirties Alexanderson and his group introduced the "amplidyne," a power amplifier based on a dynamo that was also capable of amplifying small signals into immense amounts of electrical power. The amplidyne was incorporated as an electric drive into navy turrets and main battery directors, while thyratrons drove motors in smaller applications.[90] G.E.'s power drives automated the articulation of the control system, converting low-power information generated by rangekeepers into the high-power signals required to move guns.

Where power drives coupled the rangekeepers to articulated outputs, G.E.'s switchboards routed gunnery data between system elements. Warships now had two conning towers, each with its own gun director, spotting telescopes, and rangefinder. Turrets also contained their own rangefinders and instruments. The plotting room usually had multiple rangekeepers, and eventually an entire second plotting room was added with wiring physically and electrically separate from the main control room. Switchboards covered the walls of the plotting room and connected individual elements to common "busses." In a director or a turret, different signals could be supplied or accessed by connecting equipment to these busses. By changing the switch settings, the system could be given a new configuration or optimized for different situations. For example, with two directors the main battery could split between fore and aft turrets and engage two targets simultaneously, an arrangement known as *divided fire*. The switchboard provided for entering spotting corrections received by radio from a spotting airplane or other forward observer, known as *indirect fire*.[91] Switching also made the system robust to the loss of any of its individual elements, whether as a result of battle damage or of simple mechanical failure (Fig. 2.10). Flexibility meant reliability. And such systems required G.E.'s synchronous data transmission: the Sperry step-by-step type would need to resynchronize every time a switch was thrown, but the selsyns came into line automatically.

Switchboards made naval fire control into what today we would call a hybrid analog and digital system. The rangekeepers, as well as the various telescopes and rangefinders, produced continuous, smooth data that represented the world as physical quantities. Switching, by contrast, is a discrete, logical operation. Switchboards, by routing the analog signals to different places, reconfigured the system for different data flows. Fire control manuals contained tables of switchboard setups that operators dialed in for a particular configuration, comparable to "programming" the system.[92] As we shall see, this hybrid system paralleled others of the same period: the tele-

Fig. 2.10. Operating a switchboard in a plotting room. Courtesy of Naval Historical Center, Washington, D.C.

phone network, which routed analog phone calls through a matrix of switching relays, and Vannevar Bush's Rockefeller Differential Analyzer, which combined analog computing elements through a set of telephone relays (and which Claude Shannon used to show the equivalence of switching and binary logic). With switchboards, fire control became a generalized set of instruments, data transmitters, calculators, and

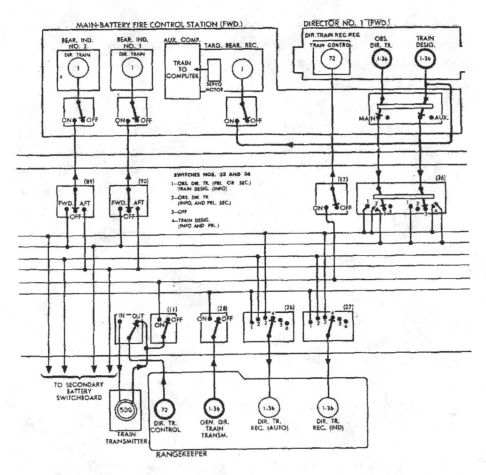

Fig. 2.11. Subset of switchboard signal routing in a fire control system. The fire control station and the director have local switchboards and connect to common wiring busses that travel throughout the ship. The main switchboard in the plotting room routes signals from the busses into and out of the rangekeeper. The subset shown here is a small fragment of the total system aboard a cruiser. Reprinted from U.S. Navy, Bureau of Ordnance, *Main Battery Fire Control System CA 68, 72, and 122 Class Ships,* OP 1387 (Washington, D.C., 1948), 47.

actuators. As a control center, the plotting room processed information, switching it and routing it to its destination (Fig. 2.11).

Drawings and diagrams reflected this system evolution. Documentation for the early machines tended to be strictly mechanical, showing the parts themselves and how they physically mounted and connected. It is difficult to find a complete system diagram of an American fire control system from World War I (none exist for the Sperry system). Through the interwar years, however, drawings and manuals began to portray flows of information. Fire control drawings of 1930 portrayed data flows yet retained images of mechanisms as icons of mathematical operations, even including human operators in the mix. By World War II fire control appeared in system- or block-diagram formats, which are so familiar today. Computing elements appear as black boxes (though sometimes in the shape of the machines themselves), relevant for their inputs and outputs but independent of their internal workings. Arrows represent data flow from one box to another. Diagrams produced during World War II pushed these graphical ideas even further, with complex, multilayered flows employing different colors for different types of data. Technologies like integrators, amplifiers, servos, and switchboards all enabled data to flow through systems and across human-machine boundaries; technical documentation both reflected and crystallized that facility.

## The Arma Engineering Company

As Ford Instrument dedicated itself to fire control, the navy still had no second supplier of gyrocompasses, so it continued to depend on Sperry Gyroscope's monopoly. BuOrd then turned to another young company formed by an ex-Sperry employee. The Arma Engineering Company was formed as a partnership between its two founders, David H. Mahood and Arthur P. Davis; the name *Arma* was formed by taking the first two letters of *Arthur* and the first two of *Mahood.* Mahood worked in the fire control division of the Sperry Company during World War I and then as the chief civilian in the Ship's Electrical Apparatus Testing Laboratory at the Brooklyn Navy Yard. This facility handled virtually all the new equipment emerging from both Sperry Gyroscope and Ford Instrument for installation on navy vessels, so Mahood's position brought him into intimate contact with the control systems of the time, as well as their problems. Davis, a young, self-educated engineer, had built switchboards at G.E. in Schenectady. Davis and Mahood formed the Arma Engineering Company on 30 January 1918 with capital of about $1,500. Like Ford, Arma was originally located in Manhattan, but it later moved to larger quarters in Brooklyn and eventually to Long Island.[93]

Like Sperry, Arma focused on gyroscopes. At the end of World War I the navy captured the design for the German Anschütz gyroscopic compass, which itself had inspired Elmer Sperry's original design. The navy gave the Anschütz to Arma, which then built a business supplying gyroscopic control and stabilization systems, searchlights, and other equipment for several navy bureaus. Soon the company applied its energies to fire control. In 1924 Arma competed with G.E. to build gun directors for the new aircraft carriers *Lexington* and *Saratoga*.[94] Arma, like Ford Instrument, became a captive supplier, fully responsive to the needs of BuOrd.

Arma resembled Sperry Gyroscope in miniature, optimized for BuOrd's secret work. Founder D. H. Mahood contrasted Arma to its larger rival. Having responded to a request by the navy "to enter fields of work which were then in the hands of a monopoly [i.e., Sperry]," Mahood wrote, "we have had no other customer but the U.S. Navy Department, have never sought any foreign or commercial contracts and have maintained the fullest secrecy which would have been impossible otherwise." As with Ford Instrument, and Carl Norden, Inc., the line between military bureau and private company became blurred. "We have considered ourselves part of the Navy Department," wrote Mahood.[95]

Arma specialized in applying gyroscopes to fire control. In 1929 it introduced a critical element to make fire control a dynamic system, the "stable element." Before its introduction a few men in the gun director had kept telescopes aimed constantly at the horizon, communicating the ship's roll and pitch to the plotting room (known as "level" and "cross level"). The Arma stable element, which was similar to modern inertial platforms, employed a gyroscope to maintain an absolute horizontal and vertical reference, effectively recreating the horizon mechanically in the bowels of the plotting room. Arma did not invent the idea. The British navy, Sperry Gyroscope, Ford Instrument, and even the National Bureau of Standards had experimented with or designed similar devices before 1920.[96] But like Ford Instrument with its rangekeeper, Arma created a workable device that filled a niche for the BuOrd, and Arma was willing to become a captive contractor. The Arma machine, located next to the rangekeeper in the plotting room, now actually fired the gun in response to a trigger pulled by a human operator. Before firing, it waited for the ship to be level or to be at the height of its roll; in light seas it could command the guns to follow the target in an automated version of continuous aim. With the stable element, the bulk of the control system moved to the plotting room. The gun director in the conning tower merely tracked the range and bearing of the target.

## Fire Control in the 1930s

The Ford Rangekeeper evolved in parallel with the Arma stable element. The two sat next to each other in the plotting room, linked by three mechanical shafts on the floor. From its introduction in 1916 until World War II the Ford Rangekeeper underwent a

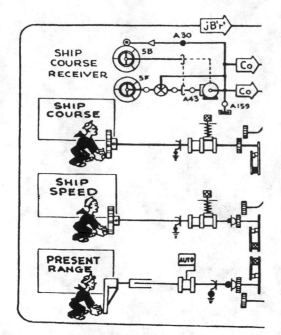

Fig. 2.12. Mark 8 Ford Rangekeeper, selection from system diagram showing user inputs. This clip represents about 5 percent of the total rangekeeper system. U.S. Navy, Bureau of Ordnance, *Rangekeeper Mark 8 and Mods,* OP 1068 (Washington, D.C., 1949).

number of variations and mark numbers, although its core function remained essentially the same: assimilating diverse data, solving for the course and speed of the target, predicting its course into the future, calculating sighting for the guns, and maintaining a plotted record.[97] An array of models and mark numbers appeared for specialized applications. Among these, the Torpedo Data Computer, or TDC, became famous during World War II for its use on American submarines (it appears in the 1958 film *Run Silent, Run Deep*).

The Mark 8 rangekeeper, introduced in the early thirties, represented the maturing of technology and served as the main battery rangekeeper for battleships and cruisers during World War II and for decades after. The Mark 8 differed little from the old Mark 1 machine in basic structure; it took initial estimates of enemy course and speed, matched them with observed data, and allowed the human operator to make corrections accordingly. It also closed that feedback loop, adjusted course and speed by itself in response to updated tracking information, extrapolated advance (future) range, and generated orders for the guns to fire.[98] The Mark 8, however, included many more variables than the Mark 1, correcting for wind, the earth's rotation, gun barrel wear, and a host of other subtle factors. The system's greater complexity allowed the machine to model the world in ever greater detail.

The original Mark 1 rangekeeper was a box on a pedestal, easily operated by a few men turning cranks. None of the variables entered automatically, except for the own ship's course, which came in via a gyrocompass repeater, though it could be bypassed and entered manually. The Mark 8, by contrast, achieved much greater automation, reflecting the navy's increasing level of comfort with electrical machinery to renew data in a system. Most information came into the unit automatically, but manual inputs were available as backups, allowing the human operator to judge the quality and veracity of the data and correct it if necessary (Fig. 2.12).[99] It also looked more like a computer in the modern sense than the older Mark 1. The Mark 8 comprised

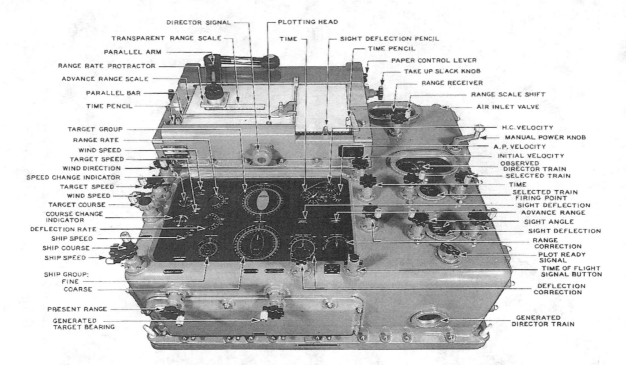

Fig. 2.13. Mark 8 Ford Rangekeeper, ca. 1930. This machine was standard on battleships and cruisers from the 1930s on, and it remained in service on battleships retired in the mid-1990s. Courtesy of Hagley Museum and Library.

five separate boxes, divided into functional units, bolted together into a single console (Fig. 2.13).

### Technical and Organizational Maturity

This maturity paralleled an organizational and technical conservatism. As the official BuOrd history put it, prewar fire control systems "were able to provide adequate fire control for main battery guns," although they were "rough in operation, unpopular with crews, and far from the ultimate" (Figs. 2.14 and 2.15).[100] After the Mark 8, Ford did not change the rangekeeper fundamentally, though it added myriad incremental improvements. The Depression and treaty limitations also meant lean years for the navy. At the time of the attack on Pearl Harbor, on 7 December 1941, all but five American battleships still used Ford's Mark 1 rangekeeper, originally designed in 1915. Even the newer systems were basically designed about 1930 and remained operational

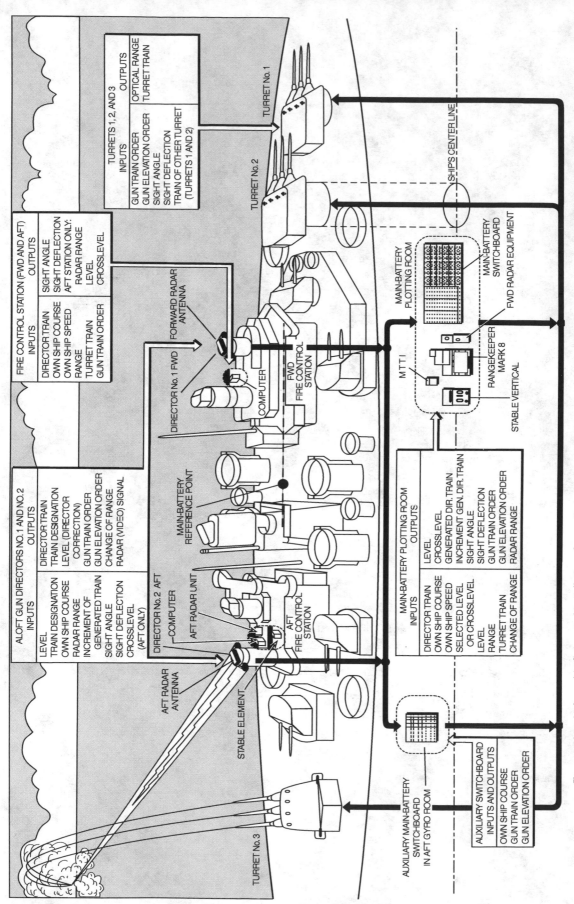

Fig. 2.14. A typical warship fire control installation of 1940. Redrawn from U.S. Navy, Bureau of Ordnance, *Main Battery Fire Control System, CA 68, 72, and 122 Class Ships*, OP 1387 (Washington, D.C., 14 June 1948), 2–3.

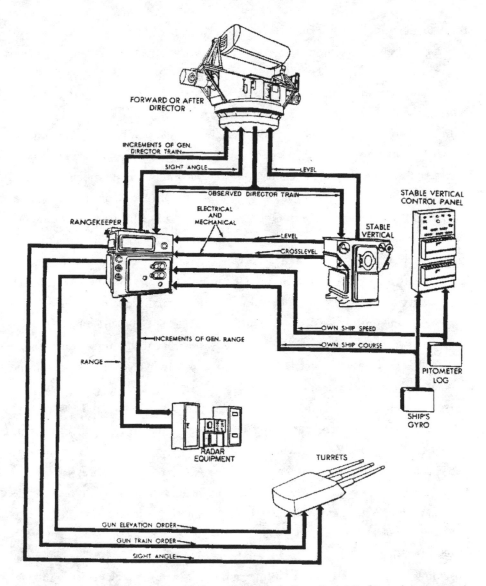

Fig. 2.15. Main battery fire control system from a cruiser including a Mark 8 Ford Rangekeeper, a Mark 6 Arma Stable Vertical, and a Mark 34 General Electric Director, ca. 1940. Both radar and an optical rangefinder are integrated into the director. Reprinted from U.S. Navy, Bureau of Ordnance, *Main Battery Fire Control System CA 68, 72, and 122 Class Ships*, OP 1387, (Washington, D.C., 14 June 1948), 82.

through the war.[101] Hannibal Ford became president of Ford Instrument in 1930, which removed him from the daily engineering concerns of the company (he would retire in 1943). Given Ford's influence over his company's technology, the change surely had practical effects, but it was symbolic as well, signaling the company's stability as an engineering firm and government contractor.

In 1933 Ford Instrument returned to its original fold. A new holding company, the Sperry Corporation, acquired Sperry Gyroscope in 1930 and then acquired Ford Instrument three years later. This acquisition, however, put BuOrd in a quandary, for its earlier animosity toward Sperry had not yet cooled. BuOrd threatened to restrict Ford Instrument's access to naval technology. "The Sperry Company has proven unmindful," it explained, "of protecting American interests of secrecy in the past and there is no assurance that they will become less careless in this respect." Probably because of these threats, Sperry Gyroscope and Ford Instrument had very little contact with each other even when they were under the same umbrella, and they retained their separate engineering cultures. As we shall see in the next chapter, Sperry Gyroscope was by this time developing antiaircraft fire control for the army. Still, the Ford Instrument engineer William Newell, who joined the company in 1926, recalled having "practically no contact" with Sperry Gyroscope.[102]

Fire control in the 1930s thus consisted of G.E., Ford Instrument, and Arma equipment, with optics from companies like Bausch and Lomb, Kodak, and Keuffel and Esser. G.E.'s synchronous data transmission system brought data from instruments of perception to its switchboard and routed those signals to different places. Power controls allowed these signals to articulate the movement of the guns. Ford Instrument's rangekeepers collected data from the system, bringing the target's motion into the machine, where it could be tracked, predicted, and sighted. Arma's stable element steadied this pitching, rolling, heaving apparatus, not by keeping it physically still but by providing minute corrections to subtract and factor out of the calculations. Still, despite its sophistication, as well as because of it, by the end of the 1930s the control system had begun to push the limits of its dynamics, especially in response to emerging technologies that threatened the battleship's very existence.

### Emergence of the Antiaircraft Problem

A new problem emerged for which engineers had few adequate solutions: antiaircraft fire control. Billy Mitchell's dramatic demonstrations of 1921, while contrived, showed that aircraft alone could sink the great capital ships. Shooting down aircraft inherited the difficulty of the surface fire control problem, but with added complexity: not only were aerial targets smaller than surface targets but they moved faster, and in three dimensions. Shells had to explode, not on impact, but after a finite time delay, introducing another variable to the system. Antiaircraft guns were smaller and more numerous than surface batteries, typically 1–5 inches in diameter rather than 14 or 16, so gun directors too had to be smaller and faster. The British were working on the problem, but they were making slow progress.[103]

In 1925 Ford Instrument built an antiaircraft gun director in response to a navy request and delivered it to the navy the following year as the Mark 19, for the USS *Maryland*.[104] This device had 55,000 moving parts. It integrated an entire control system into a single unit that sat on the deck, including a rangekeeper, a stable element, and tracking telescopes. It used the same calculations as the rangekeepers did for surface fire control but added an altitude variable for the target. While it was an impressive solution to a difficult problem, the Mark 19 was only a first step, and it embodied the limitations of the secret engineering culture: it merely adapted its existing techniques to a third dimension. It could track neither high-speed aircraft nor dive bombers and seems to have been intended not to provide an exact solution but to guide a barrage of fire to the general area of a target.[105] By 7 December 1941, 42 of these devices had been installed in the fleet, and they eventually served at Midway, in the Coral Sea, and at Guadalcanal. Several equally ungainly successors from Ford, G.E., and Arma also entered service.[106] These transitional solutions reflected the priorities of the Depression and the difficulty of the antiaircraft problem, especially when treated with traditional surface-fire techniques.

After several additional models had been developed, prewar naval antiaircraft fire control culminated with the Mark 37 director, introduced in 1939 and state of the art in 1940. It directed the new 5-inch, 38-caliber (5″/38) "dual-purpose" gun that could fire at both airborne and surface targets. The Mark 37 sat atop a cylindrical foundation, or barbette, that provided strength and mechanical stability in a heavy sea. The ends of a 15-foot stereoscopic rangefinder protruded from the sides of the unit.[107] The Mark 37 system could track targets moving at a speed of up to 400 knots horizontally and 250 knots in the vertical dimension, the latter being a special adaptation to counter dive bombing. The Naval Gun Factory in Washington, D.C., built the barbette and physical enclosure, while the instruments and rangekeepers came from Ford, Arma, and other private contractors. About 30 Mark 37s had been produced by 1940, but the Naval Gun Factory eventually made more than 800 units in 92 separate modifications (mostly minor alterations for different ballistics). Typically, destroyers had 1 unit, cruisers had 2, and battleships, 4. The Mark 37 became the most prominent American naval antiaircraft director of World War II and a visual icon on nearly all U.S. warships of the time (several on the battleship *Missouri* looked down on the signing of the Japanese surrender in 1945).[108]

### From Rangekeeper to Computer

The world of main battery control allowed solutions that were comparatively slow and steady, whereas antiaircraft fire control presented a more rapidly changing problem. The Mark 37 represents the engineering culture's move to accommodate these dynamics. The earlier antiaircraft directors had been self-contained, but the Mark 37 divided its functions between the director itself and a plotting room belowdecks. Reflecting this new complexity, BuOrd changed the designation of the Mark 37 from the older "director" to "gun fire control system." The system employed a Ford Rangekeeper, but this machine's designation now became "Mark 1 computer," recognizing the increasing ability of the machines to track more information than simply range (Fig. 2.16). Changing nomenclature reflected a new conception of fire control, with separate units organizing perception and articulation and the "computer" integrating data (Fig. 2.17).

High above the deck, inside the director itself, six or seven operators tracked the target and communicated with the plotting room, performing a mix of visual, auditory, and manual tasks (Fig. 2.18). There, the computer received tracking information from the director, calculated ballistics, and integrated to predict future position. The result of the computations included train and elevation, as well as parallax corrections to account for horizontal and vertical offsets between the director and the guns. These data were transmitted electrically, via a switchboard, to the articulated gun mounts. Four men ran the rangekeeper, and three ran the stable element. Because of the system's automated features, these operators served primarily to set up the problem on the machines and monitor them for problems. Officers in the plotting room oversaw the setup and stood by.[109] For this high-speed system, control was returned back to the men up in the director tower, who could see the enemy. Even their work would soon change: the Mark 37 was hailed as the first gun director specifically designed to accommodate radar. Still, this distinction apparently refers only to the unit's flat top,

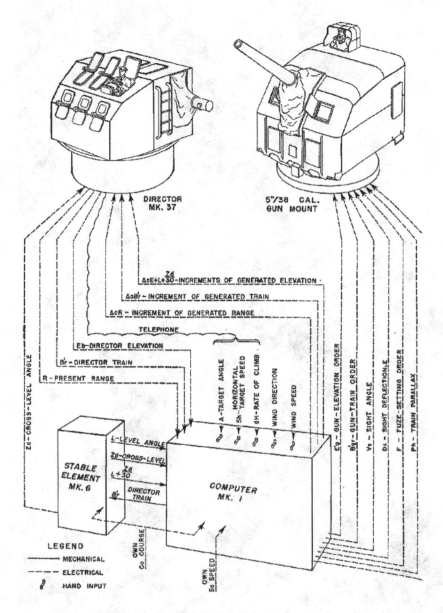

Fig. 2.16. Perception, integration, and articulation: the Mark 37 gun fire control system, its name replacing "director" because it incorporated a complete system of information. The Ford Rangekeeper became the Mark 1 "computer." By the late 1930s BuOrd was representing its fire control systems by showing information flows in addition to simply physical equipment. Reprinted from U.S. Navy, Bureau of Naval Personnel, *Naval Ordnance and Gunnery*, 428.

Fig. 2.17. The Mark 37 director with radar mounting and optical rangefinder protruding from the side of the director. This director became a common sight on U.S. warships during World War II. Courtesy of Naval Historical Center, Washington, D.C.

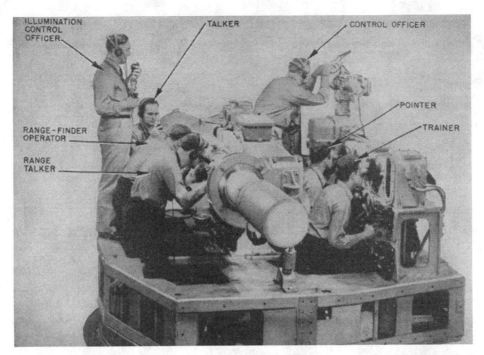

Fig. 2.18. Inside the Mark 37 director. Operators look through optical instruments and articulate their commands either into a telephone or through a handwheel on the machine. The long tube of the rangefinder protrudes into the foreground. Reprinted from U.S. Navy, Bureau of Naval Personnel, *Naval Ordnance and Gunnery*, 432.

for mounting an antenna, and its sturdy barbette. BuOrd, which focused on practical solutions from its contractors, declined to support long-term research in radar during the 1930s.[110]

Even the task of correcting estimates of course and speed, which Ford originally left to the gunnery officers, now returned to the machine. Belowdecks the Mark 1 computer converged on a solution and commanded servos to move the entire director and its scopes to track the target. With a feature called "automatic rate control" the telescope operators watched the motorized tracking, adjusting the rate of motion to keep the targets centered while the computer converged on a solution. The director and the computer formed what one user manual called a "regenerative group," a feedback loop that eased the job of tracking fast-moving targets.[111]

However useful and innovative, this feature exceeded the limits of BuOrd's or its contractors' technical knowledge. The feedback loop on the Mark 37, "a previously untried closed-circuit servo," had a stability problem: the output of the computer moved the director, which in turn affected the input to the computer. Both the computer and the power drives had time lags, so the two could push and pull each other and make the system oscillate. How these loops interacted and fed back on each other was poorly understood and caused severe problems in operations.[112] When radar was added in 1940, the noisy, often erratic signals instigated a complete breakdown. But the equipment was already in production, and the navy was preparing for war.

The stability problem in the Mark 37 was the most prominent example of a complication arising throughout naval fire control. Wherever sensitive instruments and intricate computers drove powerful servos on heavy gun mounts the systems could become unstable. "It is well known," a 1937 report read, that "the guns, because of their enormous inertia, do not respond instantaneously to a signal from the director . . . for similar reasons the guns tend to swing too far when coming to alignment with the director after such motion, giving rise to 'hunting' or oscillations."[113] Adjustments could make these systems stable, but only at the cost of unacceptably degraded performance. A series of tests identified an "inherent weakness" in the Ford system of control and in the hydraulic speed gear with which it was used. BuOrd's precious machines suffered from "insufficient 'stiffness' or 'rigidity' or a lack of prompt response to the director system."[114] Naval fire control systems, for all their precision, ruggedness, and sophistication, had run up against a problem the engineering culture could not solve: how to make a feedback loop move a large mass at high speed without making it unstable. Solving this problem required more theoretical analysis than the engineering culture of fire control could provide, so the navy turned to institutions that had developed different types of knowledge about feedback, including Bell Labs and MIT.

### Naval Fire Control in 1940

Less than half a century after Percy Scott introduced continuous aim firing, naval gunnery became an exercise in tracking and data flow; the actual firing of the guns was now but one part of a highly technical and highly distributed control system. On average, the big guns of 1940 could hit their targets once every two minutes, with a mean error of just a few percent. Measured this way, accuracy had improved by a factor of 3.5 since World War I.[115] As the foresighted Fire Control Board of 1905 had anticipated, naval gunnery now became a "system of information," joining the airplane and the machine gun in displacing people from the immediacy of combat.

Such displacement could not proceed in isolation, for it necessitated parallel shifts in industry. Changes in technology went hand in hand with changes in production, as the critical problems shifted from gun manufacturers to instrument manufacturers, then to electrical and electronics companies. BuOrd, with its fire control section, supervised and directed these shifts. An esoteric technology like fire control had few commercial applications, so the navy had to educate and train each new company it brought into its secret fold. The engineering culture surrounding BuOrd developed a set of fire control technologies that depended heavily on feedback, communications, and human-machine interaction. These precise mechanical, electrical, and optical machines combined into robust, reconfigurable systems that passed information around the ship through a series of reconfigurable switches. The engineering culture demanded secrecy, military orientation, and precision, values epitomized in companies like Ford Instrument and Arma Engineering, with their focus on low-volume precision manufacturing and naval applications.

These companies developed components and systems that allowed officers to model the field of naval warfare and manipulate it remotely, as data and machinery. Within these systems of abstraction, innovations displaced human operators with automated measurement or transformed human tasks into simple operations like pointer matching. The Arma stable element replaced the human task of following the horizon with the mechanical reference of gravity and the earth's rotation. The Ford integrator added power to mechanical representations as it calculated, enabling data to pass on without decay. G.E.'s selsyns and switchboards interchanged shaft angles and electrical quantities.

This engineering culture, developing in secrecy and peace, also invisibly accepted a set of assumptions that it embodied in its machines. The Ford Rangekeeper built a model of the world to do its calculations, transforming target data from polar coordinates into a Cartesian scheme inside the machine and then back to polar as output. These translations had costs in complexity, accuracy, and speed, but they made sense for battleships and cruisers because they assumed that the target moved along some course best represented in terms familiar to navigation. These ships' trajectories were separate from the shells they fired, so they oriented their courses for clear broadsides rather than heading straight for their targets. Classical fire control worked, then, when weapons were large machines with intricate human organizations, where people fought from afar without much direct contact with the enemy, and where the projectiles were purely ballistic and nonhuman.

As the 1930s ended, however, these assumptions began to erode. At first aircraft seemed like warships, flying straight and level, though in a third dimension, in order to drop their bombs; indeed, they had to in order to line up their bombsights. But new weapons and tactics shattered those assumptions, and with them the precise, abstract world of the battleship. Dive bombers, torpedo planes, and eventually kamikazes forged tighter links between platforms and projectiles. Now human operators directed the weapons much closer to the target, and the line between platform and weapon blurred, right up to the point of impact. A large warship could no longer afford to translate its enemies into Cartesian terms. Now they oriented around the ship itself, attacking along lines of force radiating from its center. BuOrd's fire control contractors, with their established, steady schemes, adapted only with difficulty.

After the early 1920s Sperry Gyroscope could not fit into the engineering culture of naval fire control. Rather, the company developed a different philosophy of control systems, adapted automatic controls to aircraft, and positioned itself to bring that approach back to BuOrd. During World War II, then, Sperry Gyroscope, with its tight coupling of human and machine and its alliances with university researchers, adapted to the angular, dynamic world of defending ships at sea.

# Taming the Beasts of the Machine Age

## The Sperry Company

### Sperry's Reining In

In 1940 the Sperry Corporation's engineering graphics department hired Alfred Crimi, an Italian immigrant fresco and mural painter, to make perspective drawings. Crimi had greater talents, however, and before long he was working out of a private studio at the company, making sketches and murals of Sperry's products and production lines. Crimi painted a pressure chamber for high-altitude flight training, portraying the pilots as cubist-like robots enveloped in the cylindrical structure. He drew the company's aircraft turrets, making the gunner's body transparent, "as though seen through an X-ray," with mechanical elements taking on the role of vital organs. *Life* magazine published his sketches at the height of the war, as part of a profile of Sperry's products.[1] Crimi captured the strange blurring of the human-machine boundary effected by Sperry's control systems better than Sperry's corporate photographers were able to do. In Crimi's artwork the human operator is surrounded by the machine, is intimate with the machine, becomes the machine, reining it in (see, e.g., Fig. 3.1).

By 1940 the company could write coherently about "the inability of the unaided man to operate his weapons": "His airplanes have become so big and fly so far that he must have automatic pilots instead of flying by hand. The machine gun turrets must be moved by hydraulic controls. The targets of his antiaircraft guns now move so fast in three dimensions that he can no longer calculate his problems and aim his gun. It must all be done automatically else he would never make a hit."[2] This statement, like

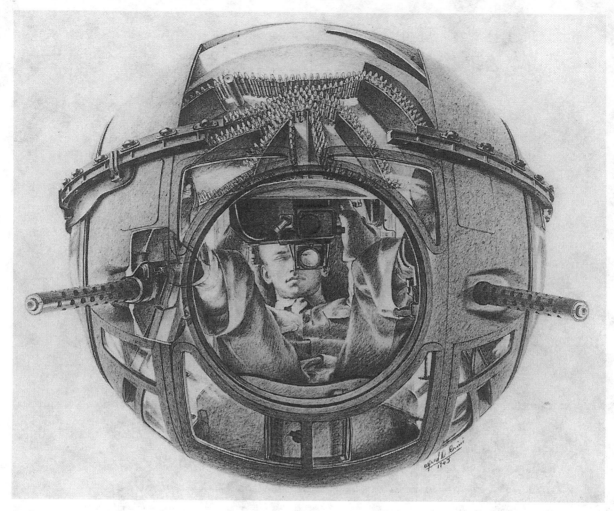

Fig. 3.1. Sketch by Alfred Crimi of a Sperry ball turret for the B-17 bomber, 1941. Note the eyepiece for the lead-computing gunsight. Courtesy of Hagley Museum and Library.

Crimi's work, epitomizes three decades of control engineering at Sperry. During much of that time the idea of controls as extensions of the human was still evolving, and the company could not articulate its goals so clearly. Rather, Sperry engineers worked out their ideas and expressed themselves by building machinery with different types of feedback, different roles for human operators, and different types of systems. They developed skills in feedback, data transmission, mechanical computers, and power drives. Sperry's engineering culture of control systems resulted from military demands, engineering research, and manufacturing constraints.

Sperry's engineering culture developed around three types of products, with varying degrees of success: automatic pilots for ships and for aircraft and antiaircraft fire control systems. As with naval fire control, Sperry's experience not only concerns the evolution of control systems but connects that history to that of American industry and its relationship to the military. Government support allowed Sperry to experiment with risky technology in peacetime, setting it up for large-scale expansion when war came. Heavily emphasizing engineering, the company relied on technology for its competitive advantage, often leaving a field when it generated competition and sometimes being forced out. Sperry built long-term relationships with officers and military organizations in the United States and abroad and frequently hired military personnel for special projects (Sperry reportedly recruited graduates of the Naval Academy who had failed the eye exam). Military services paid Sperry Gyroscope to make intricate, precise mechanical devices in small numbers at relatively high costs, but the company also excelled at manufacturing these controls for commercial customers around the world. These skills paid off handsomely during World War II, as Sperry Gyroscope's sales for simple military devices skyrocketed. In the critical area of computing, however, manufacturing challenges exceeded the company's abilities, opening the door for other approaches.

Elmer Sperry referred to airplanes as "beast[s] of burden . . . obsessed with motion." This suggestive metaphor captures the vision of control systems he imparted to his company. The phrase projects animism and autonomy onto machinery—not an animism of intelligence like the "thinking machines" of later decades but an animism of the body, more akin to horses than to calculators. Machines needed control systems, in Sperry's words, because of their "constant unstable equilibrium."[3]

## Compass and Pilot

As we have seen, Sperry Gyroscope first achieved commercial success with the gyrocompass for ships. By the end of World War I the gyrocompass had become an accepted and reliable device. Despite its frustrations in naval gunnery, the company profitably busied itself outfitting the world's fleets, both civilian and military, and Elmer Sperry won acclaim as a great American inventor and engineer.[4]

The gyrocompass included a phantom to transmit the position of the spinning gyro without interfering with its motion, but overall the machine remained a sensor and a reference. The data from a gyroscope proved steady and solid, however, so it was an

The Repeater Compass.

Wheel for changing course or controlling rudder.

Weather yaw adjuster.

Initial and meeting rudder adjuster.

Secondary rudder application adjustment.

Clutch lever for disconnecting Gyro-Pilot.

Case enclosing sprocket and chain for driving ship's steering wheel.

Steering motor is contained in this pedestal.

Fig. 3.2. The Sperry gyropilot, "Metal Mike," showing the clutch for disengaging the unit physically from the ship's wheel. Courtesy of Hagley Museum and Library.

ideal candidate for incorporation into a larger control system. Such a system made marketing sense as well, for customers who already owned the sensing device could buy additional equipment to complete the control loop.[5] In 1922, resuming a project it had started before the war, Sperry Gyroscope introduced the "Gyro-Pilot" (Fig. 3.2). This device connected the gyrocompass back through a ship's wheel, in a feedback loop that kept the ship automatically on course. It used an electric motor connected to the wheel via a chain to actually steer the ship in response to changes in the gyro reading. Elmer Sperry anthropomorphically called it the "iron quartermaster," and it quickly acquired the nickname "Metal Mike" both within industry and among the public. No human mediated the feedback loop; those who saw the device turning a ship's wheel often commented on its uncanny nature.

Sperry sales literature played on this novelty and mystery. One promotional pamphlet, *A True Story of the Devil*, tells a racist tale of a ship captain sailing in the Mediterranean. This white man invites an Arab captain, an experienced and able seaman, aboard the vessel to see it being steered automatically by the Sperry gyropilot. The

Arab sees the wheel operating by itself and stands in awe. After much searching for hidden ropes or some other source of the trick, he remains incredulous, convinced that the ship must be possessed by the devil. The American captain explains that the ship is being driven by an angel that only Christian believers can see. If the Arab were to convert from his "godless" ways, he too would see how the ship drives itself.[6] The ad makes the point that although ships may have been "beasts of burden," the gyropilot made them seem intelligent and domesticated (and Christian).

Because the gyropilot did not require a person to operate it, questions arose about its relationship to human operators. Some believed it could never replace the tacit knowledge embedded in an expert helmsman. Elmer Sperry countered by arguing that the helmsman's skill inhered in the ability to "anticipate" the ship's motions and that he had incorporated such a feature into the gyropilot.[7] Sperry Gyroscope offered as an add-on product a paper recorder to plot minute changes in course over time. The traces, the company claimed, would display in graphical form the limitations of a human operator:

> These records clearly show that even the best of men are not constitutionally adapted to perform this purely mechanical task [steering]. The man's powers of attention quickly become fatigued, he fails to detect small deviations from the course, these small deviations quickly become large deviations, too much or too little rudder is applied and the ship performs a sinuous course. The result is waste of power, both in the steering engine and the main engines.[8]

The course recorder also exposed differences between individual helmsman, as well as any variations in the quality of steering during a particular watch. In a tone that cannot have been comforting to working mariners, a common Sperry ad portrayed the gyropilot as a man looking for work:

> Wanted—a permanent position on board ship as a wheelsman. Have had experience in steering every type of merchant ship, can steer courses more accurately than others and use less rudder. Am sober, intelligent, strictly attentive to business, never ask for time off, do not talk back, am not affected by bill of fare or poor cooking, in fact do not eat at all. Wages wanted, only 54 cents per day for 24 hours service.
>
> [Signed]
> Sperry Gyro Pilot.[9]

"Helmsman regarded the course recorder as a kind of mechanical company spy," *Fortune* magazine reported, "and the marine gyropilot as a wicked device meant to send them into technological unemployment."[10] As Stuart Bennett noted in his study of instruments in the process industries, feedback devices were accompanied not only by claims of improved accuracy but also by social questions about obedience and reliability.[11]

These advertisements portrayed the human steersman as a weak link in the system. Yet customers in the traditional maritime businesses hesitated to relinquish such an important function to a machine. The company therefore hastened to assure prospective users that if they ceded control, they could grab it back anytime. The gyropilot included a large lever to physically disengage the unit, returning it to manual operation, assuring that "the regular ships *control is instantly available for emergency.*"[12] While it usurped control of the steering function, the gyropilot gave the human another task: control of the gyropilot itself.

Initially the human pilot had only one way to steer the ship: by turning the wheel. The gyropilot increased the number of inputs to seven. With the autopilot disengaged, the original wheel still worked as before. Metal Mike itself had a smaller wheel, used for setting the heading to hold (in another mode this wheel could also operate the rudder directly). The "weather adjustment," in today's parlance a *deadband,* allowed the ship to yaw a certain amount without initiating a correction. "Initial rudder adjustment" provided a means for "meeting" the helm as it returned to course, easing off on the rudder as it approached the proper course. Sperry called this "anticipation"; today it is termed *derivative gain.* "Rudder ratio adjustment," or *gain* in today's terms, determined the amount of rudder required to bring a particular ship back on course. Some of these tended to be permanent settings, varying only from ship to ship, while others needed frequent tweaking. In fact, the control system (like all closed-loop systems) required proper "tuning" to perform most efficiently, and sometimes even to operate at all.

This is not to imply that Metal Mike did not save labor; by keeping the person out of the feedback loop it kept the ship on course and relieved the operator of significant workload. Yet more than eliminating the steersman, the gyropilot altered the character of his job. He set the desired course and changed it in accordance with navigation. He also adjusted the instrument for changing weather conditions and different speeds (as often as once per hour).[13] And most important, he controlled when the automatic steering was put in effect; entering a harbor or avoiding an obstacle, for example, would not call for automatic control. The helmsman engaged and disengaged the autopilot according to the circumstances, literally trading or exchanging control between human and machine. Person and beast worked together, each making up for the other's limitations.

As with the earlier gyrocompass, the navy's interest in the gyropilot extended beyond that of commercial users. The navy cared less about labor problems and barely at all about labor costs, instead emphasizing accuracy and precision. Automatic piloting allowed better course tracking for maneuvers; more accurate courses meant more constant speeds and better firing solutions. It also had different implications for human operators. The company pointed out that in naval settings the ideal of a robotlike helmsman broke down under conditions of extreme stress: "It has been assumed that many of the helmsman's reactions under the stress of battle conditions will be mere automatic reflexes, inculcated by previous training until the familiar tasks may be performed without conscious thought. While it is a fact that predictions of human reactions must be based on averages, the man at the wheel is unfortunately not an average but an individual." Here Sperry's literature displays a certain sophistication, balancing the need to design machines for average users with the need to accommodate variation. The unique individual, with all his judgment and skill—and possibly panic, uncertainty, and mistakes—reemerges from the average in battle, potentially invalidating the machines. The solution to this quandary was not less machinery but more, to shift the operator's burdens to less stressful time periods. According to the company, once set up and adjusted, the gyropilot would literally sail through battle: "The consistent, machine-like precision of the Gyro-Pilot cannot fail to enhance the qualities which are all-essential in battle. True, man can never be displaced by the machine, but his function may well become that of the stand-by observer, rather than prime-mover in the action where perfection in every detail must ever remain the objective."[14] Combat, that most chaotic and unpredictable human situation, called for the highest precision and certainty that technology could provide. The human steersman need only stand by and watch.

Sperry's automatic pilot for ships was a classical feedback loop. It connected a sensor (the gyrocompass), through an amplifier (an electric motor that drove the ship's wheel), to an actuator (the ship's steering gear). It represented the world in a single dimension, heading, and used that representation to directly control the ship; the human intervened only to set the loop in motion and to adjust its parameters. Hence, the gyropilot raised social questions about the operator's nature and role. Was the helmsman a skilled technician? An unreliable element in a labor system? A steady warrior or an unstable observer? On ships these questions surrounded a traditional task that had been accomplished for centuries without mechanical intervention—that of the steersman, namesake and icon of Wiener's cybernetics. Soon Sperry automated a new kind of operator, the airplane pilot, addressing tasks that increasingly could not be accomplished by humans alone.

## Automatic Pilots for Aircraft

In aviation, as in the marine market, Sperry first achieved commercial success with instruments. To make the transition from a technical curiosity to a mainstream technology, the budding aviation industry in the 1920s needed to demonstrate reliable operation. Commerce needed to bring the wild airplane into the world of acceptable risk and repeatable scheduling, and the military wished to bring the chaotic aerial domain into its quantified purview. During World War I Sperry conducted a difficult development program in gyroscopic aircraft instruments for the navy. (Elmer Sperry's navy liaison for this program, the MIT graduate Luis de Florez, would later be the primary sponsor for the Whirlwind digital computer.) Sperry introduced a gyroscopic turn indicator in 1918, then a "Directional Gyro" and a gyroscopic "Artificial Horizon" in 1930. Sperry's first high-volume products, these instruments became standard on airplanes produced in the United States through World War II. Sperry's instruments, many of them based on gyroscopes, gave pilots feedback on the state of the airplane that they could not get from other sources, helping expand the operating envelope of aviation into adverse weather conditions. The directional gyro and the artificial horizon became part of the standard suite of airplane instruments and remain so today.

Sperry's aircraft instruments also proved stable enough to drive feedback loops. Just as it had introduced Metal Mike to close the loop around marine gyrocompasses, Sperry closed the loop around its aviation instruments with an autopilot. The product culminated a long series of experiments. As early as 1914 Elmer Sperry and his son Lawrence demonstrated the ability to stabilize aircraft with large gyroscopes. During World War I they built an autonomous flying bomb, or "aerial torpedo," under government contracts.[15] This device failed because the engineers could not achieve reliably stable flight without a human operator to monitor the control loops, but Sperry Gyroscope fared better with a control system that worked in conjunction with a human pilot

Lawrence Sperry died in an airplane crash in 1924, so it remained for Elmer Sperry

Jr. to develop a fully automatic autopilot in cooperation with the army in 1925–29. The army wanted the autopilot because it could keep an airplane stable during a bombing run, and automatic pilots became critical elements of precision bombsights. The company introduced its first autopilot in 1931, designated A-1. Like earlier attempts, the mechanical device employed electrically driven gyroscopes as sensors and electric motors as actuators. It could stabilize an airplane in pitch and yaw, but it also coordinated the aileron and rudder to stabilize heading, which earlier devices could not do. The innovative A-1 had reliability and maintenance problems and required frequent adjustments, but several were delivered for military and commercial use.

Sperry's refined model, the A-2, used air-driven gyros and pneumatic-hydraulic actuators for better reliability. It could also hold altitude, by amplifying the signal from a delicate pressure sensor through pneumatic actuators. Like the gyropilot for ships, the A-2 autopilot had a disengagement clutch for emergencies and several knobs that enabled the pilot to "control movement in all three planes without disengaging the autopilot." The pilot now controlled the autopilot, which controlled the plane. The company and the army followed up with an intensive program for pairing bombsights with autopilots. This pairing allowed a bombardier to look through the sight and actually control the aircraft while lining up the sight, an intimate coupling that brought the operator inside a more complex control loop.[16] The A-2 made the automatic pilot a practicable device and a commercial product. It did not eliminate the human operator, but it relieved the pilot of work and reduced fatigue. Nobody illustrated the changes for pilots wrought by the new controls better than Wiley Post.

## Wiley Post, Early Cybernetic Hero

Ironically, the man who popularized the automatic pilot was himself uniquely skilled, for Wiley Post made his mark as a pilot particularly close to his craft. In 1926, while working as a laborer in the Oklahoma oil fields, Post lost his left eye in an industrial accident. He literally replaced it with a machine, using his workmen's compensation to buy his first airplane. As one early colleague recalled, "He didn't just fly an airplane, he put it on." Another remembered Post as "as near to being a mechanical flying machine as any human who held a stick."[17] When flying, Post wrote, "I tried my best to keep my mind a total blank. I do not mean that I paid no attention to the business of handling the ship. I mean that I did it automatically, without mental effort, letting my actions be wholly controlled by my subconscious mind."[18] For Wiley Post, the autopilot merely replicated feedback loops inside himself.

Fig. 3.3. Wiley Post and the *Winnie Mae,* in which he made his around-the-world solo flight with a Sperry autopilot. Reprinted from the sales pamphlet *Round the World with the Sperry Pilot,* SGC Papers.

After leaving the oil fields, Post found a job ferrying Lockheed Vega airplanes from the factory to customers. Here he gained experience with the problems of "blind flying" through clouds and bad weather, relying on early Sperry gyroscopic instruments to keep his bearings. In 1931 Post and his partner, Harold Gatty, who had trained Charles Lindbergh in navigation, made headlines by piloting a Vega named *Winnie Mae* around the world in record time (Fig. 3.3).[19] For the round-the-world flight Post grouped his instruments right in front of his one eye and modified the cockpit so that he could fly with one foot on the rudder pedals and one hand on the wheel. As Preston Bassett, president of the Sperry Corporation, later remarked, "All combined, the setup was that of a man flying around the world with one eye, one arm, one leg, and two instruments. You will see that we are building toward a very good servomechanism."[20]

In 1933 Post repeated the trip by himself. Like Lindbergh six years earlier, Post struggled with fatigue, and he fell asleep on several occasions. Yet the sleeping Post did not spiral fatally to earth as Lindbergh had feared. The *Winnie Mae,* though absent its operator's steady hand, kept flying straight and level. Post had one of the first Sperry A-2 automatic pilots (he bought the device, but after his flight the company would not accept payment) (Fig. 3.5). The machine was not perfect; sometimes it failed and Post

Fig. 3.4. Aviatrix Amelia Earhart *(left)* inspects a Sperry automatic pilot of the type she would use on her fateful around-the-world flight. At right is Sperry president Thomas B. Lea. Courtesy of Hagley Museum and Library.

Fig. 3.5. Sperry gyropilot for aircraft installed in Wiley Post's plane: rectangular console in the dashboard, pneumatic actuators below. Courtesy of Hagley Museum and Library.

had to fly manually. The two worked together, trading control, playing on each other's strengths and making up for each other's weaknesses. Post's machine freed him for other functions, such as navigation, but primarily it allowed him to nap, considerably reducing his fatigue, the greatest obstacle on his record flight.[21]

Like Lindbergh's trip six years earlier, Post's flight was an accomplishment of both the human operator and the engineers and craftsman who had made his machinery. The blue-collar Post never achieved the acclaim of Lindbergh, but public exposure of the round-the-world flight brought the Sperry A-2 and automatic control of aircraft into public view. Amelia Earhart installed a Sperry autopilot in her flying laboratory and used one on her ill-fated around-the-world flight (Fig. 3.4). "The days when human skill and an almost bird-like sense of direction enabled a flier to hold his course for long hours through a starless night or over a fog are over," commented the New York Times, predicting the end of the era of heroic pilots. "Commercial flying in the future will be automatic."[22] Trans World Airlines (TWA) equipped its fleet of new DC-3 aircraft with the device, and by 1940 Sperry had sold 2,600 of them. As with the marine gyropilot, the coupling between perception and articulation was direct and intimate. The human operator tuned and supervised the feedback loop, trading control.

## The Sperry Corporation in the 1930s

Post's flight demonstrated the great potential of aviation controls in an exciting and rapidly changing industry that swept up Sperry Gyroscope. In 1929, the year before his death, Elmer Sperry sold his stake in the company for $4 million. In a heady, chaotic time for the airplane industry Sperry Gyroscope became part of North American Aviation, a conglomerate that included several airlines and aircraft manufactures.

Tom Morgan, the man who worked with Hannibal Ford on the first trials of the gyrocompass in 1912, became president of Sperry Gyroscope, and soon thereafter he became president of North American. General Motors acquired North American in 1933, and the fallout from that transaction created the Sperry Corporation. With Morgan as president, Sperry now became a holding company for smaller firms brought in through acquisitions (including Sperry Gyroscope Ltd., the English subsidiary).[23]

Reginald Gillmor, who had overseen the installation of the earliest gyrocompass, aboard the *Delaware,* and who had transferred British fire control technology to the United States, now headed the Sperry Gyroscope subsidiary of the Sperry Corporation. The company's business was distributed among the following markets: 30 percent naval, 30 percent military, 30 percent marine, and 10 percent aeronautical. The aeronautical sector grew most quickly. Morgan committed the company to the field and dedicated a substantial portion of its research and development budget to aircraft. By 1937 aviation accounted for half of Sperry Gyroscope's business.[24] Through the depression, Sperry Corporation's research and development budget rose steadily, averaging 2.5 percent of income from sales from 1933 to 1940, mostly supporting its large laboratory in Garden City, Long Island.[25]

Yet for the Sperry Corporation overall, control encompassed more than aircraft. During the early 1930s the Sperry Corporation took shape as an integrated control-system company. It purchased Ford Instrument in 1933, which remained under the leadership of Hannibal Ford, who thus found himself once again a Sperry employee. Ford Instrument still exclusively made naval fire control computers for BuOrd, operating out of three separate plants in Long Island City. Though owned by the same holding company, Sperry Gyroscope and Ford Instrument shared remarkably little technology well into the 1940s.

Sperry Gyroscope and Ford Instrument emphasized the perception and integration aspects of control systems, but several acquisitions during this period brought the corporation expertise and products for articulation as well, and tied the company to other domains of American industry. In 1935 Sperry acquired the Waterbury Tool Company of Connecticut, a maker of large, variable-speed hydraulic transmissions. Waterbury's hydraulic gear moved turrets for large naval guns and shell hoists, cranes, and numerous other shipboard machines; one ad called them the "'nerves' and 'muscles' for superhuman tasks."[26] In 1937 the Detroit-based Vickers Inc., the country's largest maker of oil hydraulic machinery, came into the fold. Vickers (not related to the British arms firm, Vickers Ltd.) specialized in small, high-powered controls for industrial applications, including paper making and cable manufacturing. Harry Vickers founded the company in 1920 with financial backing from Fred Fisher, of the automotive industry's Fisher Brothers. Fisher now became the single largest stockholder in the Sperry Corporation, with about 2.5 percent of its shares. (He remained on the board until his death in 1941, when he was replaced by his brother, Charles.) In 1940 Waterbury combined with Vickers to create a product line that covered the full range of hydraulic power devices.[27] Other than this consolidation, Morgan chose not to integrate his companies as divisions under the Sperry Corporation but rather to keep them as subsidiaries so that they would retain their separate cultures.

With these subsidiaries devoted to instruments, computers, and actuators, the Sperry corporate structure itself now mirrored a feedback system.[28] Indeed, the company often sold systems that included components from its multiple subsidiaries. Within the Sperry Corporation, however, fire control for the navy remained the exclusive domain of Ford Instrument. But this field was itself starting from scratch on the new problem of antiaircraft defense, and to attack this new challenge Sperry Gyroscope teamed up with the army.

### Antiaircraft Fire Control

While Sperry Gyroscope was improving airplanes with flight controls, it also built a business destroying them. In fact, the company's interwar work in control systems culminated in antiaircraft fire control. Because the problem of antiaircraft fire control required data of diverse origins and varying quality, unlike in marine or aviation controls, human operators became intimate parts of control loops as interpreters of the data. Antiaircraft fire control thus enables a useful comparison with marine and aviation automatic pilots because it shows how problems of machine representation affected both the role of human operators and the techniques of manufacturing. Also, Sperry's antiaircraft projects laid the groundwork for research into problems of prediction, computing, and human-machine interaction during World War II. When Norbert Wiener began thinking about human-machine interaction, he was addressing the problem as defined in the 1920s by Sperry and the army.

Like hitting a distant battleship, shooting an airplane out of the sky is essentially a problem of leading the target. Aircraft developed rapidly in the 1920s, and their increased speed and altitude rapidly pushed the task of computing this lead out of the range of human reaction and calculation. Fire control equipment for antiaircraft guns helped human gunners to accomplish a task that was beyond their natural capabilities.[29]

Measuring instruments had long been part of an artilleryman's tool kit. During World War I the Army Ordnance Department procured a broad array of fire control instruments for land artillery, including optical rangefinders, gunsights, periscopes, plotting boards, and traditional gunner's quadrants. American industry geared up for large-scale production, but the war ended before American-made fire control devices reached the front in large numbers. Nevertheless, the efforts did build up domestic capacity in precision optics, bringing a number of companies into the fold that would continue to build fire control instruments for decades: Eastman Kodak, Bausch and Lomb, Keuffel and Esser, Kollmorgan Optical, Leeds and Northrup, and National Cash Register (NCR).[29]

Amidst this buildup, fire control for antiaircraft guns underwent some preliminary development. Artillery officers used slide rules to calculate lead angles based on optical sighting of targets. These slide rules were incorporated into mechanized boxes; an operator would dial in data with knobs, read out an answer on a dial, and telephone azimuth and elevation to those operating the guns (*azimuth* was the term used by the army for the gun's rotation, while the navy used *train;* both services used *elevation*). Elmer Sperry, as a member of the Aviation Committee of the Naval Consulting Board, was familiar with this technology, and he did some work developing bombsights, guided bombs, and aircraft gunsights. The Army Ordnance Department knew of Sperry's work in naval fire control and invited him to submit a proposal for directing antiaircraft fire. Sperry came up with two instruments, but his company was unable to produce these devices in quantity during the war.[30]

When the war ended in 1918, the army undertook no new work in antiaircraft fire control for several years. In the mid-1920s, however, it began to develop components for antiaircraft systems, including stereoscopic height-finders, searchlights, and sound location equipment (Fig. 3.6). The latter two involved Sperry Gyroscope. Sperry had made its first searchlights in 1916 and sent them to war in 1917.[31] After the war, searchlights came to account for a significant portion of Sperry sales, for both military (navy and army) and commercial applications, and would continue to do well into World War II.

Within the army, antiaircraft came under the purview of the Instrument Department of Army Ordnance, located at the Frankford Arsenal in Philadelphia. There, in 1925, Major Thomas Wilson began developing a calculating machine for antiaircraft

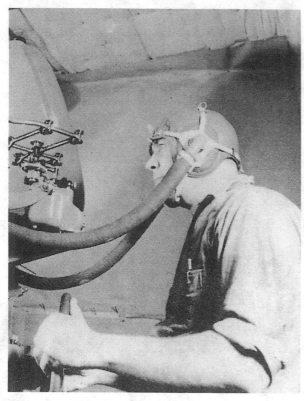

Fig. 3.6. Sperry sound-ranging equipment and human operator. Courtesy of Hagley Museum and Library.

fire control based on the system of director firing in naval gunnery. Wilson's device accepted data as input from perception components, performed calculations to predict the future location of the target, and articulated direction information to the guns.

With Wilson's director, the components of an antiaircraft battery remained independent, linked only by voice telephone. "No sooner, however, did the [antiaircraft] components get to the point of functioning satisfactorily within themselves," recalled Sperry's then chief engineer, Preston Bassett, "than the problem of properly transmitting the information from one to the other came to be of prime importance."[32] Tactics and terrain considerations often required that different fire control elements be separated by up to several hundred feet. As in the early naval systems, observers telephoned their data to an officer who manually entered the data into the calculating machine, read off the results, and telephoned them to the gun installations. From its experience in making gyrocompass repeaters aboard ship, Sperry Gyroscope knew how to automate these communications, so the army approached the company for help.

To Elmer Sperry it looked like an easy problem: the calculations resembled those in a naval application, but the physical platform, unlike a ship at sea, stood on solid ground. The army system also had its own electrical supply and stood physically separate from the guns, precluding the synchronization problems that plagued Sperry's system aboard ships. In 1925 Sperry engineers visited Major Wilson at the Frankford Arsenal and expressed interest in working on the problem. Sperry stressed his company's experience with the navy, as well as its recent developments in bombsights, which he described as "work from the other end of the proposition."[33] Indeed, bombsights had to incorporate numerous parameters of wind, groundspeed, airspeed, and ballistics, so antiaircraft directors really were reciprocal bombsights. (One reason that antiaircraft fire control equipment worked at all was that it assumed attacking bombers had to fly straight and level to line up their bombsights.) Elmer Sperry's advances to Wilson were warmly received, and in 1925 and 1926 Sperry Gyroscope built two data transmission systems (one traditional step-by-step, the other synchronous) for the army's gun directors. Major Wilson died in 1927, and Sperry Gyroscope took over the entire director development from the Frankford Arsenal with a contract to build and deliver a new director to the army.

Beginning with this project, Sperry undertook a small but intensive development program in antiaircraft fire control that would last more than fifteen years. The company set up a separate department with its own facilities and personnel and gradually developed a cadre of experts. Heading the effort was Earl W. Chafee, an engineer whose strong personality and free managerial hand allowed him to dominate Sperry's fire control work into the 1940s. The company financed its engineering internally, selling directors in small quantities to the army, mostly for evaluation, for only the cost of production. The army called these "educational orders," intended "to provide means for at least a few industrial facilities to familiarize themselves with the technique and special skill required to produce the material."[34] Sperry never sold more than twelve of any of the nearly ten models it developed before 1935; the average order was five. Sperry Gyroscope offset some development costs by sales to foreign governments, especially Russia, with the army's approval—the very type of arrangement that had so annoyed the navy.[35]

For most of the twenties and thirties, Sperry's antiaircraft work concentrated on aiming large guns (3–4 inches in diameter) firing exploding shells to relatively high altitudes to reach attacking bombers. The shells were not intended to hit the target directly but rather to explode nearby at some predetermined time after firing. Since this scenario represented the most difficult antiaircraft situation at the time, its techniques diffused into other applications, including coastal defense and traditional artillery.[36] The antiaircraft problem proved difficult enough, but aeronautic technology itself was rapidly changing. In 1925 bombers flew at 100 mph at relatively low altitude. This speed more than tripled in the next ten years, and the bombing altitude rose to well above 15,000 feet. In more ways than one Sperry was shooting at moving targets.

Increasing aircraft speeds and altitudes created a number of problems. Tracking and prediction were the same as in naval gunnery, but with a third dimension and with different distances and times. Once fired, shells travel with ever-decreasing velocity due to gravity and air resistance. Typical shells of the 1930s would take 15 seconds to reach 5,000 feet, and double that to reach 8,000 feet. A plane traveling at 100 mph at 5,000 feet would travel about 750 yards horizontally (toward the target) dur-

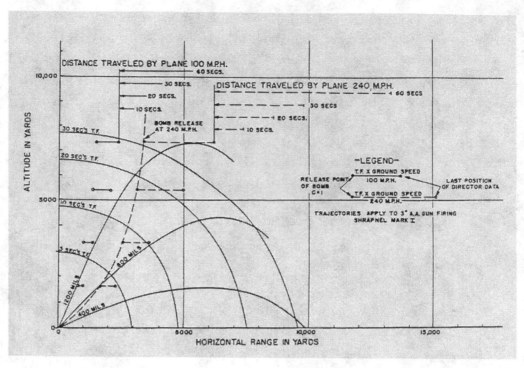

Fig. 3.7. Antiaircraft trajectories and lead times for different bomber speeds. Reprinted from Sperry Gyroscope Company, *Anti-Aircraft Gun Control.*

ing the shells' time of flight. Thus the lead for the gunner would be 750 yards. A plane traveling at 250 mph at 8,000 feet would travel 3,660 yards (more than two miles) during shell flight, requiring a lead nearly five times greater than that for the slower, lower plane. In either case, the shell would need to be fired, not at the plane itself, but at its expected location after the time of flight. The longer the time of flight, the more difficult this prediction. Furthermore, the problem had an inherent feedback loop because prediction could only be accomplished when the time of flight was known, but time of flight depended on the aiming point, itself the output of prediction.

Tactics further complicated prediction. For antiaircraft fire to have real defensive effects, it needed to shoot down (or at least scare off) attacking planes before they released their bombs. This limitation reduced the time available for the director to produce a firing solution: tracking could begin only when the attacking aircraft came into visual instrument range, and the shell had to be fired at least one "time of flight" before the bomb release point (Fig. 3.7). For a bomber flying 100 mph, if the antiaircraft guns were placed right at the bomber's target, the aircraft would be within visual range for 6 minutes and within the effective range of the guns for 2.5 minutes. For a bomber flying 250 mph typical of the late 1930s, gunners could only see the target for about 2 minutes, and they could effectively fire at it for 1 minute.[37]

One way to improve the situation would be to move the antiaircraft director and battery forward of the target, allowing it to engage attacking planes well before their bomb release points. A successful prediction depended, however, on an assumption of straight and level flight, which only held during a bombing run, when the plane needed to fly steady to align its own bombsight. An antiaircraft battery too far ahead of the target would catch the bomber before its bombing run, when the straight and level assumption was not yet valid.[38] Nevertheless, if the antiaircraft system could completely solve the problem for a given zone, it could force attackers to maneuver or climb to a higher altitude, making their job more difficult and achieving a partial tactical advantage.

### Inside a Sperry Gun Director

After producing several prototypes, in 1930 Sperry developed a gun director, designated T-6, which the army accepted and "standardized" (i.e., put into operation).[39] The T-6 was the first American antiaircraft director to be put into production, as well as the first one the army formally procured, so it is instructive to examine its operation in detail. Such an analysis also clarifies the basic features of the antiaircraft problem that would drive the development of control systems through World War II (and, in modified forms, through the Cold War and even into the era of ballistic missile defense). A technical memorandum written by Sperry's Earl Chafee in 1930 lays out the function and purpose of the machine.[40]

"The heart of the gun control system is the Computer," writes Chafee, articulating his sense of the director as the center of a distributed control system. Historians have frequently noted that before 1945 the term *computer* referred to human operators, usually women, who ran calculating machinery. Not so in fire control, where Chafee and others used the term at least fifteen years earlier to refer to these mechanical, analog gun directors (though they were not stored-program computers in the modern sense).[41] In his report, Chafee describes in detail the workings of a mechanical analog computer that connected up to four 3-inch antiaircraft guns and an a rangefinder into an integrated system (Fig. 3.8).

As in Sperry's naval fire control system, the antiaircraft director used data trans-

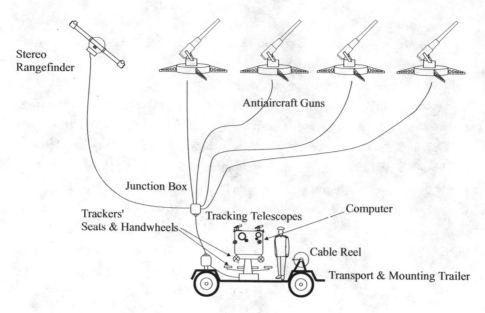

Fig. 3.8. Layout of the Sperry T-6 antiaircraft director. The human is shown only to indicate the scale.

mitters similar to those that connected gyrocompass repeaters aboard a ship to provide its three primary inputs. First, the target range came from a stereoscopic rangefinder (with one operator) similar to those used in naval fire control. The second and third variables came from two human operators using telescopes mounted on the director itself to track the target's azimuth and elevation. Each sighting device had a data transmitter that sent its angle reading to the computer. The computer received these data and incorporated manual adjustments for wind, muzzle velocity, air density, and other factors. It calculated three output variables for the guns: azimuth, elevation, and the shell's time of flight. The last, manually set into a fuze before loading, determined the amount of time after firing it would take to explode.

The computer performed two major calculations. First, *prediction,* or leading the target, modeled its motion and extrapolated it to some time in the future. Second, the *ballistic* calculation figured how to aim the gun to make the shell arrive at the desired point in space and explode; it corresponded to the traditional artilleryman's task of looking up data in a precalculated firing table. To perform the prediction, the Sperry director, like the Ford Rangekeeper, converted its input information from polar to Cartesian coordinates. The mechanisms projected the movement of the target into a horizontal plane and derived the velocity from changes in position. The director then multiplied that velocity by the prediction time to determine a future target position and then converted the solution back into polar coordinates for output. Sperry called this approach the *plan prediction method* because it represented data on a flat plan as viewed from above. (The familiar radar target display, introduced years later, in which a beam rotates sweeps around a round tube to reveal targets, became known as the *plan position indicator*, or *PPI*, an appellation inherited from this method of computation.)

Chafee did not use the term *analog*, but his machine represented the world with a physical model. "The actual movement of the target is mechanically reproduced on a small scale within the Computer," he wrote, "and the desired angles or speeds can be measured directly from the movements of these elements."[42] Once the machine had a good model, the computation became relatively straightforward. Generating the model and ensuring that it faithfully represented the world posed the major technical challenge and required human intervention.

The Sperry director also used a mechanical representation for ballistics, employing a *mechanical firing table*. Traditional firing tables were numerical lists of gunnery solutions that indicated how to set the gun for specified values of range, wind, temperature, and other factors. Sperry replaced the firing table with a "Sperry ballistic cam." This three-dimensional, cone-shaped device effectively stored the table mechanically and used a pin to look up answers (Fig. 3.9). Two independent variables were input by the angular rotation of the cam and the longitudinal position of the pin, which rested on top of the cam. As the pin moved up and down the length of the cam, and as the cam rotated, the height of the pin reflected the solution to part of the ballistics problem. The T-6 director incorporated eight ballistic cams, each solving for a different component of the computation. To adapt a director to a different type of gun, one simply replaced the ballistic cams with a new set, machined according to different firing tables.[43] Foreign governments, for example, would supply Sperry with firing tables for their own guns, and the company then machined custom cams and produced special directors. The ballistic cams constituted the permanent memory of the computer, roughly comparable to what today we would call *ROM*, or read-only memory.

Together, ballistic and prediction calculations formed a feedback loop (Fig. 3.10). Operators entered an estimated time of flight for the shell when they first began tracking. The predictor used this estimate to perform its initial calculation, which fed into the ballistic stage. The output of the ballistics calculation then fed back an updated estimate of the time of flight, which the predictor then used to refine the initial estimate.

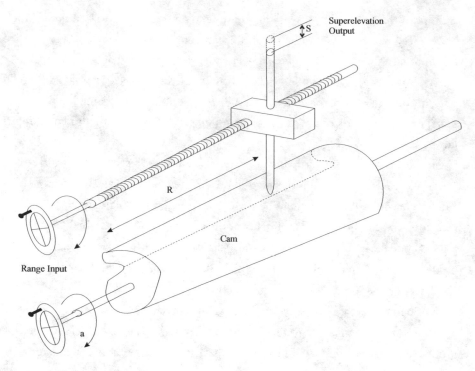

Superelevation
Output

S

R

Cam

Range Input

a

Angular Height Input

Fig. 3.9. Operation of the Sperry ballistic cam. Two variables are input by rotating the handles at the left. The pin rides along cam as it rotates, the height of the pin, *S*, providing the output value to be fed into another mechanism. In this example, for a given range and angular height of a target, the output, *S*, is the firing angle of the gun. Shaft inputs would replace handles in a real machine.

Thus "a cumulative cycle of correction and recorrection . . . brings the predicted future position of the target up to the point indicated by the actual future time of flight."[44]

A square box measuring about 4 feet on a side, the T-6 director mounted on a pedestal that allowed it to rotate (Fig. 3.11). Three crew members sat on seats, and one or two stood on a step mounted on the machine, revolving with the unit as the azimuth tracker followed the target. This arrangement provided comfortable, stable positions for the tracking operators. As the unit and the trackers rotated, however, the remainder of the crew, who stood on a fixed platform, had to shuffle around with it. While the rotation angles were small for any given engagement, it must have been awkward. Moreover, the T-6 computer required only three inputs—elevation, azimuth, and altitude—and produced only three outputs—elevation, azimuth, and fuze time—

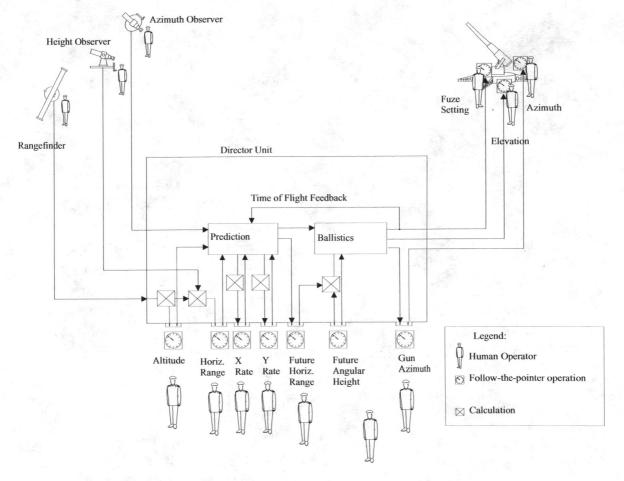

Fig. 3.10. A simplified diagram of the system layout and data flow for the Sperry T-6 antiaircraft gun director computer.

yet it required *nine* operators. These nine did not include the operator of the range-finder, which was considered a separate instrument, or the men tending the guns, but only those operating the director itself. What did these nine men do?

## Manual Servomechanisms and the Sperry Director

The army's specification for the T-6 director required "minimum dependence on the 'human element.'" Sperry Gyroscope explained that "all operations must be made as mechanical and fool proof as possible; training requirements must visualize the con-

Fig. 3.11. The Sperry T-6 director: *A*, spotting scope; *B*, north-south-rate dial and handwheel; *C*, future-horizontal-range dial; *D*, super-elevation dial and handwheel; *E*, azimuth tracking telescope; *F*, future-horizontal-range handwheel; *G*, traversing hand-wheel (azimuth tracking); *H*, fire control officer's platform; *J*, az-imuth-tracking operator's seat; *K*, time-of-flight dial and handwheel; *L*, present-altitude dial and handwheel; *M*, present-horizontal-range dial and handwheel; *N*, elevation tracking handwheel and operator's seat; *O*, orienting clamp. Courtesy of Hagley Museum and Library.

ditions existent under rapid mobilization."[45] The memory of World War I rings in this statement. Even at the height of isolationism, with the country sliding into depression, design engineers considered the difficulty of raising large numbers of trained personnel in a national emergency. Designers also considered the ability of operators to perform their duties under the stress of battle. Thus, nearly all the work for the crew was in the follow-the-pointer mode inherited from naval systems: each man concentrated on an instrument with two indicating dials, one showing the actual value for a particular parameter and one showing the desired value. With a hand crank they adjusted the parameters to match the two dials and bring the error to zero. Sperry called the operators "manual servomechanisms."[46]

Figure 3.12 shows the crew arrayed around the T-6 director in an arrangement that today seems almost comical. Strange as this configuration seems, it reveals Sperry engineers' conception of the human role in the system. In this early machine one man corresponded to one variable, and the machine's requirement for operators corresponded directly to the data flow of its computation. The men literally supplied the feedback that made the system work, although Sperry's idea of feedback was rather different from today's: "In many cases where results are obtained by individual elements in the cycle of computation it is necessary to *feed these results back* into the mechanism or to transmit them."[47] Turning a shaft to eliminate the error between two dials is a classic servo problem, and Chafee acknowledged that "servo-motors" could do the job. Still, he claimed that "it has been found in many cases to be much easier to rely on a group of operators who fulfill no other function than to act as servomotors. . . . This operation can be mechanically performed by the operator under rigorous active service conditions."[48] The term *manual servomechanism* itself is an oxy-

Fig. 3.12. The Sperry T-6 director, mounted on a trailer, with operators. Courtesy of Hagley Museum and Library.

moron, for it acknowledges the existence of an automatic technology that might replace the manual method.

Indeed, servos already replaced two operators from the previous model. While the Sperry literature proudly trumpets human follow-the-pointer operations, it barely acknowledges the automatic servos, and even then it provides the option of manual follow-ups "if the electrical gear is not used."[49] Indeed, there was more to the human servo-motors than Sperry wanted to acknowledge; men still had to exercise some judgment, even if unconsciously. The data were noisy, and even an unskilled human eye could eliminate complications due to erroneous or corrupted data. The mechanisms themselves were rather delicate, and bad input data, especially if they indicated conditions that were not physically possible, could lock up or damage the machine.[50] The crew that operated the T-6 director corresponded exactly to the algorithm inside it, and at each stage they renewed the data and verified that they faithfully represented the world.

Because of these human interventions, Sperry could develop its machines to the point that the greatest uncertainties in the system stemmed not from integration but from perception and articulation. Readings from the stereo rangefinders depended greatly on the skills of the human operator, which were highly variable, even from day to day for a single person. Setting the fuzes introduced even greater errors, because the gun crews set them by hand as they loaded the shells, which introduced variations in time. But Sperry engineers conducted time and motion studies of the crews to standardize this operation and experimented with automating the actual gun pointing. Automatic control of the guns proved a difficult problem because significant power amplification was required to make the small signals produced by the computer drive the massive guns. Sperry's acquisition of Vickers and Waterbury provided the corporation with the skills it needed to design and manufacture hydraulic drives. By the end of the thirties the company had an electro-hydraulic remote-control system to move antiaircraft guns under remote control, which it produced by the thousands.[51]

## Producing Computers in the 1930s

Throughout the 1930s Sperry continued to work with the army to develop antiaircraft computers. In later models servomechanisms replaced human operators in the computation cycle, reducing the number of human operators to four. Chafee designed the machines to be lighter, less expensive, and "procurable in quantities in case of emergency."[52] By the start of World War II the primary antiaircraft director available to the army was the Sperry M-7. It incorporated an altitude predictor for gliding targets, could accept electrical inputs from radio rangefinders, and implemented full power control of the guns.[53] This computer, culminating 15 years of work at Sperry, was a highly developed machine, optimized for reliability and ease of operation and maintenance. Its design capitalized on the strengths of Sperry Gyroscope: data transmission, intimate involvement with technical officers in the armed services, human mediation of the computation cycle, and manufacturing of precision mechanisms. It was an elegant, if intricate, device, weighing 850 pounds and including roughly 10,000 parts.

Still, producing the M-7 was not easy, and the difficulty limited its usefulness. The much-touted ballistic cams best illustrate the manufacturing difficulties of Sperry directors. These strangely shaped parts originated in the numerical firing tables provided by the army's Aberdeen Proving Ground. From the data in these tables a machinist would fabricate the cams directly, without going through the intermediate stage of blueprints. First, a rough cam would be cast, and then the machinist would drill hundreds of small holes, working from numbers on the artillery firing table. He would then file the cam and polish it, both smoothing the cam mechanically and smoothing the data mathematically. These operations required a great deal of time and skill, and ballistic cam manufacture proved a major bottleneck in Sperry's production of directors.

The process, with its flow of information from ballistics to machine control, gradually approached what would later be called *numerically controlled* machining. The historian David Noble has argued that numerically controlled machine tools emerged later as an effort to eliminate reliance on skilled machinists.[54] But the problems of cam cutting point to a different motivation: these parts were defined numerically from the beginning. Drawings for them never existed; their shapes came from firing tables and traveled through a human operator on their way to a mechanical part. The highly skilled machinists themselves worked from numbers, converting numerical data into a smooth physical representation (Figs. 3.13 and 3.14).

By 1941 manufacturing the ballistic cams, along with the other precision parts required for the gun directors, was seriously inhibiting Sperry's ability to meet its orders. Under pressure from the Army Ordnance Department the company subcontracted production of the M-7 to the Ford Motor Company (though cam production remained in-house). What followed was a remarkable episode wherein the precision manufacturing practiced at Sperry Gyroscope came into contact with classic mass production pioneered by Ford. The results revealed the limitations of both mass production and mechanical computers and spurred the later development of electronic computers for military control systems.

In 1943 Ford began to produce the directors in its Highland Park plant, the very space where the company first installed assembly lines to produce Model Ts. But Ford's techniques simply did not work when applied to mechanical computers.[55] Ford did not even begin large-scale production of the M-7 directors until three months after the original contract was to have been completed. The program eventually produced less than half of the original order of 1,856 directors and rarely exceeded half of the planned production rate of 200 per month.[56]

Sperry and the army explored two solutions to these production problems: simplifying the mechanisms of the computers and moving to electrical computations. Toward the first, the army adopted an English machine known as the "Kerrison director," named after its designer, renaming it the M-5 director for light antiaircraft guns. It had simplified (though less accurate) computations, resulting in simpler mecha-

Fig. 3.13. Numerical control: manufacturing ballistic cams. First, holes were drilled into a cylinder, each hole corresponding to a data point from a ballistics table. In the background is such a numerical table. Courtesy of Hagley Museum and Library.

nisms that required only one ballistic cam. Sperry redesigned the M-5 for high-volume production in 1940 but passed responsibility for manufacturing it on to Singer Sewing Machine, Delco, and the Ford Motor Company.[57]

The other solution to the production problems—electrical computations—Sperry investigated but never pursued. In 1936 Sperry let a contract to Professor Nicholas Minorsky, of the University of Pennsylvania, to study the possibility of replacing the calculation mechanisms of its mechanical directors with electrical components. Minorsky had worked for Charles Steinmetz at G.E. and had done pioneering work in the 1920s on the theory of control systems for ship steering. He proposed a design for an electrical director, and Sperry asked one of its engineers, Bruno Wittkuhns, to evaluate it. He found Minorsky's plan "entirely too complicated and impracticable" but came up with a scheme of his own to convert a Sperry director to electrical compu-

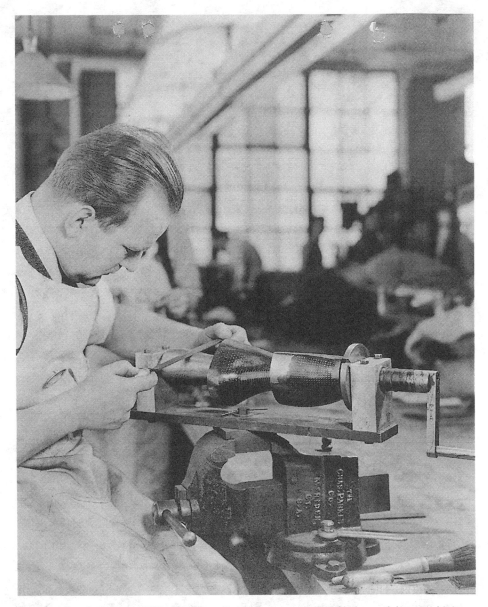

Fig. 3.14. After the holes are drilled into the cylinder, the cams are then filed smooth by a machinist, who also serves to interpolate between data points, a digital-to-analog conversion. Three cams are crafted out of a single cylinder, and the data points are still visible as holes. Courtesy of Hagley Museum and Library.

tation, noting that electrical equipment was "well suited to mass production."[58] Wittkuhns employed follow-up motors with feedback to do the calculations, although his scheme still involved mechanical cams.

The company took no further action on an electrical director, but the episode makes an important point: electrical computers first appeared on Sperry's agenda because they were easier to manufacture. As we shall see, in 1940 an engineer at Bell Labs would "invent" the device Wittkuhns described based on an electrical replacement for the ballistic cams. How machines represented the world affected not only the design of the mechanisms and the role of their human operators but also the types of skills and resources required to produce them.

As industrial products and military instruments, Sperry Gyroscope's antiaircraft gun directors were only partially successful. When the NDRC was formed in 1940, among its first projects was the creation of standardized testing regimes for antiaircraft directors. These tests proved the Sperry machines to be seriously flawed in their firing solutions because their design criteria included only static and not dynamic operation. By 1943 an electronic director developed at Bell Labs superseded the M-7, which ceased production. A decade and a half of development at Sperry Gyroscope had not produced machines that could negotiate the fine line between performance and production imposed by the national emergency.

Still, Sperry's antiaircraft directors of the 1930s were early examples of technology that would assume a critical role in the 1940s. They also illustrate the subtle interplay between computation, human-machine interaction, and manufacturing. In Sperry's systems men were the glue that held integrated systems together. As human servomechanisms, they also acted as amplifiers, renewing the data so that they could make their way through complicated manipulations without losing accuracy. In this incarnation, the "computer" was neither the machine nor its human operators but rather the assemblage of the two. And technical decisions about how to represent the firing data in the machine had concrete effects for the industries required to produce it. When building the electronic and radar-controlled antiaircraft directors of World War II, engineers at Bell Labs, MIT, and elsewhere incorporated and built on Sperry Gyroscope's experience. They too grappled with feedback, control, and the augmentation of human capabilities by technical systems.

## The Transition to War Production

In 1940 Sperry Gyroscope listed the following as its product line, along with the dates when the products were introduced:

Aircraft gyropilot (1931)

Automatic (radio) direction finder (1938)

Directional gyro (1918)

Gyro horizon (1930)

Incandescent searchlight (1924)

High-intensity searchlight (1916)

Course recorder (1918)

Ship gyropilot (1922)

Rudder indicator (1920)

Electromechanical steering system (1930)

Gyrocompass (1914)

Sound locator (1928)

Antiaircraft searchlight (1923)

Universal (antiaircraft) director (1936)[59]

All of these products, as well as the secret bombsights (not listed), were components of control systems. Only two were introduced after 1931. Despite Sperry Gyroscope's emphasis on new technology and its consistent engineering efforts, most of the company's catalog in 1940 did not represent important new inventions. The products had matured in the previous ten years, as had production methods, but antiaircraft fire control and bombsights represented the company's only significant new products in 1940. This stagnation reflects, in part, the effects of the Depression and the passing of Elmer Sperry. Also, several development programs also did not produce lasting products: Sperry naval fire control lost out to Ford and G.E., Sperry bombsights lost out to Norden, Sperry antiaircraft directors lost out to Bell Labs, and Sperry's aerial torpedoes and gyrostabilizers proved impracticable.

For every one of its product lines that stayed in production, Sperry tried several that failed; the company had great difficulty developing distributed, high-performance control systems and deploying them in the field. In fact, the company's history with automatic machinery is as remarkable for its difficulties as for its successes. The major prewar product line, antiaircraft fire control, was discontinued at the height of the wartime boom because of manufacturing complexities. Groups with no experience in fire control were able to learn the field quickly and build better systems than Sperry's. Still, during the 1930s, when military funding declined and government arsenals could not keep pace with new technology, Sperry sustained and developed control systems that otherwise would have stagnated. When the time came to ramp up production for war, the company was ready.

In 1940 the company introduced a number of new products that assured its success during the war. These included klystrons (oscillator tubes used in radar) developed by Russel and Sigurd Varian, whom Sperry supported, giving the company an advantage when radar growth exploded during World War II.[60] Sperry also excelled at simple, easily manufactured controls for fire control aboard aircraft. Unlike battleships, most World War II bombers did not use an airborne equivalent of director fire to coordinate their guns. The machine gunners who defended Flying Fortresses from attacking fighters worked individually, coordinating their fire through voice intercoms. Beginning in 1940 the Sperry Corporation produced these individual controls—as hydraulic turrets for machine gun defenses of B-24 and B-17 bombers. These devices allowed gunners to rapidly and smoothly swing their guns around to fend off attacking airplanes (and automatically prevented them from firing at parts of their own planes).

Sperry Gyroscope also built on its strength in aviation instruments and its corporate tradition, going back to the original gyrocompass, of reference and measuring devices. The company built instruments of perception. Gyroscopic sensors coupled to visual indicators called *lead-computing sights* imposed scales on the gunners' vision and indicated where to aim. "The automatic sight made possible by the simplicity and accuracy of its operation the training of more efficient gunners in shorter periods of time," the company wrote.[61] Vickers made the system's articulation component: small, electro-hydraulic power controls to move the turret. Subcontractors made the glass and steel structure.[62]

These machines, especially the famous ball turret, contributed to a popular image of mechanized air combat during World War II. Their production occupied much of Sperry's wartime resources (Fig. 3.15). At Sperry, at least, the vision of machines as "beast[s] of burden . . . obsessed with motion" survived. Only the B-29, operational late in the war, incorporated "central station" control of its defenses. Sperry Gyroscope developed a prototype of this system, but its design lost out to a G.E. design, partly

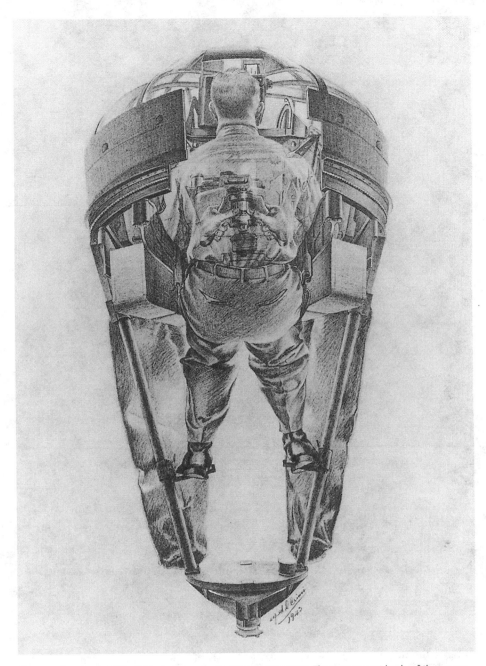

Fig. 3.15. Drawing by Alfred Crimi of an operator in a Sperry turret. The transparent body of the operator makes the machine controls visible. Courtesy of Hagley Museum and Library.

TABLE 3.1    Sperry Corporation Sales and R&D Expenses, 1933–1945

| Year | Sales ($Million) | R&D Expenses | R&D as % of Sales |
|------|------------------|--------------|-------------------|
| 1933 | 3.5716 | $112,451 | 3.15 |
| 1934 | 7.8310 | 216,370 | 2.76 |
| 1935 | 8.6901 | 254,194 | 2.93 |
| 1936 | 14.6841 | 291,033 | 1.98 |
| 1937 | 15.2773 | 352,433 | 2.31 |
| 1938 | 25.3992 | 546,527 | 2.15 |
| 1939 | 24.8561 | 789,437 | 3.18 |
| 1940 | 47.5145 | 1,049,046 | 2.21 |
| 1941 | 99.8195 | 2,211,313 | 2.22 |
| 1942 | 216.2819 | 3,483,221 | 1.61 |
| 1943 | 429.0160 | 4,902,265 | 1.14 |
| 1944 | 420.1860 | 6,783,536 | 1.61 |
| 1945 | 288.9337 | 6,663,513 | 2.31 |

*Source:* Sperry Corporation annual reports, Annual Reports Collection, Baker Library, Harvard Business School, Harvard University.

because of Sperry's overburdened production lines.[63] During the war Sperry made, not the most advanced or intricate products, but rather those that effected simple, tight assemblages of mechanical and human functions and could be produced in large numbers. Even BuOrd needed these devices. As chapter 8 illustrates, Sperry sponsored a university researcher, Charles Stark Draper, to apply flight instruments to defending ships, and his work brought Sperry back into naval fire control after a twenty-year hiatus.

Close, human-centered controls produced great rewards for the Sperry Corporation when the company devoted itself exclusively to war production. Sperry Corporation sales doubled between 1941 and 1942, and they doubled again the following year. The company relied on subcontractors for more than a third of its work, in order to meet increases in production, and in 1942 sales peaked at 17 times the 1939 figures, equivalent to sales in the nine previous years combined (Table 3.1). The government built the company a $20 million plant measuring 1.35 million square feet at Lake Success on Long Island, which opened in early 1942, and by 1943 Sperry employed 50,000 people, ten times the 1939 number. The profits were so high that the company voluntarily returned money to the government.[64]

## Conclusion: Survival of the Beast Vision

Elmer Sperry's greatest contribution may have been the very notion of a company that specialized in control systems as a discrete technology. Based on his vision, Sperry Gyroscope built a broad business stabilizing ships, guiding airplanes, and directing guns, all of which achieved higher performance than human operators could control unaided. The company rarely designed the machines themselves; rather, it added feedback loops to those designed elsewhere.

Sperry Gyroscope's control systems reined in machinery, adding precision to bring technological power into the range of human perception and reaction. Aboard ships, the company's controls closed a feedback loop between the gyrocompass and the ship's wheel, leaving the helmsman to adjust its parameters, monitor its performance, or exchange control, depending on the circumstances. In aircraft, Sperry established a similar feedback loop between gyroscopic instruments and the airplane's control surfaces. Here the human operator was a newer breed, and automatic controls extended pilots' range by reducing their fatigue. In both cases the military services valued the regularity the feedback loops provided, and Sperry built automated aiming systems around the stabilized vehicles. In antiaircraft fire control, the human operators became part of the feedback loops, amplifying and interpreting data at each stage in a complex computation. Each of these control systems called for different methods of representing the world, which had implications not only for the human operators but for manufacturing processes as well.

In 1942 the Sperry Corporation articulated the relationship between its unique approach to control systems, its organization of research, and the critical role of manufacturing:

There has come into being a whole new field of scientific accessories to extend the functions and the skill of the operator far beyond his own strength, endurance, and abilities. . . . The importance to the Government of having these organizations [the Sperry companies] carrying on continuous research along these highly technical lines independent of governmental authority or even popular support is borne out by the fact that now the products of this twenty years of Sperry development must be produced in quantities much greater than the companies can handle.[65]

Sperry argued that its control systems made the critical link between the wartime mobilization of manpower and the mobilization of industry. "Over a billion dollars of this material [control systems] must be produced by us within the next two years. But this billion dollars' worth of technical equipment will fill the vital gap between the one hundred billion dollars' worth of weapons and the thousands of men who must operate them. Without this equipment, neither men nor weapons would be effective."[66] Sperry's control systems united the beasts procured by the military with the men who would ride them into battle.

# Opening Black's Box

## Bell Labs and the Transmission of Signals

> The engineer who embarks on the design of a feedback amplifier
> must be a creature of mixed emotions.
>
> Hendrik Bode, 1940

Like any modern episteme worthy of the name, the theory of feedback has a myth of origin. One sunny August morning in 1927 Harold Black, a 29-year-old systems engineer, rode the Lackawanna ferry to work at the Bell Labs. Many Bell engineers lived in New Jersey, and on the early morning ferry rides across the Hudson to the Manhattan laboratories they frequently gathered on the forward deck for informal technical conferences. On this particular morning Black stood alone, staring at the Statue of Liberty, and had an epiphany: "I suddenly realized that if I fed the amplifier output back to the input, in reverse phase, and kept the device from oscillating (singing, as we called it then), I would have exactly what I wanted: a means of canceling out the distortion in the output." Black sketched his idea on a page from the *New York Times,* "a simple canonical diagram of a negative feedback amplifier plus the equations for the amplification with feedback."[1] He rushed into work, asked a technician to wire up a prototype, and gave birth to a foundational circuit of modern electronics. This story has become enshrined as one of the central "flashes of insight" in electrical engineering in this century and is periodically retold as an inspiration for engineers. A current textbook on control engineering, for example, prints the story of Black's vision verbatim in the first chapter.[2]

At Bell Labs from 1927 to 1940, the legend goes, Black, Harry Nyquist, and Hendrik Bode laid the foundations of control theory, which engineers then applied to all types of closed-loop systems, from servomechanisms to thermostats, fire control systems to automatic computers.[3] More than Sperry Gyroscope or BuOrd's control systems, this story of feedback earned a place in engineering legend and college textbooks. It produced design methods and graphical techniques that carried their authors' names—the Bode plot, the Nyquist diagram—and earned telephone engineering a claim to priority in feedback history. This story of Black's invention follows a cleaner, more intellectual lineage than the military and industrial tales of the previous two chapters.

But this origin myth effaces its sources. It reveals little about the concrete problems these men worked on when they produced their solutions. It skips over the relations between the men and how their backgrounds and prior experience influenced their work on feedback. It does not account for the relationship of the feedback amplifier to the long tradition of governors and self-regulating machinery that preceded it. The story also removes feedback theory from its engineering culture, the landscape of telephone engineering between the world wars.

Thus, a reexamination of the sources is in order, retelling Black's legend, not as a heroic tale, but as the story of an engineer solving the technical problems of a particular place and time and trying to convince others to support his solutions. Black did not understand as much about feedback as he later recalled. To make his idea credible, he needed Nyquist's reformulation of the problem of stability and Bode's analysis outlining the tight constraints that a feedback amplifier had to meet. He also needed "the System." Negative feedback amplifiers emerged from efforts to extend the Bell System's telephone network across the continent, to increase the network's carrying capacity, and to make it work predictably in the face of changes in season, weather, and landscape—the efforts, that is, of building a large technical system and operating it profitably over a diverse and extended geography. Black, Nyquist, and Bode all worked for a company that sought to translate ever more of the world into transmissible messages. This translation required, among other things, ever closer couplings of human and mechanical elements through the medium of sound, couplings that left a discernible mark on feedback theory. For telephone engineers, the network listened, and it spoke.

Black published his amplifier design in 1934, the same year that Lewis Mumford, in *Technics and Civilization,* noted technology's ability to abstract the world. Indeed, retelling Black's story has greater significance than simply correcting the origin myth, for it concerns the historical emergence of electrical signals as representations of the world, the technologies developed to manipulate and transmit them, and the economic and organizational conditions that made those technologies possible. Black, Nyquist, and Bode contributed to an understanding of telephony as the transmission of abstract signals, separate from the electric waves that carried them. AT&T engineers' increasing facility with creating, manipulating, and switching such signals prompted them to rethink the network not simply as a passive medium but as an active machine. Then the Bell System became not merely a set of voice channels but a generalized system capable of carrying any signal as a new currency: information.

## Network Geography

The Bell System was an engineer's dream: geographically expansive, reaching into all types of difficult terrain and climates, and yet always in control, tied to the central office. Still, in the first decade of the century American Telephone and Telegraph (AT&T) did not yet have its later hegemony; it controlled only about half the telephones in the country. Long distance was the key to expanding that share, since competing local operators could not offer the service. The Bell System thus followed its own frontier on a western expansion, often along the tracks of the railroads and the telegraph. From the turn of the century until the 1930s, AT&T expressed its technical milestones in geographical terms: the New York–Chicago line represented carrier frequency transmission; the New York–San Francisco transcontinental line represented vacuum tube repeaters; the Morristown trial simulated the entire country and represented the negative feedback amplifier. "People assimilated telephony into their minds as if into their bodies," wrote the telephone historian John Brooks, "as if it were the result of a new step in human evolution that increased the range of their voices to the limits of the national map."[4]

Despite its continental ambitions, the telephone network at the turn of the century remained a passive device, as it had been since the time of Alexander Graham Bell. Carbon microphones enabled the weak acoustic signal from a speaker's voice to modulate power from a battery, but once the wave entered the line, it traveled to the receiver without further gain. In fact, resistance in the wire imposed considerable losses, known as *attenuation*. Increasing the thickness of the wire could reduce attenuation, but the additional copper proved heavy and expensive. Around the turn of the century, then, the telephone network ran up against the limits of transmission, both in extension, which determined the furthest distance a signal could travel, and in economy, which determined the cost of more moderate distances.

Weather exacerbated the problem. The standard type of transmission, even for long distances, was *open wire*, which meant that each circuit literally had its own wire, sep-

Fig. 4.1. Open-wire long-distance transmission *(top)*, with a cable added to the towers as well. Reproduced, by permission, from AT&T Archives.

arate from the others by a few inches (Fig. 4.1). This separation minimized *crosstalk,* where one conversation leaked to an adjacent wire, and also kept losses to a minimum. Telephone poles with dozens of wires, familiar in turn-of-the-century urban scenes, distinguished this technology. But the lines cluttered the landscape and proved particularly vulnerable to snow and ice storms. As an alternative to open wire, cables bundled numerous smaller wires together. Because they could be buried, cables were more immune to weather and cheaper to install than open wires. But the small diameter and tight packing of wires in cables made for higher losses—twenty to thirty times more signal attenuation than open wire—so cables further pressed against the limits of transmission.

Solving the transmission problem required rethinking telephone theory, transcending the direct-current model that engineers used for the telegraph. In the late nineteenth century, the Englishman Oliver Heaviside conceptualized telephone signals not simply in terms of Ohm's law of voltages and currents but as electric waves traveling down the line. Heaviside observed that over certain frequencies and distances, increasing the inductance of the line could actually reduce attenuation. Such passive inductors placed at discrete intervals along the wire became known as *loading coils* and increased transmission distance. Michael Pupin, of Columbia University, and George Campbell, of Western Electric, working simultaneously, made the loading coil a practicable electrical device.[5] AT&T began commercial installation in 1904, and loading coils rapidly proliferated through the network, especially on cabled routes. Still, the loading coil remained passive; it facilitated the propagation of the wave down the line but added no additional energy.

Heaviside's contribution, however, went beyond spurring this important invention. His *operational calculus* reduced the solution of complex differential equations to simpler algebraic manipulation. He introduced a *step function* (which still bears his name) to analyze a circuit, network, or system in terms of its response to a sudden shock. How a system received the shock—what he called "indicial admittance," today termed *impulse response*—determined the response of a system to any arbitrary input. The technique was analogous to hitting the system with a hammer and watching vibrations as they died out. John Carson, of AT&T, showed that the shape, frequency, and decay of the vibrations provided sufficient information to calculate the response of the system to any input. Heaviside's work was formalized, simplified, and applied to practical problems by Carson and Vannevar Bush at MIT, among others.[6]

This *transient-response* approach described short, instantaneous events. It found wide application in telephony since a voice signal, semi-random in character, could be seen as a long succession of these events. In contrast, *steady-state* methods described systems in their long-term, stable conditions. Much of Charles Steinmetz's work on power systems, for example, used complex algebra to describe steady-state phenomena of alternating current.[7] Early in the twentieth century, however, engineers, including Steinmetz, became increasingly comfortable with describing electricity from both points of view. The translation between transient, time-domain representations and steady-state, frequency-domain representations was greatly aided by Fourier methods—Fourier series, the Fourier integral, and the Fourier transform—which expressed signals as sums of sine waves.

Engineers now saw telephone conversations as simultaneously transient and steady-state phenomena and described and manipulated them in both the time and frequency domains. Modulating a signal, for example, shifted it up or down the spectrum; a radio transmitter modulates a voice from audio frequencies up to radio frequencies for transmission, and then the receiver modulates it back down to audio. Another frequency-manipulation technique, the electric wave filter (also invented by George Campbell), selects a particular set of frequencies and excludes others. Fourier analysis and operational calculus provided the intellectual tools for attacking the transmission problem and the backdrop against which feedback theory developed at Bell Labs. These techniques allowed engineers to manipulate signals, both on paper and in actual circuits.

### Telephone Repeaters: Linking Geography, Technology, and Corporate Goals

Not only techniques but also organization and policy supported the network's expansion. John J. Carty, chief engineer of the Bell System in 1907, had a clear vision of the social role of the telephone network, which he described as "society's nervous system." He and his engineers vigorously pursued the goals of AT&T President Theodore Vail's famous motto, "One policy, one system, and universal service." Carty's vision of industrial research translated corporate goals into technical problems to be solved in the laboratory, sometimes as much for protection against competition as for advancement. One of Carty's longtime associates recalled that for Carty, the design, operation, and economics of the networks were intimately connected.[8] On all of these facets, Carty believed, science could be brought to bear.

And science he needed. By 1911 the state of the transmission art had hit its practical limit: "loaded" lines reached the 2,100 miles between New York and Denver, but attenuation and distortion so mangled voice signals that they were barely understandable after traveling the distance. Yet in 1909, AT&T initiated a project to extend the Denver line to California, making a complete transcontinental line. This geographical problem had a technical core. Bridging the distances required an amplifier, or *repeater,* an active device that added energy to the signal, as opposed to the passive loading coils, which merely stemmed its decay. To solve this problem, in 1911 Carty organized a special Research Branch of the Western Electric engineering department, with E. H. Colpitts as its head.[9] The new department sought a repeater that would renew the signal at intervals along the line to counter the energy dissipated by the wire.

The solution to the repeater problem emerged from a new alliance of corporate interests and the latest academic science. Carty gave technical responsibility for the transcontinental line to a young physicist, Frank Baldwin Jewett. Jewett started at Western Electric in 1904, after a stint as an instructor in electrical engineering at MIT. He had earned his doctorate in physics at the University of Chicago, where he had worked under Albert A. Michelson and became friendly with Robert Millikan. In 1910, when faced with the problem of making repeaters for the transcontinental line, Jewett imagined that a solution, "in order to follow all of the minute modulations of the human voice, must be practically inertialess."[10] Mechanical repeaters had existed for some time, but they were impracticable because the inertia of their elements introduced significant distortion. Jewett thought the secret to "inertialess," hence high-quality, repeaters lay in the electron physics he had studied at Chicago. At his request, Millikan sent several recent Ph.D.'s to AT&T to work on the project, and these men formed an important axis of the company's research for years to come. In the coming decades Jewett was to become an important figure in American science, but within AT&T Frank Jewett's name was intimately associated with long-distance transmission. When he retired in 1944, Bell Labs published an "implicitly biographical" tribute, not a description of the man's life but a detailed technical history of the transcontinental line.[11]

After Jewett, Harold D. Arnold was the first of the Chicago group to arrive at AT&T, where he joined Colpitts's new Research Branch. AT&T had recently purchased the rights to Lee de Forest's audion, or triode, tubes. Arnold, with fellow Millikan disciple H. J. van der Bijl, analyzed electron behavior within the audion, characterized its behavior as a circuit element, and engineered it for mass, interchangeable production. By 1913 Arnold's "high vacuum thermionic tube," later known simply as the "vacuum tube," could amplify signals in telephone repeaters.[12]

This electronic repeater made possible the transcontinental line, which opened at the Pan American Exposition in San Francisco in 1915 with great fanfare. From the East Coast, Alexander Graham Bell repeated his famous first conversation with Thomas Watson, now in California. Vail and President Woodrow Wilson both chimed in as well. The line comprised 130,000 poles, more than 99 percent of them on open wire (the few cables forded streams and rivers). It had loading coils every eight miles, and eight vacuum-tube repeaters amplified the signal in both directions. Still, cross-country calls were far from routine; a three-minute call cost more than twenty dollars and delivered only a third of the bandwidth of standard lines, which meant greatly reduced quality.[13] Its scratchy tone notwithstanding, the transcontinental line brought the entire country into the scope of Vail's unifying vision.

Amidst the fanfare, however, the transcontinental line also marked a less-noted but equally profound conceptual shift: the network became a machine. No longer was the network a passive device, for repeater amplifiers actively added energy along the route. This change decoupled the wave that represented the conversation from its physical embodiment in the cable. Electricity was no longer the conversation itself but simply a carrier, "useful only as a means of transmitting intelligible sounds . . . [with] no appreciable value purely from the power standpoint."[14] Like Sperry's human servomechanisms, a repeater amplifier could add power and renew the signal at any point and hence maintain it through complicated manipulations, enabling long strings of filters, modulators, and transmission lines. Electricity in the wires was now merely a carrier, separate from the messages or signals it carried, opening the door to new ways of thinking about communications.

Now voices became signals and could be specified and standardized. No longer did the system merely deliver conversations according to some vague notion of clarity. Now the telephone company delivered products: signals within a specific frequency range, at a specified amplitude, and with a specified amount of noise. This transformation required standard measures. To measure attenuation, for example, engineers had previously used the "mile of standard cable." This measure became the *transmission unit*, was then renamed the *bell*, and eventually was standardized as the *decibel*, smaller by a factor of ten (still today the standard measure of attenuation). Noise itself became a measurable quantity, and the limiting factor in quality.[15] The message was no longer the medium; now it was a signal that could be understood and manipulated on its own terms, detached from its physical embodiment, effecting what Mumford saw as the "dissociation" inherent in twentieth-century technics.

## The Establishment of Bell Labs

To Carty and AT&T, the transcontinental line proved the value of Jewett's alliance of physics, electronics, and telephone engineering.[16] Duplicating this success in other arenas, however, would require an organizational solidity as well. On 1 January 1925 the AT&T and Western Electric engineering departments combined to form the Bell Telephone Laboratories Incorporated. Bell Labs was responsible to AT&T for fundamental research and to Western Electric for the products of research, and the two companies funded the lab accordingly. The new lab, at 463 West Street in Manhattan, had 3,600 employees, including 2,000 scientists and engineers. Carty served as chairman of the board, which also included vice presidents of Western Electric and AT&T. Frank Jewett became president, and Harold Arnold was named director of research.

While the founding of Bell Labs represented an important milestone for corporate research, it is easy to overestimate its initial importance. The new organization resembled the old Western Electric engineering department, with only moderate differences.[17] Research conducted at Western Electric carried on largely unaltered, as did the careers of the engineers. The lab included departments for inspection, apparatus, and systems development. A special mathematics group under the leadership of Thornton Fry provided calculating services and mathematical consulting for other projects.[18] Most of those working at Bell Labs engaged in the creative, if routine, work of designing telephone equipment and making it work.

Only the Research Department performed "fundamental" industrial research. Headed by Harold Arnold and comprising five hundred people, its mission was "to find and formulate broadly the laws of nature, and to be concerned with apparatus only insofar as it serves to determine these laws or to illustrate their application in the service of the Bell System." The work covered nine main areas: speech, hearing, conversion of energy between acoustic and electric systems (i.e., speakers and microphones), electric transmission of intelligence, magnetism, electronic physics, electromagnetic radiation, optics, and chemistry.[19] Yet even within the Bell System the Research Department did not have a monopoly on fundamental exploration because the Development and Research Department (D&R) of AT&T, with a similar charter and 1,100 engineers and scientists, remained separate from Bell Labs for the labs' first ten years.

Bell Labs also had a Systems Development Department, where Harold Black worked. Yet it did not do systems engineering in today's sense of the term. The Systems Development Department did not formulate an abstract vision of the overall system but in fact had a rather concrete view of the network. This department designed the actual telephone circuits, including equipment structures, office layouts, and the electric power systems required to run the equipment.[20] "Systems engineers" served as liaisons with the operating companies, determined their needs, and translated them into engineering requirements. The 800 people working in the Systems Development Department also studied the growth of the system, projected future needs, and spawned research or development programs accordingly. While it had the widest scope of the engineering departments, Systems Development's vision of the system remained concrete, consisting of the actual groups of wires and switches that made up the network. No one at Bell Labs specifically addressed the system as an abstract entity; rather, all focused on particular pieces of the overall problem, with no systematic integration of all the activities. The overall work at Bell Labs, however, did represent "systematized research," in Director of Research Arnold's words, a concerted attack on a related set of problems.[21] The negative feedback amplifier emerged from the interactions and even the conflicts between the concrete, technical culture of systems development and the more theoretical research world growing within Bell Labs.

## The Technical Agenda

Beyond the New York–San Francisco line in 1915, wires could not go much further (crossing the oceans was considered a problem for radio). But it was one thing to span the continent and quite another to offer high-capacity, economical service over that distance. Furthermore, just meeting demands for growth proved a constant problem: about 800,000 new subscribers were added in 1925 alone. Such expansion required planning and forecasting future requirements based on the rate of growth and detailed cost analysis to determine when new technologies would be required.[22] Engineering studies considered a series of trade-offs between the diameter of the wire, the number of repeaters, the cost of the terminal equipment, and the number of available channels. Increasing the capacity and hence cutting costs over existing long-distance routes began to drive technical development at Bell Labs.

Bringing down the costs of connections meant distributing the capital of the line over several channels. The most promising method, *carrier multiplex*, modulated several voice signals onto high-frequency carrier signals. If these modulations occurred in distinct frequency bands, they could all travel over the same line, in much the same way that separate radio stations occupied the single electromagnetic spectrum. In-

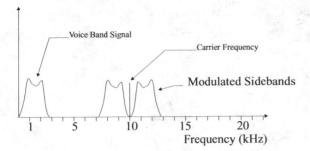

Fig. 4.2. Spectrum of a voice band signal modulated onto a carrier

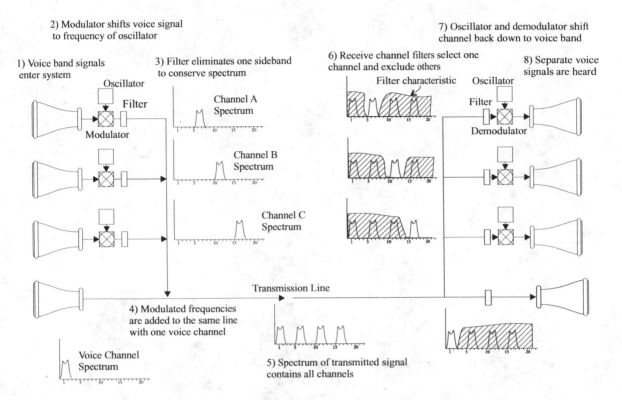

Fig. 4.3. Carrier modulation on cable

deed, the technique became known as "wired wireless" (Fig. 4.2).[23] At the receiving end, a wave filter separated out the voice channels. The idea had been around for a long time: both Elisha Gray and Alexander Graham Bell had investigated carrier techniques in their telephone research.[24] But only when vacuum tubes could cleanly modulate, filter, and amplify the signals did carrier telephony became practicable. The first commercial carrier system, type "A," installed in 1918, put four two-way channels on open-wire pairs.[25] Still, carrier systems had their problems: because of the high frequencies, carrier signals faced greater attenuation than traditional voice band signals and hence required more repeaters (Fig. 4.3).

Another means of increasing capacity was transmission through cables. Cables, which consisted of tightly packed bundles of wire pairs, carried ten times as many circuits as open wires, but at the cost of high attenuation. In October 1925 a cable opened between New York and Chicago, but with delicate and precise construction pushing the limits of the medium. Success came at great cost in machinery and material, requiring a costly, low-resistance cable and extensive loading and repeater equipment.[26] Making long cables practicable and economical required numerous repeaters and massive manpower distributed along the route to maintain the delicate devices. A simple comparison clarifies the difficulties of both carrier and cable transmission: the original (open-wire) transcontinental line used fewer than 10 repeaters across the continent; a carrier system over the same distance needed 40, a cable would require 200, and more still would be needed when carrier and cables were combined.[27] Hence carrier and cable transmission required amplifiers of extremely high quality.

### Searching for the Linear Amplifier

An ideal amplifier is a pure multiplier, taking an input signal and multiplying it by some number (called *gain*) to produce an output. On a graph of output versus input the amplifier's response is literally a straight line whose slope is the gain: a perfect amplifier has a linear relationship between input and output. A real amplifier, however, has phase shift, a time delay from input to output that varies with frequency (it is measured as a difference in angle between a sine wave input and its output). Also, in a real amplifier the output-versus-input curve tends to be, not linear, but more S-shaped (Fig. 4.4). This nonlinearity introduces distortion that causes two problems. First, if the signal is modulated on a carrier, the nonlinearity produces extraneous harmonics outside of the desired signal band. When several signals are modulated onto the same wire, the harmonics from one channel overlap the bands of others, causing crosstalk, one conversation bleeding into another (Fig. 4.5). Second, since each amplifier adds some distortion, a long line with numerous repeaters can garble speech beyond recognition. In the 1920s, as the line became longer and longer and more and more signals were squeezed onto a single wire, the amplifiers had to become corre-

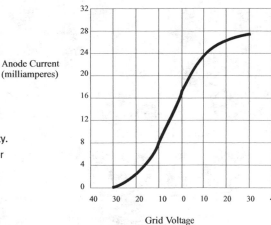

Anode Current
(milliamperes)

Grid Voltage

Fig. 4.4. Typical vacuum tube nonlinearity.
The output, anode current, is not a linear
function of the input, grid voltage.

Fig. 4.5. Nonlinear amplifier causing distortion and crosstalk in a carrier system.

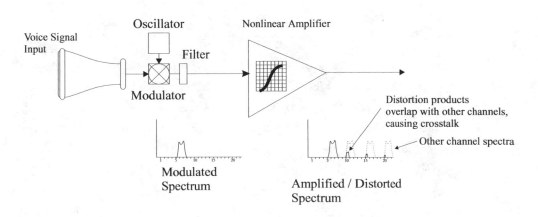

spondingly higher in quality. This problem was a high priority for Bell Labs at its
founding.

When he joined the Systems Engineering Department of Western Electric in 1921,
it was to this problem of linear amplifiers that Harold Black turned his energies. A
Massachusetts native, he had graduated that year from Worcester Polytechnic Institute with a degree in electrical engineering. At that time, the new type "C" carrier systems, which had not yet entered service, were having problems with distortion and
crosstalk. The first approach, and the logical starting point, was to make the vacuum
tubes themselves more linear. Toward this goal, Black worked with Mervin Kelley and
his vacuum tube engineers, but with little success. Despite their utility as circuit ele-

ments, vacuum tubes remained complex, unruly—and nonlinear—devices (hence Kelley's efforts, years later, overseeing the development of the transistor).[28]

Black began to rethink the problem in terms of signals. He now conceptualized the output of the amplifier as containing a pure, desirable component—the signal—and an impure, unwanted component—the distortion. The problem, then, was to somehow separate the two and keep only the pure signal. He came up with a clear, if inelegant, solution: a *feed-forward* amplifier that generated its own distortion and subtracted it from the output signal. Black built a laboratory prototype that achieved the desired result and applied for a patent in 1925.[29] This setup proved that a low-distortion amplifier was possible, but it was far from practicable. Black's overly complex new amplifier required careful attention and continuous adjustment, which engineers could do in a testing lab but not for a system deployed in the field.

### Stabilizing Black's Box

For three years, then, Black struggled to simplify his solution. Finally, in 1927 he had the epiphany on the ferry: if the gain of the amplifier were reduced by some amount, and that amount were fed back into the input, the linearity could be vastly improved. In fact, the distortion was reduced—that is, the linearity was improved—by the same factor that the gain was reduced. A simple explanation of the idea appeared in Black's 1934 paper (Fig. 4.6). Black showed that the gain of the amplifier depends only on the feedback network, $\beta$, and not on the gain, $\mu$, of the amplifier itself. This assumption holds to within $1/\mu$, so if the amplifier gain is 100, then 1 percent of the gain is determined by the vacuum tube, and 99 percent by the feedback network. Yet it was possible to make the feedback network with only passive elements, such as resistors, capacitors, and inductors, which are both more linear than vacuum tubes and more stable with respect to temperature and other changes. For example, a feedback amplifier with a vacuum tube gain of 100,000 could be enclosed in a feedback loop that reduces its gain to 1,000. The linearity of the amplifier overall thus increases by a factor of 100, an incredible improvement. The price, of course, is to throw gain away and settle for a much reduced level of amplification.[30] On 29 December 1927 Black succeeded in making a feedback amplifier whose distortion was reduced by a factor of 100,000 (and whose gain was reduced accordingly).[31]

Still, Black had no easy time convincing others at Bell Labs of the utility of his idea. He recalled that Jewett supported him in his research but that Director of Research

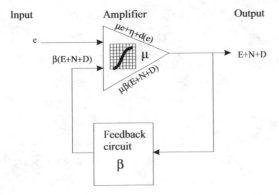

**Input**      **Amplifier**      **Output**

e

$\mu e+\eta+d(e)$

$\beta(E+N+D)$    $\mu$    $E+N+D$

$\mu\beta(E+N+D)$

**Feedback circuit**

$\beta$

**Amplifier System with Feedback**

e   Signal input voltage
$\mu$   Propagation of amplifier circuit [gain]
$\mu$e   Signal output voltage without feedback
$\eta$   Noise output voltage without feedback
d(E)   Distortion output voltage without feedback
b   Propagation of feedback circuit
E   Signal output voltage with feedback
N   Noise output voltage with feedback
D   Distortion output voltage with feedback

The output voltage with feedback is E+N+D and is the sum of $\mu e+\eta+d(E)$, the value without feedback plus $\mu\beta(E+N+D)$ due to feedback.

$$E+N+D=\mu e+\eta+d(E)+\mu\beta(E+N+D)$$

$$(E+N+D)(1-\mu\beta)=\mu e+\eta+d(E)$$

$$E+N+D=\frac{\mu e}{1-\mu\beta}+\frac{\eta}{1-\mu\beta}+\frac{d(E)}{1-\mu\beta}$$

$$\text{If}\ |\mu\beta|\gg 1,\ E=-e/\beta$$

Under this condition the amplification is independent of $\mu$ but does depend on $\beta$. Consequently the over-all characteristic will be controlled by the feedback circuit which may include equalizers or other corrective networks.

Fig. 4.6. Harold Black's negative feedback amplifier. Redrawn by author from Black, "Stabilized Feedback Amplifiers," 3.

Harold Arnold refused to accept a negative feedback amplifier and directed Black to design conventional amplifiers instead.[32] Black had similar difficulties with the U.S. Patent Office. His application for a "Wave Translation System," originally filed in 1928, was not granted until 1937.[33] To a generation of engineers who had struggled to make the vacuum tube amplify at all, throwing away the hard-won gain seemed absurd.

More important, no one could understand how an amplifier's output could be fed back to its input without a progressive, divergent series of oscillations. In fact, Bell engineers at the time found it difficult to make high-gain amplifiers *without* feedback. Subtle, uncontrolled feedback would arise through unintentional effects such as stray capacitance between wires or even between elements within the tube itself and cause the amplifier to go into *parasitic oscillation,* or *singing* (much like the whistling in a poorly tuned public address system). In 1924, for instance, two Bell Labs engineers, H. T. Friis and A. G. Jensen, studied what they called "feed-back or regeneration" as it occurred through a tube, noting that it "makes the total amplification vary irregularly in a very undesirable manner and also makes the set 'sing' at certain frequencies."[34] Black's work ran counter to that of experienced amplifier designers: they sought to eliminate feedback, not to incorporate it.

Black interpreted the resistance to his ideas as evidence of their radical nature. Yet he was an engineer in the Systems Development Department with only a bachelor's degree; he did not possess the analytical sophistication, the communications skills, nor the prestige of the research scientists at Bell Labs. His lab assistant during this period, Alton C. Dickieson, recalled that Black was in constant conflict with his own management, as well as with the rest of Bell Labs.[35] Such conflicts were one thing for a lucid genius, but Black was far from self-explanatory. "A compulsive, non-stop talker," Dickieson recalled, Black "was inventive and intuitive, but not particularly clear at exposition." His negative feedback circuit was only the latest in a series over a period of several years, all of which Dickieson wired up and built, but as Dickieson recalled, "None of the schemes we tried showed any real promise." Dickieson also recalled "quite a bit of rivalry" between the Ph.D.-trained researchers and the systems people. "There seemed to be some feeling that *exploratory* development was the exclusive province of the research people. Mathematicians like Thornton Fry [head of Bell Labs' Mathematics Department] found Black's mathematics beneath contempt."[36] Black—restless, creative, and a bit arrogant—was traversing the established boundaries of the organization, running headlong into the cultural differences between the Research Department and his own, lower-status Systems Development Department.

Credible as Dickieson's recollections seem, no contemporary accounts exist to support or refute them. The documents do allow, however, a thorough analysis of Black's ideas and how Black himself had to transform them (and enlist others to transform them) in order to win their acceptance. A key point surrounds his claim that the epiphany on the ferry included a concern for dynamic stability: that if he "kept the device from oscillating (singing, as we called it then)," it would work. He implies that he understood the stability of the amplifier to be the central problem. But a look at Black's conception of stability at the time reveals it to be different from today's meaning, that is, freedom from oscillation. In fact, Black's conceptions of both stability and negative feedback differed markedly from those of much of the engineering community at the time, although they would have been familiar to engineers working on the telephone network.

## Two Cultures of Feedback and Stability

Today, the adjective *negative* in the term *negative feedback* means that the feedback signal subtracts from the input signal rather than adding to it (i.e., the sign of the feedback signal is reversed). In analogy to James Watt's flyball governor on a steam engine, when the engine speeds up, the spinning balls slow it down, and when the balls spin slower, they speed up the engine. Hence the feedback is negative.

In Black's time, however, even the specific-sounding term *negative feedback* had yet to acquire a stable definition. The idea of positive feedback had become current in the 1920s with the introduction of the regenerative amplifier. Positive feedback, or *regeneration,* in a radio amplifier increased the sensitivity of a receiving tube by sending a wave back through an amplifier many times. Black insisted that his "negative feedback" referred to the opposite of regeneration: gain was reduced, not increased. Yet in the analogous steam-engine governor Black's sense of *negative* means that the power required to spin the balls reduces the power output of the engine (as opposed to the balls' triggering an action that slows it)—hardly a significant effect for a steam engine. In their 1924 paper, Friis and Jensen had made the same distinction Black used between "positive feed-back" and "negative feed-back," based not on the sign of the feedback itself but rather on its effect on the amplifier's gain.[37] In contrast, Nyquist and Bode, when they built on Black's work, referred to negative feedback as that with the sign reversed. Black had trouble convincing others of the utility of his invention in part because confusion existed over basic matters of definition.

Misunderstanding also arose over the critical idea of stability. Dickieson recalled why those concerned with singing in amplifiers did not take Black seriously: "Harold did not even approach the question of stability—he simply assumed that it did not sing."[38] Actually, Black was deeply concerned with stability: his first published paper on his amplifier appeared in 1934 with the title "Stabilized Feedback Amplifiers." But for Black, *stability* referred not to freedom from oscillation but to the long-term behavior of components in the telephone network.[39] Life in the network exposed a telephone repeater to a harsh world, and Black sought to insulate the signal from the brutal reality. He wanted to use feedback to stabilize the characteristics of the amplifier over time. Weather, aging of components, changes in the power supply, and any number of other factors could affect the performance of an amplifier. Rain and temperature, for instance, changed the resistance of the wire and caused significant variations in attenuation, sometimes by a factor of 100 or more over the course of a single day and comparably with the change of seasons.[40] These fluctuations could greatly alter the physics of transmission, a potentially disastrous effect for a system operating close to its physical limits.

Yet to the scientifically trained engineers at Bell Labs, another sense of *stability*, freedom from oscillation, posed the main difficulty for the feedback amplifier. The Bell Labs engineer Homer Dudley, discussing Black's paper in the journal *Electrical Engineering* in March 1934, listed achieving freedom from singing as one of two major difficulties for the amplifier (the other was achieving sufficient gain). Yet this type of stability was not Black's concern. His original patent application, filed in 1928, makes no mention of even the possibility of "singing" or oscillation.[41] When he resubmitted the application in 1932, he added this clarification: "Another difficulty in amplifier operation is instability, not used here as meaning the singing tendency, but rather signifying constancy of operation as an amplifier with changes in battery voltages, temperature, apparatus changes including changes in tubes, aging, and kindred causes. . . . Applicant has discovered that the stability of operation of an amplifier can be greatly improved by the use of negative feedback." Black even acknowledged the other meaning of *stability* but assigned it unequivocal second billing: "Applicant uses negative feedback for a purpose quite different from that of the *prior art* which was to prevent self-oscillation or 'singing.' To make this clearer, applicant's invention is not concerned, *except in a very secondary way* . . . with the singing tendency of a circuit. Its primary response has no relation to the phenomena of self-oscillation."[42] In the patent, Black "simply assumed" that the amplifier did not oscillate.

Black's conception of stability, strange as it may seem, derived from his position in the Systems Development Department as opposed to the Research Department. Where a researcher might focus on the theoretical behavior of the system, Black was concerned with its concrete, daily characteristics. To systems engineers like Black, "stable" amplifiers were those that retained consistent performance in the face of the varying conditions experienced by equipment in the telephone network. Consistency, regularity, and stability of the circuit elements themselves were critical to transmission systems. Black employed this operational conception of stability in the analysis of his amplifier. He used the term *stability* as an engineer who saw the system as a concrete, operational entity, not as one who thought in abstract diagrams.

Nevertheless, systems engineers, despite their emphasis on transmission stability, should also have been familiar with the other meaning of stability. A full repeater requires two amplifiers, one for each direction of transmission. The amplifiers would sing if the signal from one direction of transmission leaked into the other. In response to such problems, telephone engineers filtered out the singing frequencies and limited the amount of gain in each repeater. Carrier systems also tended to sing, either locally or through the transmission line.[43] The now familiar telephone handset, introduced in the late 1920s, depended on understanding and preventing the singing, or "howling," that resulted when the mouthpiece picked up sound from the earpiece.[44]

Moreover, the stability of motion had been a popular topic in physics in the late nineteenth century. Physicists such as E. J. Routh addressed dynamic stability as the absence of oscillatory behavior in a mechanical system and provided mathematical tools for analysis. Stuart Bennett observes that at least some telephone engineers in the 1920s were aware of this work, although they were unsure how to apply it to vacuum tube circuits.[45] So this sense of *stability* would have been familiar to electrical engineers at Bell Labs, possibly even to Black. Those building on Black's work commonly used *stability* to refer to oscillation or singing, and not the stability of transmission.

Multiple, overlapping conceptions of negative feedback and stability thus surrounded the introduction of Black's amplifier. The Bell Labs research culture was not monolithic but comprised at least two engineering cultures: Ph.D.-level mathematicians and scientists interested in fundamental questions and systems engineers like Black, concerned with building the network and keeping it running. Their differing backgrounds and differing notions of ideas like stability help explain why the Research Department did not take Black seriously. As Nyquist and Bode's contributions make clear, it would take both approaches to make the feedback amplifier a practical reality.

When Black "invented" the negative feedback amplifier, he invented a different machine from the one he remembered and from the one it eventually became. Especially in light of his claim that he recognized feedback as a unifying principle across different types of systems, these clashing visions raise the question of whether Black drew on the long tradition of regulators and governors that preceded him.

## Singing and Hunting

Feedback techniques had of course been in common use for a long time in governors, regulators, thermostats, automatic pilots, and numerous other devices. Black later indicated that he understood his feedback amplifier as part of that technological trajectory. The significance of the origin myth rests on Black's supposed recognition that feedback was isomorphic across diverse types of systems. Indeed Black's patent, as issued, states that the negative-feedback principle applies to more than electronic amplifiers: "The invention is applicable to any kind of wave transmission such as electrical, mechanical, or acoustical . . . the terms used have been generic systems." But the patent does not specify what those other applications might be, and a steam-engine governor, an automatic pilot, or a servomechanism hardly fit into the category "Wave Translation System," the title of Black's patent. Black likely had in mind the numerous electro-acoustic translations required in telephony. Neither the patent, nor any of Black's early writings, nor any Bell Labs feedback theory for at least ten years mentions regulators, governors, automatic pilots, or any of the myriad devices we now understand as employing negative feedback.

Nonetheless, such devices were in wide use within the telephone network. Telephone repeaters needed regular adjustment because of environmental effects on the properties of transmission lines. In the late 1920s Bell Labs installed automatic regulators in repeater stations. These devices adjusted amplifier gain by using a feedback loop that sensed the wires' characteristics. The 1929 New York–Chicago line, for example, included 6 regulating stations among its 20 repeaters. These transmission regulators, like Black's "stable" feedback amplifiers, relieved network maintenance personnel of adjusting the delicate amplifiers.[46] In this light, Black's amplifiers effected a kind of automation.

Regulators and governors could also be found within Bell Labs' engineering culture. Sound movies, for example, required tight control lest variations in film speed change the pitch of the sound and become annoying to the listener. Similarly, television systems in development at Bell Labs in the 1920s employed large mechanical disks to scan the picture (instead of the later electron beams). Keeping these disks exactly aligned required precise regulators. In a series of papers published in 1927–29, H. M. Stoller, of the Apparatus Department, explicitly compared his speed controls to steam-engine governors and even discussed the phenomenon of "hunting" (equivalent to singing in an amplifier).[47] He included a drawing of a flyball governor in the *Bell Laboratories Record* and used *stability* in the sense of freedom from oscillation. Stoller even used the words "feed back" for the electrical speed regulation in his own circuits.[48] Had Black looked, he would have found discussion of traditional mechanical regulators in his own organization and its publications.

In fact, the analogy between a mechanical regulator and an electronic one would not have been a great leap for Black, as Stoller had made the connection clearly but without much fanfare. But Black did not take that step. He did not see his negative feedback amplifier as analogous to regulators and governors, and he did not see "hunting" in those devices as comparable to singing in an amplifier.

This critical look at Black's conception of his amplifier provides some perspective on the origin myth. Black's flash of insight, however much it enlightened him on the structure of negative feedback, did not give him an artifact he could sell. Nor did it give him the modern conception of a negative feedback amplifier or a broadly applicable notion of feedback. But it would be wrong to suggest that Black would have found a more receptive audience for his invention had he realized that the amplifier's stability was a key problem, that negative feedback worked similarly to regulators, that singing resembled hunting. These judgments we can only make with hindsight. The important historical point must be made positively: Black saw the amplifier as a means of throwing away gain to achieve linearity in a vacuum tube, a way of stabilizing the repeaters in the telephone system subject to variation and hazard. On these points he was always clear, consistent, and determined.

In his 1934 paper, "Stabilized Feedback Amplifiers," Black presented his amplifier to the world. He attributed the time delay between his 1927 insight and the 1934 paper to corporate secrecy, but that can account for at most five of the seven years. Black's paper, in fact, was not the first word from the telephone company on the negative feedback amplifier; that one, which Black cited and discussed, had appeared two years earlier. It was the work of an ally to whom Black had turned for help but who remade Black's box: Harry Nyquist rethought negative feedback by redefining stability.

### The Morristown Trial

Harry Nyquist, a Swedish immigrant with a Ph.D. in physics from Yale, brought negative feedback from Black's curiosity into the network. Nyquist belonged not to Bell Labs but to the Development and Research Department of AT&T, and he stabilized Black's box by bringing it into the frequency domain.[49]

In May 1928 Nyquist asked Black to join in developing a new carrier system and to include the negative feedback amplifier in a trial of new transmission techniques. This project, known as the Morristown trial, tested carrier transmission on a long-distance cable. It employed 78 repeaters of Black's design spaced 25 miles apart. The cable folded back on itself, so all the amplifiers were located in the same laboratory in Morristown, New Jersey.[50] During this project, Nyquist attacked the stability problem of Black's amplifiers.

Nyquist had already worked on both transmission stability and regulation. He patented a method of "constant current regulation," without feedback, for smoothing out fluctuations in power supply voltages, and he devised ways to use transmission regulators to compensate for phase shifts.[51] Nyquist brought this experience to negative feedback, in his 1932 paper, "Regeneration Theory." There he defined stability in terms of transient disturbances and outlined measurable conditions that determined an amplifier's stability. "For the purpose of studying the singing condition," he wrote, "it is permissible to regard the feed-back phenomenon as a series of waves."[52] If all disturbances impressed upon a circuit died out after a finite period of time, the circuit was *stable*. If the disturbances went on indefinitely, the circuit was *unstable* (Fig. 4.7).[53]

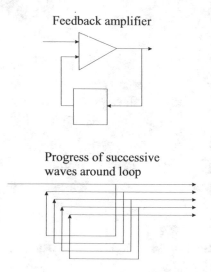

Feedback amplifier

Progress of successive waves around loop

Fig. 4.7. Harry Nyquist's frequency domain interpretation of feedback and stability: "For the purpose of studying the singing condition, it is permissible to regard the feed-back phenomenon as a series of waves" (Nyquist, "Discussion of Black, 'Stabilized Feed-Back Amplifiers,'" 1311). Redrawn by author from Nyquist.

In light of this definition, it became clear to Nyquist that two conditions were necessary and sufficient to make an amplifier unstable and sing. First, the magnitude of the wave coming around the feedback loop must be equal to or greater than that of the input to the amplifier; that is, the gain must be equal to or greater than 1. And second, the feedback wave must be inverted compared with the input wave; that is, its phase shift must be 180°. If, for any frequency, both of these conditions were met, then the amplifier would be unstable and would oscillate.

Nyquist turned these conditions into a simple, empirical method for determining stability. First, the loop must be broken so that the amplifier would not feed back on itself. Then its open-loop characteristics must be analyzed, plotting two easily measured quantities, gain and phase shift, on a polar plot as they vary with frequency. If the resulting curve enclosed the point representing a gain of 1 and a 180° shift, the system was unstable. If the point lay outside the curve, the system was stable (Fig. 4.8).[54]

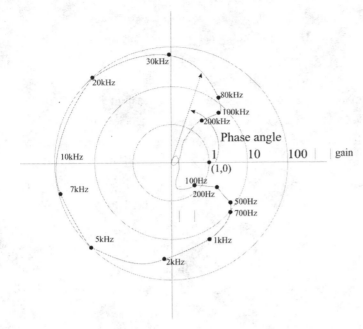

Fig. 4.8. Original-style Nyquist diagram, showing gain versus phase angle plotted on a polar plot. Since the curve does not enclose the point (1,0), the system is stable; if it did, the system would be unstable. Redrawn by author from Bode, "Feedback: The History of an Idea," 114.

This plot became known as a *Nyquist diagram,* and the technique is still referred to as the *Nyquist stability criterion,* or the *Nyquist criterion.* The Nyquist diagram reduced a significant amount of calculation to a simple procedure, a graphical technique, and a tool for engineers to think with. It is still used today.[55]

### Feedback as a Network Problem

It remained for one more Bell Labs engineer, Hendrik W. Bode, to complete telephony's prewar phase of feedback theory. Bode came to Bell Labs in 1926, fresh from a master's degree at Ohio State, where he had also received his bachelor's degree (he received a Ph.D. in physics from Columbia in 1935). Bode's expertise was not in feedback nor even in amplifiers or vacuum tubes but in the useful but esoteric *network theory.*

The theory of electrical networks treated resistance, capacitance, and inductance as *impedances,* complex numbers representing magnitude and phase. *Network analysis* described the behavior of existing networks and derived from George Campbell's early work on wave filters. *Network synthesis,* developed at Bell Labs and elsewhere in the 1920s, formulated a network based on a specified response or behavior.[56] These methods compressed a great deal of algebra into standardized building blocks for making complicated filters with a minimum of components, optimized for a variety of parameters.

As the Bell System adopted carrier transmission and began to manipulate signals in the frequency domain, electrical networks became increasingly critical to telephony. Filter networks, for example, separated specific frequencies out of the spectrum. Similarly, equalizer networks compensated for the distortion in a transmission line by returning frequencies to their original proportions. With these networks, as with repeaters, each element required proportionally more quality as the size of the system increased. Bode recalled that in the early 1930s he "plodded through a long program intended to reformulate certain areas of network theory related to equalizers."[57] In 1934 Bode developed and published a general theory that accounted for all types of filters. Like Nyquist, Bode created graphical tools, calling them "a sort of algebra" that allowed designers to manipulate network designs as diagrams without solving their tangled equations.[58]

Bode's network theory merged with feedback amplifiers in the context of yet another new transmission medium, coaxial cable. So-called "coax" cables had only one conductor surrounded by a conductive shield and could carry much higher frequencies on a single wire than those composed of twisted pairs. Coax operated in the frequency range of millions of cycles per second (MHz), which allowed several hundred conversations to be multiplexed together. In addition, coaxial cables improved transmission stability because their losses varied with temperature simply and uniformly. Still, as with the jump from open wire to cable, the move to coaxial cables placed heavier demands on repeaters, equalizers, and system performance overall. A transcontinental line made of coax would require 600 repeaters.[59]

In 1934 Bode was asked to design an equalizing network for the feedback path of an amplifier for a coax repeater with a bandwidth of 1 MHz that would carry about 400 phone conversations. Here he made novel use of the feedback path. He realized that the overall amplifier behaved like the *reciprocal* of its feedback elements: when the feedback path divided, for example, the amplifier multiplied. An equalizer network, to cancel out the distortion of the transmission line, modeled the inverse of the line. A network to model the transmission line itself could easily be built, but creating a network as the inverse of the line might not be easy, and it might even be impossible. But if the equalizer network were in the feedback path of an amplifier, then the amplifier itself would create the inverse. The feedback network had only to mimic the transmission line exactly, a much more straightforward task (and physically realizable by definition).[60]

The trouble was, Bode had to design the equalizer network for an existing amplifier, and such *post hoc* modification made it unstable. Bode recalled that he "sweated over this problem for a long time without success" and finally, "in desperation," redesigned the entire amplifier by applying his procedure for designing an equalizer to the entire closed-loop system. Where Nyquist had provided a way to determine whether an existing amplifier was stable, Bode now aimed to design a stable amplifier to meet specified parameters for performance.

## Feedback Amplifiers and Mixed Emotions

Bode's 1940 paper, "Relations between Attenuation and Phase in Feedback Amplifier Design," remains his best-known and most succinct contribution to feedback theory. The opening pages have a decidedly pessimistic tone, as Bode comments that the stability of a feedback amplifier "is always just around the corner." He begins,

> The engineer who embarks upon the design of a feedback amplifier must be a creature of mixed emotions. On the one hand, he can rejoice in the improvements in the characteristics of the structure which feedback promises to secure him. On the other hand, he knows that unless he can finally adjust the phase and attenuation characteristics around the feedback loop so the amplifier will not spontaneously burst into uncontrollable singing, none of these advantages can be actually realized.

Bode likens a feedback amplifier to a perpetual motion machine, which always works, "except for one little factor," a little factor that never quite goes away, despite all the tweaking. Elsewhere Bode compared designing a feedback amplifier to "a man who is trying to sleep under a blanket too short for him. Every time he pulls it up around his chin, his feet get cold."[61]

Bode then elucidates the relations between gain and phase shift around the loop, "which impose limits to what can and cannot be done in a feedback design." The conditions for stability, he continues, exact a cost for feedback that "turns out to be surprisingly high." It "places a burden on the designer," and if he does not understand the mathematics "he is helpless." Bode seems to be addressing Black himself, and his uncritical exuberance for the benefits of feedback, regardless of stability problems. "Unfortunately, the situation appears to be an inevitable one. The mathematical laws are inexorable."[62]

Where Nyquist recognized broadly that stability was a function of phase shift and attenuation, Bode defined a specific relationship and expressed it with an equation. Like Nyquist, Bode did not leave a great deal of mathematics for design engineers. Rather, he laid out a simple, graphical technique: plot gain and phase shift, based on observed and analytic quantities, on a logarithmic graph. These graphs approximate exponential response curves with easily drawn straight lines. They survive to this day as "Bode plots," used to determine stability.

Nyquist's stability conditions produced an answer: the amplifier was stable or it was not. But real amplifiers, even if they passed Nyquist's test, still faced variations in temperature, manufacturing differences, and a host of other factors that could alter their characteristics and make them unstable. To deal with these uncertainties, Bode's technique assessed just how close to being unstable an amplifier was. He introduced the concepts of *phase margin* and *gain margin,* which in effect answer the question, When the gain reaches 1, how much phase shift is left before 180° (and instability)? or conversely, When the phase shift reaches 180°, how long before the gain reaches 1 (and instability)? These measures relate the formulas to real amplifiers, "whose ultimate loop characteristics vary in some uncontrollable way." Bode also refined Nyquist's graphic to its modern form, rotating it 180° to erase the effect of a single vacuum tube amplifier, which is inherently inverting and thus adds a phase shift of 180°.

Nyquist's criterion also suggested that to avoid instability, it would be beneficial for the gain to cut off as quickly as possible outside the useful band of frequencies. Bode showed that the gain had to cut off more gradually, for a too rapid cutoff could alter the phase shift in such a way as to induce instability. Because of Bode's limited cutoff rate, the amplifier actually needed to work in a range higher than the useful frequency band in order to be stable. Practically, this meant that "we cannot obtain unconditionally stable amplifiers with as much feedback as we please."[63] The amplifier Bode had originally been asked to examine, for example, only needed to amplify signals up to 1 MHz. Paradoxically, however, it would need the ability to amplify up to 30 MHz just to stay stable, even though it might never see such signals in practice. By imposing limits on the possible performance of the feedback amplifier, Bode brought the sophistication of the network designer to the problem, and he brought the negative feedback amplifier fully into the frequency domain.

Between 1934 and World War II Bode refined his work and taught courses in it at Bell Labs. During the war, the lab distributed an unpublished manuscript to other laboratories working on control systems. Servomechanism designers adapted Bode's technique for mechanical and hydraulic systems. While Bode's name is permanently associated with feedback, he always linked it to its network roots: "It is still the technique of an equalizer designer. . . . I can imagine that the situation may well seem baffling to someone without such a background."[64] Bode's 1945 book, *Network Theory and Feedback Amplifier Design,* reflects his primary experience in networks, with secondary application to amplifiers. Bode acknowledged a certain amount of "unnecessary refinement" of the design methods in the book but explained that they were required for telephone repeater amplifiers, which had unusually high standards for performance.[65] Even today, through Bode's contributions and the plots that bear his name, feedback techniques retain traces of the network theory of the 1920s.

## The Network Machine

With their work on negative feedback, Black, Bode, and Nyquist brought telephone signals into the realm of signals, frequencies, and networks. Their work embodied Carty's idea of the telephone network as "society's nervous system" or, as others in the Bell System called it, "the spinal cord of a nation."[66] Repeater amplifiers exemplified how, within the engineering culture of Bell Labs, machines and people talked to each other in novel ways.

Engineers in other realms of the Bell System also developed technologies that expanded the horizons of perception, integration, and articulation to tightly couple human beings to the network. In the 1920s, for example, automatic switches began to replace human telephone operators. The *Bell Laboratories Record* called the new automatic switch a "mechanical brain" because it could select from thousands of possibilities to make the right connection. Users could now dial their telephones themselves, without resorting to a human operator (though they now had to call "information" desks, where operators looked up numbers in phone books). One Bell observer made a direct analogy between these relays and neurons, suggesting that "by the proper assembly of relays there could be constructed electrical equivalents of human brains."[67] Similarly, "talking pictures" replaced textual representations (subtitles) with anthropomorphic ones (recorded sound). Mechanical reproduction of the moving image had its own silent mystery, but adding the mechanical voice made the film come alive, as any number of popular reactions to the new technology attested.[68] The Bell Labs engineer Hugh Stoller even suggested that the governors that regulated the new soundtracks might also regulate silent films, allowing the orchestra conductor to better stay in synchrony with the image—rationalizing the orchestra as well as the machine.

Research engineers at Bell Labs also studied the borderline between the network and its users, the translation between electrical and acoustic energy, site of perception and articulation. During World War I, AT&T engineers worked on using sound to detect attacking airplanes.[69] They developed microphone detectors and binaural direction finders for antiaircraft systems, loudspeaking intercoms for battleships, and telephone sets for fire control applications for both the navy and the army.[70] After the war, they focused on perception, studying listening and the nature of speech. Harvey Fletcher, another of Millikan's Ph.D.'s from Chicago, examined noise, intelligibility, and the structure of the human ear. He adapted electrical transmission theory to electro-acoustic systems, creating hearing aids and an artificial larynx. Others analyzed articulation, looking at acoustics in auditoriums and the pitch sensitivity of the ear. New notions of sound as a signal were not limited to transmission networks but redefined a broad range of human communications.[71]

From this and related work in electronics emerged the new field of high fidelity audio, also connected to the network. On 27 April 1933, for instance, the Philadelphia Orchestra performed a remote concert. From their home hall, they played through a high-bandwidth line to an audience in Washington, D.C. The orchestra's director, Leopold Stokowski, did not conduct his orchestra in person but rather operated the volume and tone controls in Washington. "Seated at his controls," reported the *New York Times,* "Dr. Stokowski superimposed his interpretation on that of the invisible orchestra." Carrying the music, taken straight from the Morristown trial, were Harold Black's negative feedback amplifiers.[72]

In a similar vein, the Bell Labs researcher Homer Dudley likened human speech to "a radio wave in that information is transmitted over a suitable chosen carrier." Dudley built two speech synthesizers, dubbed Vocoder and Voder, which combined skilled operators with electrical models of the human vocal tract to synthesize mechanical speech through the telephone network. "Communication by speech," he wrote, "consists in the sending by one mind and the receiving by another of a succession of phonetic symbols with some emotional content added."[73] AT&T displayed the Voder with much fanfare at the 1939 world's fair. The AT&T booth offered the public free long-

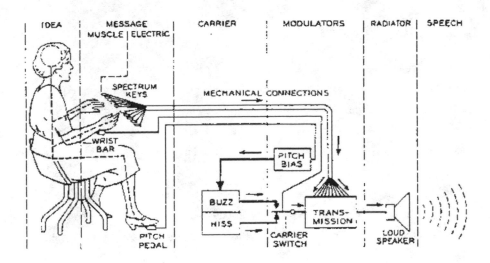

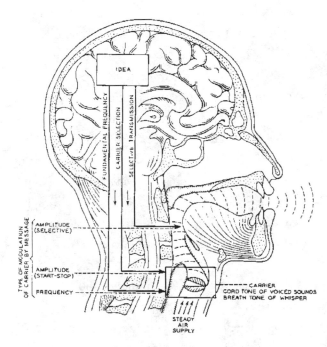

Fig. 4.9. Voder diagrams showing physiological analogy of telephone messaging with information flow across human-machine boundaries. Reprinted from Dudley, "Carrier Nature of Speech," 497, 509.

distance calls, from which the company collected data on human hearing and on the content of phone conversations.[74]

Stereophonic hearing, artificial organs, "invisible orchestras,"—each furthered, in its own way, the integration and extension of human activity by the telephone network. Each also reflected the increasing facility for generating, recording, and reproducing sound, abstracted as audio signals represented by electricity (Fig. 4.9).

## Measuring Text and Speech

Electronic manipulation of signals need not be limited to audio. Theoretical work at Bell Labs showed, in fact, that text, speech, and images could be treated as equivalent when transmitting them down a wire. This approach first coalesced around the problems of telegraphy, for the last word in the name American Telephone and Telegraph remained far more than vestigial. Telegraphy represented a significant source of revenue, and in the early 1930s AT&T introduced teletypewriter service (later "telex"), proudly promoting the service at the Century of Progress Exposition in Chicago in 1933–34.[75] Just as they did with voice lines, Bell engineers pressed to increase the capacity of telegraph lines, searching for ways to overlay both types of signals onto common carriers.

In attacking this problem, Harry Nyquist studied how to represent the world with electrical signals. In a series of papers in the late 1920s he worked out the interchangeability of telephone and telegraph signals, the effect of noise on transmission rates, and the relationship between the bandwidth of a transmission channel and the number of signals it could transmit. In his analysis Nyquist chopped a telegraph signal into discrete bits, or "signal elements," transmitted them individually, and then used them to reconstruct the original signal. This technique showed how to modulate telegraph signals onto carrier frequencies, transmit them alongside speech signals, and separate the signals out again at the receiver. Nyquist showed that the bandwidth required for a telegraph signal was equal to half of the pulse rate of the telegraph.[76] Years later Claude Shannon drew on this "Nyquist rate" for his sampling theorem, which still determines the rates at which our continuous world is sampled and converted into digital form.[77]

Where Nyquist worked out the relationships between continuous and discrete signals, the Bell Labs researcher Ralph Hartley proposed quantitative measures for signals that were independent of the signals' nature or content. In a paper titled "Transmission of Information," Hartley compared the capacity of different channels and measured information as "the logarithm of the number of possible symbol sequences," a definition Nyquist hinted at as well.[78] Hartley grouped transmission media—wire, radio, even direct speech—under the common rubric "line" and characterized each by its bandwidth. Similarly, Hartley defined as "messages" the sequences of symbols sent through those lines. What mattered, Hartley argued, was the range of frequencies inside a message, not whether it represented text, sounds, or pictures. The problem of communications, then, became matching the frequency range of the message to the bandwidth of the line. For example, if the bandwidth of the available transmission channel was narrower than that of the message, then one could record the message on a tape and then play it back more slowly through the line, effectively trading time for bandwidth. Any message could thus be matched to the bandwidth of any channel by playing the tape at a different speed. Applying this notion to the novel television signals, Hartley showed that bandwidth required for transmitting images over telephone lines depended on the picture resolution and repetition rate of the pictures. Anyone used to sending data over a modem today will recognize the significance of Hartley's work relating bandwidth, time, and information content.

Nyquist's and Hartley's work epitomized the engineering culture of the telephone company in the 1920s and '30s, as it began to conceptualize the telephone network as a transmitter of generalized signals, not simply of telephone conversations. Where Black's feedback amplifier enabled the network to renew and manipulate telephone signals, Hartley and Nyquist reformulated those signals as generalized representations of the world. When analyzing the number of symbols required to represent a continuous signal and send it down the line, Nyquist employed only *two* symbols, which he represented by the numbers 0 and 1. Indeed, Nyquist's and Hartley's notions began to resemble digital representations, as their techniques analyzed the subtleties of converting discrete pulses to and from the continuous world. These men laid the groundwork for the theory of information that Claude Shannon would articulate in 1948.[79] In the first paragraph of his famous work, Shannon cited Nyquist's and Hartley's work on transmission and information theory: each related the abstraction of signals to the extension of human activity by the telephone's spreading network.

## Conclusion

Harold Black, Hendrik Bode, and Harry Nyquist brought negative feedback and the vacuum tube within the realm of signals, frequencies, and networks. They furthered the separation of the message inherent in the telephone signal from the energy required to transmit it down the line. Black's feedback amplifier aimed to regulate transmission and insulate the performance of the technical network from its physical and meteorological environment. Nyquist and Bode addressed the immediate problems of frequency response and dynamic stability. Because self-regulation could rapidly turn to oscillation, avoiding instability became a primary concern of feedback amplifier design. The development of the feedback amplifier connected at every point to problems of the telephone network, including long-distance transmission, interchangeable signals, and the role of fundamental research in the overall system. Others working within the system developed similar methods for dealing with speech, music, text, and pictures.

Communications theory did not simply reflect and facilitate converging technical systems. It also supported AT&T's corporate goals. Frank Jewett, speaking to the National Academy of Sciences in 1935, rejected the distinctions between types of signals: "We are prone to think and, what is worse, to act in terms of telegraphy, telephony, radio broadcasting, telephotography, or television, as though they were things apart." Jewett argued that these technologies merely represented different embodiments of a common idea of communication: "They are merely variant parts of a common applied science. One and all, they depend for the functioning and utility on the transmission to a distance of some form of electrical energy whose proper manipulation makes possible substantially instantaneous transfer of intelligence."[80] In Jewett's opinion, government regulation persisted in making distinctions between media (radio, telephony, etc.), each controlled by its own vested interests. If policy followed science, Jewett argued, it too would treat all signals as equivalent. Then AT&T, with its natural monopoly, would emerge as the unified communications company: a builder of transmission, a carrier of long-distance signals, and a switcher of information.

To Theodore Vail's "One policy, one system" motto, Jewett might have added "one signal" to update it with the advances in technology, theory, and the organization of research at AT&T in the 1920s and '30s. Feedback theory at Bell Labs contributed to the rapidly converging ideas about signals and communications that Jewett articulated. It was in this environment that Harold Black had his vision of feedback on the Lackawanna ferry in 1927.

This analytic work, and the graphical techniques it spawned, had no parallel in the world of naval fire control or Sperry's manufactured controls. Neither of those engineering cultures had physical, social, or financial networks extensive enough to support the fundamental research carried out at Bell Labs. As we have seen, the role of theory in the telephone network derived not only from technical problems but also from monopoly position and geographical extent. Sending signals far and wide required that they be abstracted from their physical substrates. Nyquist and Hartley accomplished that abstraction by defining information capacity as bandwidth and establishing the equivalence of diverse types of signals: telegraph messages, voice signals, and television images.

Still, despite burgeoning theories of information and Frank Jewett's call to unify communications as "merely variant adaptations of common physical phenomena," Bell Labs engineers kept their ideas within the existing network. They did not see their contributions to feedback theory as significant to governors, regulators, servomechanisms, or automatic controls. Contrary to Black's recollection, the realization that feedback behaved similarly in a variety of settings did not crystallize until World War II, when new institutions brought engineers from diverse backgrounds together to construct military control systems. Only then were the techniques developed at Bell Labs to deal with feedback, frequencies, and noise applied to mechanical and hydraulic systems, as well as to the human operators themselves. Only then did feedback become prominent as a general principle in engineering, and only afterward, with the work of numerous other engineers and theorists, did Black's, Bode's, and Nyquist's ideas move beyond amplifiers into a broad range of disciplines. Feedback is indeed fundamental to our technological world, but Harold Black's epiphany, more than being a foundational moment, was one of a series of technical insights that allowed engineers to separate human communications from their electrical substrates, send them through geographically extensive networks, and represent the world in a common language of signals.

# 5

# Artificial Representation of Power Systems

## Analog Computing at MIT

Business office practice has been revolutionized by the advent of computing machines.
These deal almost entirely in terms of numbers, as indeed does the business man. . . .
Applied physics, and in fact many other branches of science, frequently deal, however,
with functions as a whole, and usually resort to figures only as a rather laborious
means of dealing with functions or the curves which represent them.
Vannevar Bush, F. D. Gage, and H. R. Stewart, "A Continuous Integraph"

### Feedback Culture

Engineers at BuOrd, Ford Instrument, Sperry Gyroscope, G.E., and the many other companies making control devices in the 1920s and '30s did not belong to a common subdiscipline of engineering. They belonged more to discrete engineering cultures, what we might call "feedback cultures," than to a discrete practice of control engineering. The latter presupposes a recognition of control as a distinct intellectual, theoretical, and professional activity, whereas the term *feedback culture* refers to techniques, tools, knowledge, and, above all, people who were skilled in applying feedback mechanisms.[1]

Feedback cultures in the 1930s had two defining characteristics. First, they primarily addressed the behavior of machines in a *steady state*. The speed of an engine or the course of a ship would be set, and the regulator or servo would "hunt" back and forth while settling on that speed or course. The feedback mechanism maintained *consistency* in the face of external disturbances, such as changes in the load on an engine or changes in wind bearing on a ship. The focus was less on the regulator's behavior while the setpoint was changing than on its ultimate stability. Hunting was acceptable as long as the system settled out in a reasonable amount of time. Automatic pilots, for example, kept airplanes straight and level but did not control them in rapid maneuvers. Feedback culture emphasized the steady state, and not the transient or dynamic behavior of machines, systems, and processes. Regularity and stability were the order of the day.

The second characteristic of feedback culture was its reliance on a set of practices, techniques, and mechanisms endemic to their applications, without a common theoretical framework. Published work discussed specific systems—engine governors, voltage regulators, automatic steering devices, autopilots. Analysts focused on particular types of machines and did not address theory or generality.

This situation began to change in the 1930s, when Harold Hazen, at MIT, published a theory that analyzed control systems with high speeds of response and addressed numerous different machines and systems. Hazen's work emerged not from any particular field of control systems but from an engineering culture at MIT centered on electric power. During the 1920s and '30s, that culture, under the leadership of Vannevar Bush, developed machinery and techniques for modeling these systems in the laboratory. As they moved beyond electric power into more general realms, Hazen and his colleagues articulated an approach to representing the world in machines that later became known as analog computing. Institutional imperatives, a technical aesthetic of generality, and a desire for efficiency drove engineers to make their analog calculators flexible and programmable. The analog approach led not only to a theory of servomechanisms but also to one of switching, routing, and binary representations.

## Early Theories of Control

When Hazen published his theory, feedback devices had existed for a long time, and some mathematicians and physicists had attempted to model them. In 1868, for instance, James Clerk Maxwell had published "On Governors," which proposed a taxonomy of regulators and governors, investigated the phenomena of dynamic stability, and classified devices as stable or unstable according to their response to external disturbances. Still, "On Governors" lacked the idea of a closed feedback loop, so fundamental to later conceptions of control, and was inconsistent in terminology and definitions. "In the world of engineering, the paper was ignored," wrote Otto Mayr, who found no significant references to it in the literature between 1867 and World War I.[2]

Still, Maxwell did have some influence. The Cambridge physicist E. J. Routh and the Swiss mathematician Adolph Hurwitz independently developed a simple method for using the coefficients of a system's characteristic equations to determine its stability without actually having to solve the equations. Indeed, the Routh-Hurwitz stability criteria are in use today by control engineers to determine the stability of a given system (with a yes or no answer). Still, a successful theory needs a receptive audience, and at the time this work remained too theoretical and mathematical for practicing engineers, who did not have broad training in science until the early twentieth century and hence could not apply theoretical tools with sophistication. Furthermore, as Stuart Bennett has pointed out, the insights afforded by Routh and Hurwitz were behind the practical knowledge of the time, so engineers did not use them widely until after World War II. Nevertheless, between theory and practice, by the end of the nineteenth century the basic issues surrounding governor design—speed of response, time constants, stability (freedom from oscillation or hunting), and accuracy (steady state stability)—were well known.[3]

A popular textbook provides a window into the state of regulation theory at the start of the twentieth century. *Governors and the Governing of Prime Movers,* by Professor Willibald Trinks, of the Carnegie Institute of Technology, appeared in 1919. Trinks noted that engineering students learned about steam-engine governors in a course on steam engineering and about pressure regulators in a hydraulics course; rarely was the subject treated as a whole. Trinks aimed to unify the study of regulating mechanisms by defining a governor as "both a measuring device and a motor," introducing the connection between perception and articulation that would characterize control systems in the twentieth century. Extensively citing European and American publications, Trinks analyzed stability, the "promptness" of the governor's return to equilibrium, the natural period of vibration, and a host of other behaviors for a variety of governors. Trinks raised modern issues in control, but he always discussed particular mechanisms, not general systems. His text articulated the approach to control prevalent at the start of the century: feedback devices were additions to prime movers, considered on their own as mechanical devices but independent of their human operators.

An early insight into the theoretical side of control systems, as opposed to regulators and governors, came from Nicolas Minorsky (the man who suggested an electrical fire control computer to Sperry Gyroscope). Minorsky, a Russian immigrant engineer, had been an assistant to Charles Steinmetz at G.E. and shared Elmer Sperry's interest in human interactions with machines. He studied the human eye's ability to perceive a ship's angular motion as it deviated from its course. Minorsky and G.E. installed a prototype automatic steering gear (a competitor to Sperry's gyropilot) for testing on the battleship *New Mexico*. This work led to Minorsky's paper "Directional Stability of Automatically Steered Bodies" (1922), which addressed steering as a feedback problem.[4]

Minorsky countered Sperry's claim that a gyropilot, or even a human pilot for that matter, "anticipated" a turn in any meaningful way. "There is not so much question of intuition as of suitable timing based on actual observation," Minorsky wrote. "Once the element of observation is removed from the helmsman, there can be no accurate steering, whatever his intuition may be."[5] He went on to discuss a number of different methods of feedback control and introduced what became famous as the *three term controller*. This technique, which remains standard practice today, feeds back not only the error signal itself but a weighted sum of the error signal, its time derivative (rate of change), and its integral (accumulation over time); it thus acquired the name PID, for "proportional plus integral plus derivative" control. Later Minorsky published a complete analysis of the Sperry gyropilot, showing it to be of the "proportional plus acceleration" type.[6]

Minorsky's rejection of anticipation took an initial step away from anthropomorphism and toward mathematics as a model for feedback controls. Still, he did not connect his work with vehicles to other work on governors or regulators, and he left numerous questions unanswered. How did the behavior and stability of an individual servomechanism relate to that of the overall system? Did Sperry's phantom in the gyrocompass, for example, share characteristics with the servo that controlled the rudder? or with the larger ship-rudder-gyropilot system? Could the closed-loop behavior of a system be predicted in a quantitative way? Engineers, increasing mathematically literate and scientifically trained, were receptive to new formulations.

## The Theory of Servomechanisms

In 1934 a young MIT professor of electrical engineering, Harold Hazen, published two papers in the *Journal of the Franklin Institute* that addressed these questions and began a transformation from empirical feedback cultures to formal engineering. These papers, "Theory of Servo-Mechanisms" and "Design and Test of a High-Performance Servo-Mechanism," overcame the two limitations of feedback cultures of the time. First, Hazen shifted the emphasis from steady-state mechanisms to dynamic systems relying on continuously varying inputs or setpoints (hence the "High-Performance" of the second title). Second, the papers proposed a theory that applied to all types of loops, from local servomechanisms to large-scale control systems. Hazen also made a critical conceptual leap: he stated that because they translated the low-power input from an instrument of perception into a high-power, articulated output, servomechanisms behaved fundamentally like amplifiers. Earlier papers appear strange to modern engineers, but Hazen's papers so changed the language of the topic that their methods and terminology look familiar to a present-day control engineer.[7]

What enabled Hazen to see that the servo was fundamentally an amplifier, that different types of machinery could be controlled according to a single theory of systems? What problems were he and his colleagues working on that led him to this formulation? Curiously, Hazen's background did not include immersion in industrial feedback cultures. He had had little experience with the control systems of the time, and certainly none comparable to the work at a leader like Sperry Gyroscope. He had had some industrial experience in electrical power at G.E., but no evidence suggests that he worked with governors there. His doctoral thesis, written only three years before his servo papers, lists not a single reference to control or regulation.[8] His papers, in fact, do not even use the term *feedback*, but rather *closed cycle*, for servomechanisms, nor do they use the term *stability*, although they do discuss freedom from oscillation as an important requirement.

No, Hazen's insight did not result from long years in the feedback culture, tinkering and adjusting governors and stabilizers. Rather, Hazen came from, and helped form, a laboratory culture of analog calculating machines. He designed and built machines to represent the physical and mathematical world; that experience enabled, indeed directed, him to develop a practice of analog computing and to understand control as a general principle. To trace the route by which Hazen came to the servo problem, then, we must first examine the trajectory of electrical engineering at MIT in the decade before his famous papers, as well as the problems that defined the intellectual climate at that time.

### Engineering in the 1920s and the Stability Problem

In the United States after World War I, tensions between technology and society pulled engineering in several directions at once, generating a strange mixture of conservatism and reform. Progressives thought engineers could apply their knowledge and objectivity to solving social problems and help make a more equitable society. Conservatives thought the profession should gracefully serve corporate America, the new repository of science and technology. Engineers were caught between social ideals and their desire to consolidate a professional identity.[9] Herbert Hoover, the most prominent engineer of his day, exemplified these tensions: he rose to prominence and power as a manager, administrator, and problem solver but rarely referred to himself as an engineer.[10]

Similar paradoxes could be found at the other end of the political spectrum as well. Charles Steinmetz, though a committed socialist, spent his career doing research for G.E. Steinmetz did more than any man of his day to bring mathematics and theory to the practice of electrical engineering.[11] His field had an inherently close alliance with physics and depended on a high level of mathematics and abstraction. During the 1920s electrical engineers actively shaped the American scene, overseeing the construction of telephone and electric power networks: systems of subtle, often invisible workings, huge capital investment, and obvious social importance.

By the 1920s regional electric power systems were proliferating throughout the country. They increasingly connected into interregional and national "grids" in order to even their load factors and broaden their markets.[12] These networks, referred to as "superpower" systems, raised difficult technical problems. They connected a number of generators—at hydroelectric, steam, and coal stations—each with devices regulating voltage and frequency. The generators drove long-distance transmission lines connected to a series of loads, such as factories, streetcar systems, and residential areas. The characteristics of these distributed networks, however, were poorly understood. By the early '20s electrical engineers recognized that the stability of these networks posed a problem, but they lacked consensus on how to approach it.[13]

A simple example illustrates the stability problem. Consider an alternating-current system with two generators feeding a transmission line with a load at the end. Because of the alternating current, the generators on the line must be synchronized so that their sinusoidal power curves reinforce rather than fight each other. When a load is applied to the line, each of the generators will have to supply additional power. This will cause them to briefly slow down and hence "fall back" in phase. After some time, the governors on the engines driving the generators will respond to the drop in speed and bring the generators back "into step." But the generators on the line may be of different sizes and have different reaction times, hence they will come into step at different rates. As one generator comes into step, another generator coming from behind could push it to overshoot the point of synchronism, and the two might oscillate, or *hunt*, about the stable power point. If these oscillations are small and decay with each successive cycle, they will die out harmlessly. If, however, they grow, the system will become unstable: progressive oscillations will cause it to shake itself apart or to fail by exceeding its power limits.

Another way to think about the stability problem is that the power being drawn from a network has to match the power being fed into it. As one 1925 analysis put it, "The problem of stability is one of securing a proper balance between mechanical input to a generator and its electrical output, and the electrical input to a motor and its mechanical output. . . . It is, therefore, an exceedingly complicated problem, involving mechanical factors such as inertia, governors, gate speeds, etc., and also electrical factors such as machine characteristics, line constants, breaker operation, etc."[14] By the mid-1920s, unstable long-distance lines were causing power surges and blackouts, calling the industry's attention to the problem.

Before Steinmetz died in 1923, he established the foundations for analyzing the AC machinery that made up power networks.[15] His techniques primarily analyzed steady-state power distribution, during normal operation. Yet the stability problem became critical in the face of short-lived events, or *transients,* such as lightning strikes, sudden applications of load, and short circuits. When a factory started up, for example, or a section of the grid tripped off, a transient moved through the network in the form of a traveling wave. Ideally, the network would damp the transient, and it would die away after a short time. If the transient initiated secondary effects that caused it to grow, however, it could increase indefinitely, or until the network was damaged or shut itself down. Steinmetz made some initial forays, but by the mid-1920s the study of transient phenomena was still in its infancy.[16]

To engineers at MIT the stability problem represented both a challenge and an opportunity, for they had a different set of priorities from their industrial counterparts. As educators, they needed to train technical talent for industry, but as academics, they also sought intellectual legitimacy within the university. The former meant working on problems of immediate relevance, while the latter meant making fundamental contributions of general importance. They also had to stay competitive with the large, well-financed research laboratories at G.E. and AT&T. Dugald Jackson, head of MIT's electrical engineering department, realized that these constraints, if creatively applied, need not be mutually exclusive.

Like Steinmetz, Jackson had a spirit of reform, but of a different sort: he aimed to educate engineers to become leaders of corporations, where they could steer technology to improve society. Jackson went to MIT in 1907 to revitalize a flagging department. He expanded the facilities, reorganized the curriculum, and established the graduate program. Jackson also extended an institutional hand to G.E., with whom he set up a cooperative educational course.[17] He moved the electrical engineering department beyond an exclusive focus on 60-cycle power to include radio transmission and high-voltage machinery. The department quickly rose to prominence, and increasingly the names of the professors in the department appeared on the key textbooks in the field.[18] Power system stability seemed ideal for the newly exciting department; it was a difficult analytic problem of fundamental interest but with great relevance to industry. This description fit not only power system stability but also calculating machines and servomechanism theory—all aspects of Vannevar Bush's research program.

## The Bush Approach

The name Vannevar Bush, of course, is among the best known in twentieth-century science, both for his work in computing and for his contributions to science policy and the organization of research, especially regarding the establishment of the Manhattan Project. Yet Bush's early career focused on circuit theory and electric power. He came to the institute in 1915 as a graduate student, working with Arthur Kenelly, who had been an electrical assistant to Thomas Edison. Bush earned MIT's fifth Ph.D. in engineering the following year. During World War I he worked on sonar submarine detection and even published on gyroscopic stabilization of ships.[19] As a young faculty member, Bush enacted Jackson's philosophy of immediate relevance and industrial relationships, although he added theoretical sophistication and an emphasis on fundamental problems. In 1923 Bush began working on the problems of transients and power system stability. When Karl Compton became president of MIT in 1930, he swung the pendulum back toward basic science and fundamental research. Bush easily adapted, moving away from industrial concerns and toward fundamental research in calculating machinery and its application to scientific problems.[20]

One mathematical technique united Bush's early work: the operational calculus developed by the Englishman Oliver Heaviside, which Bush applied to power system transients. Heaviside dealt with short-lived transient phenomena by using mathematical operators to reduce the intricate manipulations of differentiation and integration to simple algebra. Mathematicians did not take the work seriously because Heaviside did not rigorously prove its validity, but engineers paid attention. The operational calculus could indeed solve useful problems, and AT&T's John Carson adopted it for simplifying analysis in telephone engineering.[21]

Building on Carson's work, Bush put the operational calculus on a rigorous mathematical foundation. Heaviside's methods revealed how a step input, analogous to a transient, would affect a system, and Bush made the mathematics more consistent and understandable. Bush's first book, *Operational Circuit Analysis,* applied the technique to practical problems. It included an appendix by Norbert Wiener on Fourier analysis and frequency domain techniques, which began to relate transient and steady-state analyses. Wiener, who was on the mathematics faculty at MIT, served as a mathematical mentor to Bush, and the two collaborated for many years.

Bush's key conceptual contribution was the observation that throughout engineering one finds the basic idea of the *circuit,* defined as "a physical entity in which varying magnitudes can be sufficiently specified in terms of time and a single dimension" (as opposed to field problems in two or three dimensions).[22] Bush applied operational calculus to circuits in hydraulics, mechanics, electricity, and acoustics. This unifying, analytical project treated engineering systems as abstractions and paralleled the work at Bell Labs that manipulated electricity as signals in both the time and frequency domains. It also carried over into Bush's work on calculating machines and influenced Hazen's extension of servo theory to a variety of systems.

### Transients and Governors

If all systems were circuits, then one could study a system by modeling one form of circuit with another. In 1925 Bush brought this approach to power system transients. With his colleague R. D. Booth, Bush noted the difficulty of the problem and three ways to attack it: mathematical analysis, tests of laboratory models, and examination of direct experience.[23] Interconnected power systems were too new for there to be much direct experience on which to draw, so analysis and tests were the favored tools. "The final check of theory is by test," Bush and Booth wrote, "and the final attack on the actual problems of system design must be by analysis." Their central question was: when analyzing a system of power stations connected by transmission lines and operating close to its power limits, "what is the degree of stability of such a network when subjected to disturbances of the types likely to be encountered in practice?"[24]

To help answer this question, Bush and Booth proposed a *point-by-point method* of calculation, starting with the steady state of the system and then calculating how it changed for an increment of time during a transient. They could then piece together how a system behaved for a time interval and then extend that solution out for a series of such intervals. This method involved tedious repetition unless one had an easy way to figure the individual points, so Bush and Booth described a *superposition method,* whereby the engineer physically overlays paper graphs of the machinery's characteristic curves in order to solve for its operating point. Bush had used graphical techniques before, for a "profile tracer" he built to survey land elevations.[25] For the laborious stability calculation, a graphical approach led to his research program in calculating machines.

Bush and Booth also recognized that "the behavior of exciters, governors, and regulators" played a critical role in power system stability. "Unfortunately complete information in regard to the behavior of all types of governors is not yet available in the form necessary."[26] Experts in the field agreed with Bush and Booth's assessment of the inadequate state of feedback devices. "A complete paper could be written on steam and hydraulic governors," one commentator wrote of Bush's work, "and such a paper would I regret to say deal chiefly with their shortcomings." More study of individual machines, their regulators, and their behavior when connected into systems was definitely in order. "There is room for a great deal of improvement, and such improvement will come by studying their [governors'] characteristics in connection with the problem of stability."[27]

## MIT's Culture of Stability

During the period 1924–31 much of the work under Bush's supervision concerned network stability or transient phenomena. By the end of the 1920s nearly one in five graduates of MIT's electrical engineering program addressed stability-related issues in their theses. During these years Bush's laboratory was a lively place, as young engineers attacked important practical problems with new techniques. "Ideas," recalled Hazen, "were just growing thick and fast all over." Bush would visit the laboratory as an overseer and problem solver. "We in that laboratory would work around for a few days. . . . Bush would drop in unannounced. . . . When he left, the air was full of new ideas to be exploited, and everyone was ready to start ahead full speed again."[28]

Bush's students included a number of men who would later become leaders in electrical engineering. Frederick Terman, for example, would fulfill Jackson's vision of academic leadership and industrial relationships as he went on to build Stanford's engineering programs and become "the father of Silicon Valley."[29] Terman's 1924 doctoral thesis, "The Characteristics and Stability of Transmission Systems," addressed the proposed superpower systems, which would span vast areas of geographic space and "must operate under conditions and near limits not approached by any of the lines now in existence."[30] Terman argued the problem of stability in electrical power networks intimately related to the behavior of governors and regulators. Not only did individual devices affect transient behavior, Terman wrote, but the network itself had much in common with these devices. He sought to understand the effects of regula-

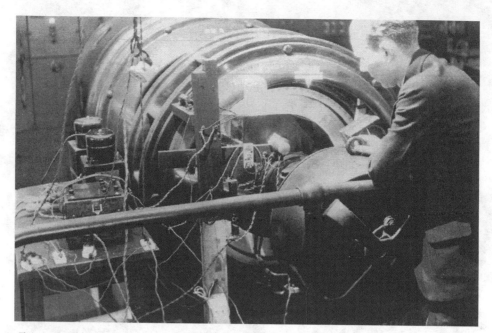

Fig. 5.1. Harold Edgerton studying power system stability with a stroboscope and frozen north and south poles on generator. Courtesy of MIT Museum.

tors and governors on the stability of the system and, in turn, how the characteristics of the network influenced the behavior of the governors. "An electro-mechanical transient develops very similar to the general phenomenon of hunting," Terman wrote, concluding that the stability of the overall system had characteristics similar to those of an individual governor.[31]

Another Bush student with a bright future, Harold Edgerton, arrived at MIT in 1926 following a period "on test" in the cooperative program at G.E. His 1927 master's thesis, "Abrupt Change in Load on a Synchronous Machine," calculated system stability using Bush's point-by-point method and showed ways to avoid instability by applying loads gradually. For his doctoral work Edgerton used a stroboscope synchronized to the AC power line to visually freeze the poles of a generator. He attached white cardboard signs marked "N" or "S," for the north and south poles, to the generator. Though the rotor was in motion, the letters would appear to stand still under the flashing light. When a sudden load appeared, the letters seemed to rotate backward, then slowly catch up, overshoot, and oscillate about a certain position as the governor acted (Fig. 5.1). Edgerton's 1931 dissertation presented similar observations made with a new mercury-arc stroboscope, work that led him to his famous work in high-speed flash photography.[32]

Other Bush students during this time included Kenneth Germeshausen, later a co-founder, with Edgerton and Herbert Grier, of EG&G Inc., and Cecil Green, later a philanthropist and the founder of Texas Instruments. Titles like "The Parallel Operation of Alternators through Long Transmission Lines" and "A Study of Synchronous Machines Not Running at Synchronous Speed" convey the flavor of the engineering culture concerned with power system stability.[33] The stability problem and the behavior of rotating electric machinery under transient conditions shaped the environment at MIT in the 1920s within which students like Harold Hazen matured.

## Modeling and Calculation

Bush adopted two strategies to study transients in power systems: modeling and calculation. After his work on transients with Booth, Bush began thinking about building laboratory models of complex power systems. In 1924 he suggested to Harold Hazen, then an undergraduate, that he write a bachelor's thesis on a small-scale circuit that mirrored the behavior of an electrical power network. Hazen, the son of a lumber and coal dealer from Three Rivers, Michigan, had been introduced to the machine shop by his Sunday-school teacher, and he built electromechanical inventions in his father's basement. He arrived at MIT in the fall of 1920 and would remain for nearly 60 years.

Small models of power networks were not new; Edison's Menlo Park laboratory built one around 1880, and by the 1920s they were becoming increasingly common for both AC and DC analysis.[34] Between 1919 and 1923, O. R. Schurig, of G.E., developed a DC calculating table for analysis of short-circuit conditions in networks. When Schurig built a more generally applicable AC model, however, the machine itself developed a stability problem, "hunting itself out of synchronism" and "shaking apart" when more than a few elements (e.g., miniature motors and generators) were connected together.[35] While electrical parameters (transmission lines) could easily be replicated in miniature, mechanical components (motors and generators) did not scale well, hence the instability. Small rotating machinery just did not have enough inertia to represent larger machines. Put another way, a small motor has little energy storage (inertia) compared with energy dissipation (friction), whereas with a large machine it is the other way around. Thus the miniature systems would not adequately model the real ones and would be even more susceptible to instability.

Hazen, at Bush's suggestion, solved this problem by building a miniature power network that substituted *phase-shifting transformers* for motors and generators. These devices, built with parts loaned from G.E., had electrical characteristics similar to those of generators but did not actually rotate. An operator could adjust them by hand, however, to vary their phase shifts, which corresponded to varying loads and torques. Although the model could emulate only steady-state problems, if one used the point-by-point method, the machine could solve for a transient by breaking it up into discrete points. Each point was then solved in steady state, and a series of points were assembled into a curve for the transient.

With the phase-shifting transformers, Hazen took a step forward in his representation of machinery: they allowed the model to replicate the behavior of the system at smaller scale. Hazen and his fellow student Hugh Spencer described the construction of their miniature power system as "a simple, compact, accurate, easily manipulated laboratory scale means of solving networks." The very title of their paper, "Artificial Representation of Power Systems," suggested their new approach.[36] Today we would call these models *simulations,* but Hazen referred to them as "miniature networks," or "network models."

## Artificial Representation of Power Systems

By themselves, models were not new to engineering; whether for patents, testing, or demonstration, engineers had long put effort into scaled-down versions of their creations. Mathematicians even built solid models of three-dimensional functions. Models of electric power networks had their own problems, however, because they were intended to model not only structure but also behavior and to provide reliable measurements. How small one could make the model depended on the accuracy needed and on the quality of the measuring devices used to observe the model. Attaching a voltmeter to a full-size power network has no effect, as the meter imposes a negligibly small load compared with the amount of power in the system. In a miniature, however, the meter introduced loading that could seriously affect the phenomena under study. As Hazen put it, "When you put a voltmeter on [a miniature network], it's like whacking a factory load onto the actual power system, and that doesn't do . . . any parasitic power requirements take a major toll on accuracy."[37] When building miniature networks, one started with the loading imposed by the meter and then scaled the size of the simulation accordingly. The problem of energy loss (and hence lost accuracy) in a simulated system would persist throughout Hazen and Bush's explorations and would shape Hazen's conception of the servomechanism.

When Hazen graduated, Jackson and Bush encouraged him to stay on and pursue a graduate degree.[38] Hazen went instead to G.E. "on test," as part of the G.E./MIT co-op program, to further pursue network modeling. (Bush himself had spent a year there.) Most of the engineers in this program conducted routine testing, but Hazen got a more interesting assignment, in the office of Robert E. Doherty, G.E.'s chief consulting engineer and a close friend of Bush's. Hazen recalled that the environment was "*the* high-brow engineering office of G.E."[39]

G.E. was then designing a 500-mile transmission line to bring Canadian hydroelectric power into New England and New York. This project posed a difficult stability problem that brought together Bush, the MIT-related consulting firm of Jackson and Moreland, and Westinghouse, in addition to G.E. In Hazen's words, "A five hundred mile line, it was soon found out by those who looked at it, provided a very soft, mushy electrical and energy connection between the generating in far off Quebec and the load center. And what will happen if you just suddenly throw on a little more load? Well, it will oscillate, and you can throw it out of step. The system will break down."[40] Elsewhere Hazen described the problem with an analogy: "Its operating characteristics resembled roughly the towing of one car by another with a long elastic cable stretched almost to the breaking point. Under these conditions, any mishap, such as a short circuit or a sudden adding of load, would in effect snap the towing cable."[41] Hazen spent several months on this problem in Schenectady. In fall 1925 he returned to MIT to continue his investigations as a research assistant and eventually as a graduate student, bringing with him equipment borrowed from G.E. Hazen's 1929 master's thesis, like his undergraduate work, approached the network problem through modeling. He built on the earlier experience to construct, in collaboration with G.E., a new, larger machine, the network analyzer.

### The Network Analyzer

Like the Morristown trial at Bell Labs, the network analyzer brought a geographically dispersed technical system into a single place where it could be studied under controlled conditions (Fig. 5.2). The machine consisted of a set of transmission lines and transformers that replicated the steady-state behavior of a complex network in a laboratory setting. It could also solve transients by running numerous points in series, "making a transient oscillation calculation step by step."[42] Whereas Hazen's earlier power system model represented a particular network, this machine could adapt to

Fig. 5.2. Harold Hazen *(seated)* and the network analyzer. Also pictured are Samuel Caldwell *(left)* and Sidney Caldwell *(right)*. Courtesy of MIT Museum.

new problems, being "sufficiently extensive and flexible to represent numerous actual systems." The network analyzer offered up to 8 generating stations, 60 lines and cables, 40 loads, and a host of transformers, condensers, and other elements. For a given problem, the user employed a plugboard from a telephone exchange to configure and connect these elements to represent any particular system. Just like the fire control systems G.E. built with flexible switchboards, the network analyzer was, to use a modern notion, programmable. Hazen referred to it, in terms that would reappear many years later at MIT, as a "network computer."[43]

The network analyzer typified the engineering approach emerging under Bush: applying mathematical methods and academic research styles to practical problems. It allowed students to experiment with power system design in ways they could not in the real world and to develop an intuitive sense for the systems under study. The network analyzer also embodied MIT's industrial relationships, for flexibility was an organizational as well as a technical feature. The machine served not only students and researchers but also commercial clients, including the American Gas and Electric Service Corporation, G.E., Jackson and Moreland, Illinois Power and Light, Union Gas and Electric, and the Tennessee Valley Authority.[44] Initiating the centralized computing facilities that would become common in succeeding decades, the network analyzer remained operational at MIT until the early 1950s.

Most important, with its programmable structure, the machine embodied Bush's technique of modeling engineering systems as circuits. The nature of the system resided not in the physical form of the object but in its abstract behavior and hence could be modeled by an analogous system. In a similar vein, Hazen used hydraulic models to study the currents in the Boston city water system and flows in the Cape Cod Canal.[45]

## The First Product Integraph

The second approach to the power network problem was calculation. Bush's point-by-point method of evaluating transients required a great deal of mathematics, much of it repetitive and tedious. Hence Bush and his students began building mechanical calculating instruments to directly evaluate the differential equations that described the networks.

AT&T's John Carson pointed out that the stability problem reduced to the problem of integrating the product of two functions.[46] In 1924 Bush, along with his associates Herbert Stewart, a graduate student, and F. D. Gage, a research assistant, built the product integraph, a machine for evaluating these integrals. This machine's components reflected the influence of electric power. Its main integrating unit was a standard watt-hour meter, not unlike the devices used today for measuring household power consumption. For Bush, Gage, and Stewart a typical calculation involved "the problem of transients in circuits due to an applied alternating voltage"—the stability problem.[47]

Like models of power networks, integrating machines had been built before. *Planimeters,* which integrated a curve or an area on paper, had been in use for navigation and surveying since at least the nineteenth century. The naval architecture department at MIT used them for calculating the stability of hull designs. The terminology of the day referred to the MIT devices as "calculating instruments" (analog, graphical devices), as opposed to "calculating machines" (which referred to "numerical" devices like adding machines). The new Bush "instrument" could integrate not only the area under a curve; like fire control computers, it could integrate a function whose upper limit had not yet been defined. The instrument was called an *integraph* since, like hand-operated devices bearing that name, it recorded the result of an integration as a plot or graph.[48] Whereas Bush's point-by-point method overlaid graphical curves to solve the networks, the integraph calculated and drew those curves.

In the product integraph, two functions were first plotted on paper and then fastened to a table or platen that moved along a track at constant speed. Above the table, fixed sliders each had a pointer that could move vertically across the paper as the graph moved laterally. Today's term *graphical user interface* applies literally to this instrument, for two human operators followed the curves with mechanical pointers as the table moved from left to right. These human operators contributed the most significant source of error for the product integraph, as much as 2–3 percent; all other mechanical and human errors did not exceed 1 percent. Since these errors input to an integrator, however, they tended to average out, provided they were as often below the proper mark as above it. Whereas the human operators of Sperry's antiaircraft computer integrated out noise in the data, here the mechanical integrator averaged out human errors. In both cases, as with continuous aim firing, tracking became a difficult problem for control.

These hand-operated pointers generated electrical quantities as input to the watt-hour meter. A power company charges for watt-hours, the product of voltage and current accumulated over time, so the watt-hour meter measures the integral of the product of two functions (Fig. 5.3). Because the output of the watt-hour meter was a delicate spinning disk, if the next stage in computation required too much driving force, it would load the watt-hour meter and cause it to slip and lose accuracy. This situation mirrored the phantom in Sperry's gyrocompass, the human servomechanisms in antiaircraft computers, and the role of repeater amplifiers in the telephone network (Fig. 5.4). In each case the signal had to be renewed as it made its way through the machine. The product integraph used a servo motor "in such a manner that the motor follows exactly the rotation of the watt-hour-meter."[49] This servo motor then drove a pen on a graph connected to the table, plotting output curve as the table moved by.

### The Second Product Integraph

Building on the experience gained with the first machine, Harold Hazen and King Gould built a second product integraph in 1927. This machine, however, had an im-

Fig. 5.3. Vannevar Bush *(left)* and the product integraph, late 1920s. Harold Hazen is second from right. Electric motors drive the plotting tables, vertical boards contain the servomechanisms, and an automobile (Model T) radiator hangs above to cool the precision resistance instruments. Courtesy of MIT Museum.

Fig. 5.4. Functional layout of the first product integraph.

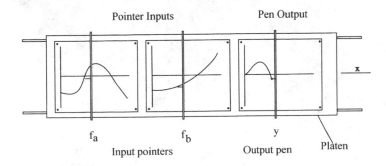

portant new feature: by "back coupling" the output of the integrator to the input, it could solve differential equations rather than just evaluate integrals. This loop enabled the machine to operate on its own results. As Bush wrote, "It is the feedback connection which 'mechanizes' the equal sign in the equation, because it applies the constraint which forces the machine to operate so as to equalize the two sides of the equation."[50]

Again the idea was not new. Similar feedback formed the key difference between Charles Babbage's Difference Engine and his more powerful Analytical Engine. Lord Kelvin realized that he could connect chains of integrators to solve differential equations by "compelling agreement between the function fed into the . . . machine and that given out by it."[51] The MIT machine's back coupling also paralleled the continuous cycle of correction Ford embodied in his rangekeepers, as they fed back the output of the calculation to make the system converge on a solution. Indeed, Hazen hoped to use the connection to solve for the behavior of a vacuum-tube oscillator, itself a feedback circuit.

Hazen also added another stage of integration, a mechanical wheel-and-disk integrator of his own design. He might have used Hannibal Ford's mechanical integrator, but he did not. Ford's fire control work remained secret in the late 1920s, when Hazen built the integraph. Bush knew of Ford's work but was unaware of the details, and secrecy prevented him from informing his students. Still, Ford patented the two-ball integrator as "mechanical movement" in 1919, so it was public information.[52] Ford's integrator could drive a heavier load than Hazen's, but the wheel-and-disk integrators designed at MIT were faster and more accurate than the Ford design, though their fragility made them useful only in a laboratory setting. Hazen would still have to contend with the critical problem of loading.

## Loading Problems

Like the measurement problem in the network analyzer, loading set fundamental limits on mechanical calculators. Hazen and Bush stated the problem and described their solution as follows:

> It is essential that these integrator shafts—in the first stage the watt-hour meter rotor; in the second, the wheel shaft [of the integrator]— be free from all friction and load torque, and hence they cannot directly furnish energy to drive the recording shafts. A servo-motor follower mechanism is therefore used to drive each recording shaft. . . . This mechanism is really the key to the success of the machine from the practical point of view.[53]

According to Hazen, Lord Kelvin understood the potential of mechanical integrators, but he could not build them into complete systems or useful calculating devices because of the loading problem, "the discrepancy between the energy available from a delicate, accurate calculating mechanism and that required to operate dependent apparatus."[54] For the MIT machines, the servomechanism coupled between the stages and abstracted the numerical data away from the machine itself. No longer were the numbers tied to the shaft positions; rather they could be renewed, or amplified, with each successive stage.

Solving the loading problem with a servo distinguished the MIT machines from previous generations of mechanical calculators and made the product integraph into what the control engineer Henry Paynter called an "active mathematical instrument."[55] Once again, as in Ford's rangekeeper or Black's repeater amplifiers, the servomechanism separated the signals, which could be manipulated on their own, from their representations in machinery, which were tied to mechanical limits. The servo made the successive stages of the integraph into modular system blocks—just as human servomechanisms in the Sperry antiaircraft director renewed the information at each successive stage, and just as repeater amplifiers in the telephone network boosted the signal as it flowed through the network. This renewal meant that energy and friction no longer limited the size of the machine. Just as repeaters allowed telephone signals to travel the length of the continent and beyond, servomechanisms allowed Bush and Hazen to build larger calculators.

### The Differential Analyzer

Bush's mechanical calculators entered their third generation with the construction of the differential analyzer in 1928–31. This machine could perform six levels of integration to 0.1 percent accuracy. Though it required a great deal of setting up, lubricating, troubleshooting, and adjustment, it succeeded as a practical calculating device and was applied to problems in a broad range of disciplines. Like the network analyzer, and at about the same time, it became a computing facility at MIT, where scientists from other departments and institutions went to run calculations.[56]

The differential analyzer represented more than a simple progression toward more powerful, general-purpose machines. Bush's research program also reflected the changing institutional priorities for research at MIT. With Karl Compton's appointment as president in 1930, the institute began a shift from industry-oriented problems toward more fundamental scientific work. Bush and his group easily adapted; indeed, Bush helped drive the change. Whereas in the 1920s they had built specialized devices to solve particular industrial problems, they now carried out fundamental research in machine calculation that applied to a broad array of scientific disciplines. Hence the differential analyzer ran problems beyond the bounds of electric power: it graphed equations for electron orbits, engineering structures, geology, cosmic rays, and electronics (Fig. 5.5).[57]

The differential analyzer also traveled around the world, connecting MIT to in-

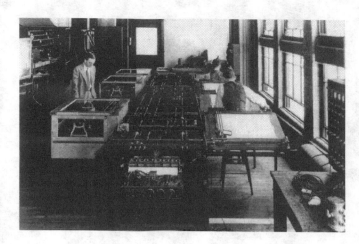

Fig. 5.5. Bush's differential analyzer. Integrating units are in wooden cases to the left, rods and gears at the center connect the units together, and input data come from tables on the right. A printer, for numerical outputs, sits on the table in front of the unit. Courtesy of MIT Museum.

dustrial, academic, and military networks. Bush's lab hosted numerous visitors, some of whom left with blueprints, though one staff member remembered being instructed by Bush not to explain the details of the machine to visitors from Japan.[58] Differential analyzers were reproduced at the Ballistics Research Laboratory of the Army Ordnance Department in Aberdeen, Maryland, at the Moore School of Electrical Engineering at the University of Pennsylvania, and at G.E. in Schenectady. Douglas Hartree and his associate Arthur Porter, of the University of Manchester, in England, supported by the arms firm Metropolitan-Vickers to study feedback systems, built a differential analyzer out of an erector set and then a full-size machine (their machine is now on display at the Museum of Science in London). Other versions appeared later in Ireland, Norway, Sweden, and Russia.[59] By this time the work had transcended the limited goals of power system analysis and focused on calculating machines in their own right. A 1935 list inventories papers derived from the product integraphs and the differential analyzer. It has 54 entries; half relate to transients or stability in power systems, but half deal with unrelated scientific problems.[60]

The differential analyzer also led Harold Hazen to do further work with servomechanisms. For this machine, he solved the stage-to-stage loading problem with Nieman torque amplifiers instead of follow-up servos (Fig. 5.6).[61] These devices employed friction belts on rotating drums (something like a rope around a capstan on a ship) to magnify the output torque from the integrators. Nieman torque amplifiers were not servos because they did not use feedback to hold a particular position. Nevertheless, they had one key characteristic in common with servomechanisms: amplification. In fact, the torque amplifiers had such high gain that, like electronic ampli-

Fig. 5.6. Nieman torque amplifiers (wheels with pulleys) and wheel-and-disk integrators, shown in integrating units with covers removed. Courtesy of MIT Museum.

fiers, they could become unstable if the slightest amount of feedback, such as mechanical vibration, inadvertently coupled the output to the input. Torque amplifiers could become feedback mechanisms by accident. Bush made an explicit analogy between their behavior and that of singing in electronic circuits:

> Now such a torque amplifier is quite analogous to a two-stage thermionic-tube amplifier, and it has many of the properties of the latter, including the possibility of self-oscillation. It was soon found, in fact, that when the amplification of such a low-input unit was raised to around 10,000 it was very prone to go into a condition of violent oscillation, usually ending in disaster. This was presumably caused by a small part of the output being *fed back* in one way or another into the input. This problem caused quite a struggle.[62]

Despite the oscillation problem, Bush made no analogy to feedback amplifiers; he was writing before Black published his work. Operational difficulties notwithstanding, torque amplifiers extended the reach and the significance of the differential analyzer. A machine that could couple stages together without losing energy, corrupting data, or compromising accuracy could make a truly general system, that would be infinitely extensible.

The earlier integraphs enacted a fixed set of equations, but the differential analyzer, like the network analyzer, had "extreme flexibility"; its very structure could change. It presented the user with a set of mechanical elements that corresponded to mathe-

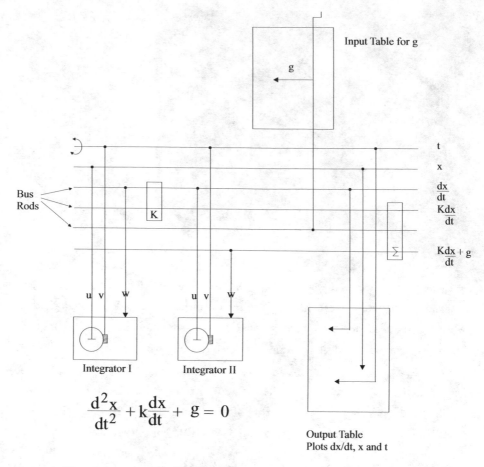

Fig. 5.7. A differential analyzer set up for solving a basic falling-body problem, using Bush's electrical-type notation. Bus rods rotate and transmit data from one unit to another. For this problem, no real-time inputs are needed; all data are input as initial conditions. An input table is provided so that the value for gravity can be easily changed. Redrawn by the author from Bush, "Differential Analyzer," 457.

matical functions and could be rearranged for each problem. Bush compared the method of setting up the differential analyzer to that of the network analyzer: "The scheme of connecting the machine for a specific problem which has been illustrated is quite general. . . . It has certain features in common with the 'plugging' of a desired circuit on a switchboard, and the resulting diagrams have something of an electrical atmosphere about them."[63] The differential analyzer embodied, in a machine, Bush's approach to circuits, for the configuration of the machine was more than a mere mechanical activity; it was an intellectual one with a degree of generality. He designed a graphical notation, with an "electrical atmosphere," for specifying the configuration so that engineers could design mathematics the way they designed circuits. "This [the layout of the machine] is more than a diagram," he wrote of the mental exercise, "it is a process of reasoning, and as such it is recommended to those who seek to import to youth the meaning, as contrasted with the formalism, of the differential equation" (Fig. 5.7).[64] The differential analyzer produced more than answers and graphs; it enacted the world's continuous mathematics in a concrete machine.

## Graphical Calculation as a Research Problem

Historians of computing have referred to the differential analyzer as the first practical, general-purpose means of machine computation for engineering and science.[65] We must be wary, however, of seeing the machine merely as a point in the progress of modern computers, and not as a component of MIT's engineering culture, lest we overlook its institutional setting and that of engineering science between the wars. The historian Larry Owens has gone further, integrating the differential analyzer into its educational environment, writing of the machine's role in making differential equations concrete for pedagogical purposes.[66] Indeed, Bush described how the differential analyzer provided "the man who studies it a grasp of the innate meaning of the differential equation." Bush recalled that the machinist in his laboratory learned differential equations, with no mathematical training, simply by working with the machine. "It was very interesting to discuss this subject with him," Bush wrote, "because he had learned the calculus in mechanical terms—a strange approach, and yet he understood it.[67] Owens argues that the differential analyzer was a tool for teaching graphical language to engineering students and that it "embodied an engineering culture belonging to the first decades of our century." Still, Owens speaks only of the culture of graphical drawing, and not of the burgeoning attempts in numerous fields to represent the world in machines. We can see the differential analyzer as still more fully embedded in the engineering culture that built machines as continuous, graphical representations of physical phenomena, what we now call analog computing (Fig. 5.8).

## Hazen's Analog Approach

MIT's practice of analog computing was no subtly embedded web of assumptions, visible only through the lens of historical analysis. Rather, it was a well-developed and

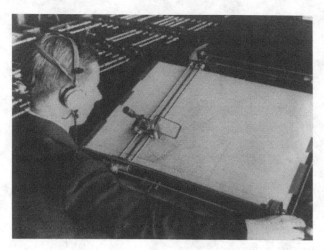

Fig. 5.8. An operator taking voice commands and using a graphical input table to enter curves into the differential analyzer. Courtesy of MIT Museum.

clearly articulated approach to engineering problem solving. In his 1931 dissertation, "The Extension of Engineering Analysis through Reduction of Computational Limits by Mechanical Means," Harold Hazen examined contemporary problems in electrical engineering and how they might be adapted to mechanical solutions. He clearly explained the approach that emerged under Bush's leadership in the preceding years:

It is well to state the sense in which mechanical referring to computations is used in this thesis. It represents the idea of expressing [an] abstract quantity as a physical quantity, such, for example, as length, electric current, light flux, or angular displacement; of applying by physical means the mathematical concepts enumerated to this physical representation of quantity; and of obtaining as a result a physical quantity which can be returned to the abstract form.[68]

This eloquent statement reveals Hazen's deep dedication to the analog art.

The history of computing as currently written tends to view analog computing as unquestionably inferior, a mere predecessor to the digital revolution. By contrast, Bush and his colleagues, steeped in the culture of electric power systems, saw analog computing as an improvement over what they called "numerical" computing (which manipulated symbols rather than physical analogs), not as a precursor to it. They were well aware of the numerical punched-card processing popular in business. But engineers studying power systems, even its transients, lived in a smooth and continuous world. They aimed to build machines that worked smoothly as well, without the messy discontinuities of numerical data. In fact, for the MIT engineers, the continuous nature of the machine was a decided innovation over the numerical methods of office machinery. Bush, Gage, and Stewart wrote that while business practice and business machines relied on numbers, "applied physics, and in fact many other branches of science, frequently deal . . . with functions as a whole, and usually resort to figures only as a rather laborious means of dealing with functions or the curves which represent them."[69] According to this view, numbers (the "resort to figures") were intermediate representations between the physics of the problem and its solution in the machine: "Where a physical problem is involved, models or analogies may replace the need for the solution of algebraic equations as such." New, advanced calculating machines would not use these abstractions but "will deal directly with the functions themselves."[70] For Hazen, numerical computation was not only "costly to apply in involved problems" but also inelegant, needlessly complex, and divorced from the physical intuition that made analogs so valuable. Numerical methods, he wrote, had "an artificiality irksome to the physically minded."[71] Analog calculation, by contrast, tightened the connection between machinery and the world, without the intervening stage of mathematics.

In this sense Hazen refused to follow Mumford's progression of ever-increasing abstraction; instead he preferred machine representations with physical likeness to the world. Of course, analog computers also artificially represented the world in a machine. Hazen's own servomechanisms segregated data from their mechanical substrates, but always by substituting one physical quantity for another. Yet when symbolic representations (e.g., numbers or punched cards) replaced physical ones, Hazen became uncomfortable. He was simply unready to plunge headlong into a world where machines manipulated symbols that had no physical analogs to their referents.

## Lightening the Load: The Cinema Integraph

In 1931 Hazen received his Ph.D. and was made an assistant professor. Although he focused more on teaching and less on machine building, he continued to work with servos. The differential analyzer became popular as a general computing facility, and pressure mounted to increase its throughput. Hazen, together with his student Gordon Brown, designed an *automatic curve follower,* which automatically tracked the curve and automated the entry of data. The device used photocells to sense a curve on paper; electrical and mechanical amplifiers then sent the signal to drive the tracking head. Hazen and Brown exhibited it at the Chicago World's Fair in 1932–33 and attracted large crowds. Although it was intended to automate the data entry for the differential analyzer, no evidence indicates that it was used for calculation (Fig. 5.9).[72]

Gordon Brown had come to MIT from his native Australia as an undergraduate. Through the 1930s, indeed through much of his career, he followed one step behind Hazen, and like Hazen, he would remain at MIT until retirement. Brown too cut his

Fig. 5.9. Harold Hazen and his automatic curve follower for entering data into the differential analyzer. Note the thick black curve being tracked by the machine's photocells. The machine employed Hazen's high-performance servomechanism. Courtesy of MIT Museum.

teeth on the stability problem, and he built a special meter for taking power measurements from the network analyzer. This device employed a negative feedback amplifier of the type that Black was developing at Bell Labs. Brown cited Black's work in a paper on the meter but did not suggest an analogy between servomechanisms and electronic amplifiers with feedback.[73]

Brown's 1934 master's thesis and his 1938 dissertation both dealt with the *cinema integraph*, a further line of research into methods of integration.[74] Norbert Wiener, who advised the Bush laboratory on calculating machines, suggested a way to speed up calculation by lightening the load, literally, on the mechanisms. Plot images of functions on film, Wiener suggested, shine light through the film and electronically integrate the light passing through it with a photocell. King Gould built an infrared version of this device in the late twenties, and Truman Gray built a visible light machine, the photoelectric integraph, in 1930. Brown's device used movie film for images of functions. Although the device anticipated the need for faster electronic integration, it proved an intellectual dead end and never became the general-purpose computing facility that the differential analyzer did. The cinema integraph did function, in Brown's words, as "a machine for producing dissertations" (Fig. 5.10).[75] It also produced Hazen's servomechanism theory.

The cinema integraph used a servo to position the film and also to operate a light shutter to accurately measure light flux through the film.[76] For these problems, Harold Hazen designed the high-performance servomechanism described in his 1934 Franklin Institute paper. Bush, again reflecting the shifting emphasis of MIT in the thirties toward more fundamental approaches, recognized the broader importance of these ideas and urged Hazen to generalize the results in a theoretical work.

## Hazen's Theoretical Papers

In light of Hazen's analog philosophy and his use of servos to renew data in calculating machines, his theory of servomechanisms was not as great a leap for him as it would have been for an engineer immersed in the feedback culture, whose primary goal was the stability of a specific machine. Hazen's 1934 Franklin Institute paper, "Theory of Servo-Mechanisms," provides not only analytical theory but also definitions and taxonomy. It begins by describing how automatic machinery will replace human operators. Hazen distinguishes between "open cycle" (without feedback) and "closed cycle" (with feedback) control, what today would be called open-loop and closed-loop control.[77]

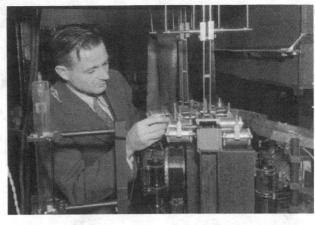

Fig. 5.10. Gordon Brown and the cinema integraph. Courtesy of MIT Museum.

Closed-cycle control is Hazen's primary concern, and he distinguishes between closed-cycle "servomechanisms" and "control systems." The servomechanism, introduced by the French engineer Jean Joseph Léon Farcot in the late nineteenth century, controls a particular device that might be part of a larger control system. In the ship-steering example the servo moves the rudder and maintains its position according to a setpoint. Such devices commonly aided the helmsman, but they did not perform automatic steering. Another loop attached the rudder servo to a compass or gyroscope to keep the ship on course. In his gyropilot Sperry closed this loop by including his gyrocompass, the rudder servomechanism, and the ship itself in the feedback loop. Hazen would call this loop a "control system."

Hazen's classification of controllers provides the consistent hierarchy of servomechanisms that Maxwell attempted without success. Hazen's "Theory of Servo-Mechanisms" defines three types of servos: relay, pulsed, and continuous. In a relay, or on-off, servo, "widely used because of its simplicity," the actuating force is constant in magnitude when present.[78] A common thermostat works this way because it only controls the binary state of the furnace, turning it on when the actual temperature is lower than the desired temperature and off when it is higher than the desired temperature. The second type, the pulsed servo, operates during regular, fixed intervals, affecting a sort of periodic correction, and is commonly used in digital control systems today.[79] Not surprisingly, Hazen's primary interest, however, lay in the third type, continuous control, "in which the restoring force, acting continuously on the output element, is approximately proportional to the deviation of the output."[80]

Hazen analyzes the three types in turn, evaluating each for oscillation and time lag

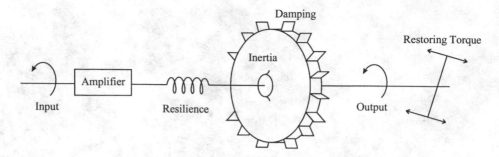

Fig. 5.11. Harold Hazen's mechanical analogy of the servomechanism as an amplifier, 1934. Note the lack of a feedback path, even though the restoring torque is a function of the difference between the input and output. Redrawn by the author from Hazen, "Theory of Servomechanisms," 317.

in response to an input. He emphasizes time lag because of his interest in dynamic performance. He cares how the servo performs while the input is changing, an emphasis that reflects his years of work on transient phenomena. Continuous controllers work best, he concludes, "where accurate, rapid following [of an input] is required" and are also easiest to analyze. He establishes a unitless "figure of merit" for servos, which could both evaluate performance and serve as a quantitative basis for design (the notion survives today as *damping ratio*). Hazen's "easy and complete analysis" shifted emphasis away from the simpler, relay-type systems and toward the higher-performance, continuous controllers.[81] He also presented a general-purpose model of the servo loop as a mechanical system, abstracted from any particular implementation (Fig. 5.11). Again, the abstraction directly inherited Bush's treatment of varying types of systems with the common notion of the circuit.

One aspect of Hazen's theory in particular enabled him to achieve this generality. He recognized that the sensors that provide the error signal (instruments of perception) are generally low-energy devices, whereas the machine to be controlled tends to require higher power (articulation). Hazen thus defined the servo as "a power-amplifying device." In an automatic pilot for a ship, for instance, "this disparity [in power level] exists between the energy magnitude associated with the measuring instrument, a compass, and that associated with driving the rudder." With no autopilot, "the helmsman serves as a human servo-mechanism."[82] Hazen's use of Sperry's paradoxical term illustrates that the formalization of the servomechanism was never far from the idea of the human operator. Hazen cited both Sperry on the automatic pilot and Minorsky on ship steering but brought the servo into a more theoretical frame. He discussed the primary behaviors of servos, "oscillation and lag." Because of friction and inertia, the output of the servo does not exactly follow the input but lags it by some time interval; if that lag were too great, "oscillation would be expected to occur if definite preventative means were not employed."[83] Paralleling Nyquist's realization that the phase shift in an amplifier could make it sing, Hazen uses the term *unstable* to describe this condition.

Unlike the telephone engineers, however, Hazen broadened his analysis beyond the situation at hand. He argued that his theory of servomechanisms applied to the speed control of steam turbines and water wheels, the stabilization of ships by gyroscopes, the operation of gyrocompass repeaters, the automatic stabilization and guiding of aircraft, and "in fact the automatic recording or control of almost any measurable or measurable and controllable physical quantity." Hazen intended his work to be the beginning of a unifying theory. "To the writer's knowledge," he added, "no systematic quantitative treatment of even the simple common types has previously been given."[84] The papers Hazen cited all discussed control mechanisms for specific machinery or classes of machinery; none mentioned generality or theory. Hazen not only proposed a theory of servomechanisms but suggested that it could be expanded to systems in general, thus breaking down his own distinction between servos and control systems. "Entire closed-cycle control systems," he observed, "are dynamically similar to servo-mechanisms and their operation is investigated by the same methods."[85]

Hazen's "Theory of Servo-Mechanisms" overcame the two limitations of the feedback cultures of the time. Whereas previous work had considered primarily steady-state response, Hazen also considered "high-speed response" as a desirable characteristic. And whereas previous work had not addressed unified theory, this paper initiated the development of analytical tools for control systems.

Hazen's work, generally considered fundamental to the field of control theory, was the subject of intense study for a generation of engineers. The profession quickly recognized its importance. In 1935 the Franklin Institute awarded Hazen's papers the Levy Medal, its highest award for technical contributions. Hazen, who was modest to the point of self-effacement, allowed in his memoirs that the paper became "a standard bibliography item in every subsequent paper or book in the field, and is still widely regarded as a classic."[86] Indeed, nearly every American publication on feedback control in the subsequent decade references Hazen's theory paper, often as the first citation.[87]

Hazen's contributions emerged after nearly ten years of designing servos and applying them in calculating machines. Yet curiously, Hazen himself did not see the overall research program on transients, electric systems, network models, and calculating machines as influences. His emphasis on dynamic behavior, he explained, responded to the need for speed in the cinema integraph and "was not associated with the Differential Analyzer work."[88] This statement reveals a difference between the engineer's conception of historical causality and that of our study. For Hazen, the influential context for his servo work was the project in which it was immediately applied. Here, however, we are concerned with the institutional environment in which he was raised and trained and the problems he and his colleagues faced. Feedback, stability, servos, and MIT's emphasis on fundamental research in engineering shaped the philosophy of analog computing that produced Hazen's papers.

### Limitations of Hazen's Vision

Despite his accomplishments, we should hesitate to see Hazen's 1934 interventions as a grand, unifying gesture. For all his insight into servomechanisms, computation, and analogs, Harold Hazen never made the leap to digital representations, nor to electronics. Hazen's transient analysis retained the time-domain legacy of power systems and did not venture into the frequency domain so familiar to telephone engineers. The limitations of Hazen's vision reveal the inchoate nature of feedback theory in the 1930s.

While defining servos as power amplifiers, Hazen distinguished them from electronic amplifiers. "The servo-mechanism," he wrote, "differs from the simple amplifier in that the responsibility for the functional relation is not placed directly on the amplifying element of the servo." This statement implies that the actual amplifying element in the servomechanism need not be a precision device (i.e., it has no "responsibility for the functional relation"). As long as it provided adequate power to bring the error signal into correspondence with the input, "such an amplifier element can be a relatively crude affair."[89] This notion is equivalent to Black's idea that a passive feedback network could make a vacuum tube into a linear amplifier because linearity did not depend on the tube itself. But Hazen did not make that connection. Curiously, when distinguishing his mechanical servo from an electronic amplifier, he compared his closed-loop example not with a feedback amplifier but with an open-loop amplifier.

Indeed, for Hazen the servo did not constitute a closed feedback loop. Rather, he saw it as an open-loop amplifier driven by an error signal proportional to the difference between the desired and the actual position. Hazen's canonical diagram included no feedback element, nor even the sensor itself. No diagram in either paper included the closed loop so familiar today. Nor did either paper even use the term *feedback*. Hazen was always clear that the input was the difference between a desired and an actual input and that "closed cycle," servos were his primary concern. Yet his omission of closed loops and the term *feedback* is not semantic or coincidental. For Hazen the servo was an open-loop amplifier in which the error signal came from an external element or from a simple subtraction.[90] Hence Hazen did not see any equivalence between his servo-as-amplifier and Black's telephone repeaters.

The conceptual importance of this divide is underscored by the fact that it did not arise from mere isolation of research groups. In fact, MIT and Bell Labs were in close touch. During the 1930s, Bode corresponded with Bush and his colleagues on network synthesis, visited MIT several times, and toured the electrical engineering laboratories. A 1936 conference on network theory brought together Bode, George Campbell, and Thornton Fry of Bell Labs and Bush, Hazen, and many others from MIT. Yet the program made no mention of feedback.[91] The divide between servo theory and feedback amplifiers was deep, not easily overcome by contact on related matters.

Hazen later recalled that he knew of Nyquist and Bode's work on feedback networks but "mentally associated [it] only with communications network theory. I did not recognize at the time the intimate and fundamental interconnection between this and the transient analysis approach [developed in the 1934 papers]."[92] In 1938 we do find Hazen commenting on a paper that described a negative-feedback electronic amplifier designed to make sensitive measurements on a network analyzer.[93] He discussed feedback in electronic amplifiers in detail but made no mention of his servomechanism theory. Hazen's failure to see the connection between servos and feedback amplifiers in 1938 traces the frontier of his vision.

One MIT engineering student did see an analogy, but the recognition went relatively unnoticed. John Taplin read Black's and Nyquist's articles on feedback and found them similar to Hazen's work. "They were all studying the same thing," he recalled, "but they called it by different names." Taplin discussed his idea with Nyquist and then designed a servomechanism using frequency domain instead of the MIT transient-analysis techniques.[94] But there was no eureka moment, no instant union of disparate fields. Taplin himself left MIT for a successful industrial career when he graduated in 1935. The theories of servomechanisms and negative feedback amplifiers began to merge, but so hesitatingly as to be nearly unnoticed. It would take a new institutional environment to solidify the union and bring it to broad attention and utility.

### The Rockefeller Differential Analyzer

Servomechanisms formed but one branch of growing work on calculation, which generated one more significant machine and one additional theoretical insight. Every time the differential analyzer ran a new problem, the machine had to be disassembled and rearranged according to the new equations, a cumbersome, time-consuming, and error-prone task. In 1935 Bush initiated a project to automate these rearrangements, making the machine a production line for calculation. Instead of rotating shafts to interconnect the calculating units, this new machine would transmit its data electrically. A central switchboard interconnected all the units, which could then be rearranged simply by resetting the switches by remote control.

This machine became known as the Rockefeller Differential Analyzer because the Rockefeller Foundation sponsored its construction. After a lengthy and troubled construction period it achieved equivocal success. At least as important as its technical capabilities, however, the Rockefeller Differential Analyzer established relationships between MIT and Rockefeller. Bush's interaction with Warren Weaver, director of the Natural Sciences division of the foundation, shaped both the research program in the 1930s and the wartime work in control. Weaver, a mathematician with a mechanical bent, had a natural interest in the differential analyzer.[95] In 1932 Weaver visited MIT to inspect the machine and gained a deep understanding of the computer, "very impressed by the power and accuracy of the machine." Still, he insisted that Rockefeller support no work in the field of engineering and told Bush not to hope for support from the Rockefeller Foundation.[96]

Undeterred, Bush gently lobbied Weaver for several years, transforming his ambitions from simply building a new machine to setting up a major international center of calculation "to which research workers everywhere will turn for their solutions of their equations."[97] The foundation's aversion to engineering accelerated Bush's move from immediate industrial problems to the fundamental problems more akin to science. Bush emphasized to Rockefeller the differential analyzer's scientific applications and its potential for opening new fields of scientific research, including the "mathematical biology" that Weaver supported.[98] The strategy worked: in mid-1935 Rockefeller gave Bush $10,000 for the early design and planning of a new differential analyzer, followed up by an additional $85,000 for construction the following year, significant sums, especially given the state of the economy at the time.[99] In the following years Weaver made numerous visits to MIT to discuss progress on the analyzer.[100] This new machine would be much more versatile than the earlier versions but also more complex to build.

Indeed, difficulties arose early. The servos had problems with stability and speed.[101] MIT evaluated two of Hannibal Ford's integrators but found that they had problems working at high speed, so MIT opted to develop its own.[102] By 1938, the year Bush left MIT to head the Carnegie Institution, in Washington, D.C., the new differential analyzer had fallen behind schedule. The institute was simply not set up to handle a project of such size. It overwhelmed Samuel Caldwell, who headed the operation, now formalized, with the help of a Carnegie grant, as the Center for Analysis.[103] By 1940 it still was not complete, and the project suffered further when engineers began to leave for the war effort. Weaver put in more Rockefeller money to finish the machine, now citing its utility for national security as a rationale.[104] In 1942 MIT warned Caldwell that he would have to close down the project if he could not complete it quickly, and it did finally go into service that year.[105]

When complete, the device had 18 integrators, could be expanded to accommodate 30, and worked to an accuracy of 1 part in 10,000. The integrators, similar in structure to Hazen's earlier wheel-and-disk type, now used glass instead of metal disks in order to achieve better accuracy. The mathematical units connected together through a compact servomechanism that employed Hazen's recent theory to achieve much higher performance than the earlier servos. The new machine incorporated a "crossbar" switch, borrowed from the telephone network, "to provide paths by which any [data] transmitter can reach any receiver." Indeed, the mathematical units had electrical outputs, so they could be routed electrically rather than mechanically. Bell Labs donated its prototype crossbar switch when the device's development had been completed.[106] The Rockefeller Differential Analyzer, with its combination of servos and telephone switches, began to combine control and communication.

Whereas naval fire control routed information from instruments through banks of switches and the Bell System manipulated signals through banks of relays, Bush's new machine manipulated mathematical quantities through its "trunking" system. A user could set up any mathematical problem merely by selectively opening and closing the switches. Punched paper tapes determined the relay switch closures, the multiplication gear ratios, and the initial conditions of the integrators. The process of setting up

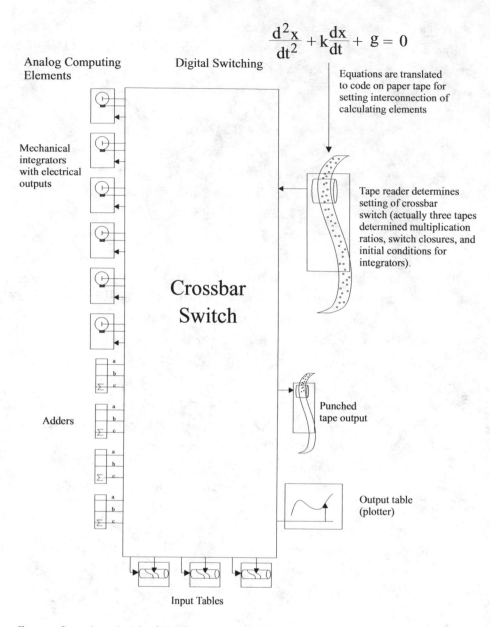

$$\frac{d^2x}{dt^2} + k\frac{dx}{dt} + g = 0$$

Analog Computing Elements

Digital Switching

Equations are translated to code on paper tape for setting interconnection of calculating elements

Mechanical integrators with electrical outputs

Tape reader determines setting of crossbar switch (actually three tapes determined multiplication ratios, switch closures, and initial conditions for integrators).

Crossbar Switch

Adders

Punched tape output

Output table (plotter)

Input Tables

Fig. 5.12. Operation principle of the Rockefeller Differential Analyzer

the machine now reduced to punching the right codes on the tape. A centralized "supervisory control panel" ran the whole process by remote control (Fig. 5.12). The new system also entailed a change in labor: curves were still entered by hand, but now by women with clerical skills rather than male graduate students (Fig. 5.13). With its switched routing of analog signals, the Rockefeller Differential Analyzer was a hybrid

Fig. 5.13. Data-entry operator for the Rockefeller Differential Analyzer. Courtesy of MIT Museum.

analog/digital machine, similar to hybrid systems in the institutions examined in previous chapters.

Setting up the Rockefeller machine entailed a conversion: translating a set of continuous equations into a set of holes on tape. When analyzing this problem, the MIT student Claude Elwood Shannon articulated the potential of the hybrid. Shannon had come to the institute as a research assistant to work on the differential analyzer in 1936, after earning dual bachelor's degrees in mathematics and electrical engineering. Shannon wrote on the mathematical theory of the differential analyzer and elaborated on Bush's circuitlike notation for differential analyzers. While working on the Rockefeller machine, Shannon also became interested in the relays themselves and their potential for computation.

Shannon's 1937 master's thesis, "A Symbolic Analysis of Relay Switching Circuits," examined the logical structure and synthesis of relay circuits "in automatic telephone exchanges, industrial motor-control equipment, and in almost any circuits designed to perform complex operations automatically." Shannon drew on electrical network theory and showed that "several of the well-known theorems on impedance networks have roughly analogous theorems in relay circuits."[107] He also applied Boolean algebra to systems of relays, demonstrating that the relay circuits could be analyzed and synthesized with binary arithmetic, classical true-false logic, and the now canonical functions of *not, or,* and *and.* Once translated to algebra, switching functions could be manipulated and rearranged according to familiar rules and reduced to their simplest form, thus enabling the designer to create the most efficient circuit possible for a given logical function. The paper closed with design examples for an electronic combination lock and a binary adder. With a simple, brilliant stroke Shannon brought the design of switching systems into the world of mathematical logic and network theory.

## Conclusion

While historians have noted the importance of the differential analyzer, Bush himself insisted that the calculating machine provided only one class of solutions and that artificial models provided a necessary compliment.[108] The differential analyzer and the network analyzer represented two distinct but similar expressions of the engineering culture Bush created for his research group. What began as an attack on the industrial problem of power system stability evolved into a program with broad applicability, significant theoretical components, and visibility within the scientific community.

Responding both to technical problems and to institutional and professional currents, the differential analyzer and the network analyzer developed analogs of the world. Bush and his students developed an approach to engineering problems using the common language of circuits, with which they examined problems of feedback, control, and computing. Harold Hazen formulated a new approach to servomechanisms based on their role in the differential analyzers. For Hazen, feedback devices became amplifiers, renewing mathematical signals within the machine and enabling servomechanisms and control systems to be grouped under a common theoretical umbrella. This conception also led Hazen to articulate a coherent approach to analog machines that captured his preference for direct, physical representations of the world. Simultaneous with that analog approach, however, related techniques emerged to manipulate and switch analog signals through matrices of relays. Whereas Nyquist and Hartley explored the relationships between continuous signals and discrete pulses, Claude Shannon formalized the translation between continuous equations and discrete switch closures.

When Shannon published his "relay algebra," it caught the attention of a young engineer in the mathematics department at Bell Labs. George Stibitz was already building calculators of his own out of telephone relays, and he read Shannon's paper with delight. Stibitz immediately adopted Shannon's network-like notation to design relay networks and soon coined a new term for these exciting new calculators that employed switching circuits, Boolean algebra, and binary arithmetic. He called them *digital*.[109]

# Dress Rehearsal for War

## The Four Horsemen and Palomar

Despite Hazen's contributions, control engineering in 1940 remained local, tied to discrete engineering cultures. No conference had brought its practitioners together, no special publications were dedicated to feedback problems, no theory or textbooks solidified their common foundations. Still, in the late 1930s connections began to form. Things proceeded hesitatingly and informally at first; nevertheless, a set of ideas began to emerge, and a cadre of men skilled in their application began to develop. It would take a war to solidify the connections, but the coalescence of control in the 1940s would depend on the continuity of these prior links.

### BuOrd's "Four Horsemen"

For Harold Hazen and MIT these links began to form in 1936. BuOrd recognized the importance of Hazen's work and asked him to develop a course on servomechanisms. The request flowed from a minor but continuous connection between the navy and MIT. Bush had long served as an officer in the naval reserve, and in the 1920s he had done reserve duty on the battleship *Texas,* which tested the Ford Rangekeeper prototype in 1916. Though Bush and Hannibal Ford never met, Bush acknowledged in his memoirs that Ford's machines could do nearly everything the differential analyzer could do, but many years earlier. Hazen too joined the naval reserve and spent only a few days on active duty in 1936.[1] BuOrd began to have trouble with the stability of the servos that articulated the output from its rangekeepers, particularly in the Mark 37 director, so they asked to send some young officers to MIT to learn the new servo-mechanism theory from Hazen.

In response to this request, in 1938 Hazen began planning a special course in controls, but he soon handed the work over to Gordon Brown, who had just joined the faculty.[2] Bush, who had been vice president and dean of the engineering school at MIT since 1932, was named president of the Carnegie Institution in Washington, D.C. His relinquishment of his MIT post in the beginning of 1939 initiated an administrative reshuffling. Edward Moreland, then head of the Department of Electrical Engineering, replaced Bush as dean, and Hazen replaced Moreland as department head, a post he would hold until 1952. Gordon Brown thus took over teaching and research in servomechanisms.

In the fall of 1939 BuOrd sent a few officers to MIT. Lieutenants Edwin Hooper, Lloyd Mustin, Alfred Ward, and Horacio Rivero, who sometimes called themselves the "four horsemen," stood at the intersection of fire control and servomechanism theory. Partly because of their fortuitous arrival and partly because of how they applied what they learned at MIT during World War II, all four of these men eventually became admirals. Graduates of the Naval Academy in 1931 and 1932, they served as gunnery officers in the fleet for several tours. Gunnery at the time represented the high-profile career for bright young officers, "before the real surge of glamour of naval aviation," Hooper recalled. Though he had been accepted as a student in mechanical engineering at MIT, he had chosen to attend the Naval Academy instead. Rivero had rejected a Rhodes scholarship in favor of his commission and then struggled to switch into gunnery from a career in communications because "ordnance was *the* thing in the navy."[3] During their early tours they learned the details of fire control. All four attended postgraduate school in gunnery at the Naval Academy in 1938, a necessary stop for a rising gun club career. After their first year of postgraduate work, they left for the new fire control course at MIT. They arrived in September 1939, unsure what to expect.

MIT hosted many naval officers as students, but most were from aviation or naval architecture, not gunnery. The Bureau of Navigation, not BuOrd, ran the postgraduate program at MIT, so when the four horsemen arrived, they didn't quite fit in. The university did not think they could earn master's degrees in just two semesters, and the navy captain in charge at MIT agreed. The four lieutenants insisted, but soon found they had taken on a bit more than they could handle. Although the previous year at Annapolis had been spent preparing for the MIT course, Hooper, Mustin, Ward, and

Rivero had to do remedial work in mathematics to keep up, making for a grueling schedule. They studied transients in linear systems and mathematical analysis by mechanical methods. What really excited them, however, was Charles Stark Draper's work, which seemed to have applications to fire control. Halfway through the year, Draper agreed to teach them about gyroscopes instead of delivering his planned lectures on aviation instruments and to credit them for the original course, without informing the navy.[4]

Also during the fall of 1939, Gordon Brown began teaching servomechanism theory, the first university course on the topic. His seminar included the four BuOrd officers, two students from Draper's lab, and a few others. They studied Minorsky's and Hazen's papers and applied their principles to naval fire control. Brown taught that servomechanisms could act as amplifiers to unburden the computational elements of a fire control system, allowing them to drive massive machinery. In the spring semester Brown and his students began setting up a laboratory, partly with equipment borrowed from Sperry Gyroscope.[5] "Everything we had for the course for the first couple of years were scavenged gifts," Brown recalled. "I would have to drive my car down to Long Island [to Sperry] to bring it back loaded with these things."[6]

Employing the material from Brown's course, Hooper and Ward wrote a joint master's thesis on controlling large turrets with small electric signals. The thesis addressed the classic problem of naval fire control: how to direct a ship's guns at long range against a target. This was the problem that the navy had originally intended them to work on when it first approached MIT in 1936.[7] Hooper and Ward applied Hazen's conception of the servomechanism as an amplifier to move guns with electrical servos. Their first two citations were Hazen's 1934 papers, and like Hazen, they did not use the term *feedback*. They did use Hazen's amplifier idea to design a servo to increase a signal from 1/200 horsepower to 8 horsepower, pointing out that it would work at up to 100 horsepower. The device employed a variable-speed hydraulic drive produced by Sperry's Waturbury subsidiary, and they borrowed much of their electronics from the differential analyzer.[8]

The other two students in the navy course examined a still newer problem that was rapidly becoming urgent. War began in Europe at the start of the fall semester in 1939, and the British navy was beginning to realize that its ships were vulnerable to fast German aircraft that were difficult to hit with antiaircraft fire. Lloyd Mustin, a pistol-shot expert, had worked in antiaircraft before coming to MIT. He and Rivero analyzed ships under attack from short-range, high-speed airplanes, especially dive bombers, strafers, and torpedo planes. "As far as is known," wrote the two lieutenants, "no control device for the short-range problem has been developed anywhere which pretends to solve the three-dimensional problem involved."[9] Even in the academic world, now, antiaircraft fire control was replacing long-range gunnery and power system stability as the driver for control systems research.

Mustin and Rivero's thesis focused on controlling light antiaircraft machine guns following rapidly moving targets. This problem was particularly amenable to the transient analysis typical of MIT's engineering culture. Their work clearly showed the influence of Hazen's emphasis on dynamic performance: "There can be no compromises as to speed; the solution must be delivered at the point of application, and in its 'steady state' within a fraction of a second after the device has gotten on its target ... at a power level sufficiently high for it to be applied automatically and directly to the point of use."[10] Because of the vibration and smoke produced by the guns, they wrote, the controllers should be located some distance away, driving the guns by "remote control."

Mustin and Rivero analyzed how a gyroscopic device based on a commercially available turn indicator for aircraft might predict the path of an oncoming airplane. The basic problem was to derive the rate, or angular velocity, of the target and then to calculate the lead. But differentiating a function is a difficult task, highly susceptible to error, because a pure differentiator amplifies noise. Mustin and Rivero built on Charles Stark Draper's work with aircraft instruments, rigging a gyro to calculate lead angles in a smooth, stable, and accurate measurement. But when Mustin and Rivero approached Draper for help, Draper "froze" and "shut up like a clam."[11] Mustin and Rivero had run into another of the prewar practitioners of control systems, also now coming to MIT for help: Sperry Gyroscope.

Draper had been consulting on aircraft instruments for Sperry Gyroscope for several years. He created his own field, aircraft instrumentation, and embodied it in his Instrument Laboratory. Draper's work, like that of Bush and his disciples, emphasized transient phenomena, models and analogs of physical systems, and graphical solutions. It also made use of industrial relationships. Before coming to MIT, Draper had worked at Sperry Gyroscope, and he had close contacts with Chief Engineer Preston Bassett, President Reginald Gillmor, and Director of Research Hugo Willis. In the mid-1930s Sperry began supporting Draper's work, commercializing the products of his research, and hiring graduates of his laboratory. In the fall of 1939, as war broke out in Europe, Draper began to build a device to compute the lead angles for guns on tanks by adapting a gyroscopic turn indicator he had developed. It was this project, nicknamed "Doc's shoebox," that he was working on when Mustin and Rivero brought their idea for a lead-computing sight for antiaircraft guns.[12] No evidence documents what caused Draper to begin developing an antiaircraft sight, but he may have been influenced by Mustin and Rivero's thesis. It is clear, however, that in June 1940, with a contract from Sperry Gyroscope, Draper turned his attention to antiaircraft fire control.

Despite these promising projects, the four horsemen completed their degrees in the spring of 1940 and returned to the navy. Before taking up their new assignments, Hooper, Ward, Mustin, and Rivero embarked for the summer on a Cook's tour, visiting industrial, military, and research laboratories working on fire control. By September the navy was mobilizing in response to events in Europe and Asia and cut short the tour. For their next assignments, Ward joined the Naval Inspector's Office at Ford Instrument, Hooper became the naval inspector at G.E., Mustin went to the Naval Gun Factory, and Rivero to BuOrd's Fire Control Section. MIT's work in control from the preceding decade now began to diffuse back into the military and its contractors through the dual conduits of industrial relations with Sperry Gyroscope and military liaison with BuOrd.

The usefulness of servomechanism theory for fire control was thus established. "By 1940 the development of rigorous methods of analysis and synthesis had reached the stage of adolescence," recalled Gordon Brown, "when suddenly the work was blocked out by the fog of military security."[13] It would remain invisible until 1945. Mustin and Rivero's thesis, with the vague and deliberately uninformative title "A Servo Mechanism for a Rate Follow-up System," was classified "confidential" when written. It remained so until 1972.

Also in 1940, Brown wrote a paper incorporating his control research and teaching experience. "Transient Behavior and Design of Servomechanisms" presents a general summary of the field to date, introduces its basic principles, discusses transient response and analysis, and presents design examples. The second footnote cites Hazen's two 1934 papers. Brown mentioned that John Taplin recognized the similarities between Black's and Nyquist's work on amplifiers and Hazen's on servomechanisms. Brown also mentioned feedback amplifiers in the context of servos—probably the first time in print. Brown did not employ any of the frequency-response methods from Bell Labs, but he did introduce a novel "system operator" equation (now called a *characteristic polynomial*) inspired by Taplin.[14] Brown also modified Hazen's canonical servo model by adding an explicit feedback element and introduced a block-diagram notation to represent the behaviors of the components. Brown planned to present this paper at the annual meeting of the American Society of Mechanical Engineers (ASME) in the fall of 1940.

Brown never presented the paper to the ASME. In fact, it would not see publication for five years.

In July 1940, a few months before the ASME meeting, Brown greeted an important visitor and explained to him the current state of servo research. Brown detailed the previous year's fire control course and the contributions the four naval gunnery officers had made, and complained that his work was being delayed by shortages of personnel and equipment. The visitor was Warren Weaver, who had been sent by Vannevar Bush, who had just formed the National Defense Research Committee. As head of the Natural Sciences division of the Rockefeller Foundation, Weaver was one of the few men in American science senior to Bush. Just a few weeks before Weaver's visit to Brown, Bush had asked Weaver to set up a special NDRC committee devoted to fire control. When Weaver's committee first met in 1940, it quickly classified Brown's paper and issued it as a restricted report. With that news, however, came a contract for Brown to extend his research and found the Servomechanisms Laboratory.

## Controlling the Palomar Telescope

Weaver's was not an unfamiliar face at MIT. He had been the program officer for Rockefeller's support of the new differential analyzer. At the Rockefeller Foundation, Weaver had brought scientists together to solve interdisciplinary problems before; he had even brought civilian researchers together with military contractors to solve a control problem. Weaver and Rockefeller, in fact, came fresh from a dress rehearsal for wartime science: controlling the Palomar telescope. Before attending to the transformations wrought by the NDRC and Weaver's wartime committee, then, it is worth tracing this earlier interaction of foundations, scientists, and the navy, for it set the stage for Weaver's moves in 1940.

In the late 1920s Weaver's boss and mentor, Rockefeller head Max Mason, had committed the foundation to supporting the new Palomar telescope, to be built in California under the auspices of the California Institute of Technology. Brainchild of Cal Tech founder George Ellery Hale, and built under the auspices of Cal Tech's Robert Millikan, the telescope was the most ambitious scientific instrument of its time. To control the massive structure, Millikan and Mason turned to the navy. "So far as we know there is no other organization, institution, or individual," Millikan wrote to the secretary of the navy in 1935, "possessed of such an experience in the design of heavy, precision instruments."[15] The navy gave Captain Clyde S. McDowell a year's leave to work as supervising engineer on the project. McDowell had been active with the National Research Council during World War I, had advocated collaborations between scientists and the navy, and had been instrumental in setting up a scientific research program.[16] He had experience with naval fire control and knew both Hannibal Ford and Vannevar Bush, the latter probably through secret work Bush had done for the navy.[17] McDowell took on the Palomar project with enthusiasm, aiming "to make it representative of the best technical knowledge available today in this country."[18]

The author Ronald Florence called Palomar "the perfect machine." Yet constructing it presented numerous engineering problems: fabricating the huge mirror, transporting it across the country, and building an enormous, delicate instrument on top of a mountain in the wilderness. A major difficulty would also be aiming the 250,000-pound telescope with watchlike precision. The astronomer Sinclair Smith set out the requirements for controlling the telescope: smooth drive at a constant speed (independent of temperature), user-settable rates, precise settings for all positions in the sky, automatic controls for the dome, a memory to return the instrument to a stored position, and controls that could be set from several stations.[19] Smith desired the control system to hold the position of the mirror within about 1 arc second (a 3,600th of a degree).

McDowell immediately set about bringing the navy's fire control technology to bear in building the instrument.[20] He visited Ford Instrument "to see his [Ford's] latest development on control for turrets and guns" and found that "the development work that [Ford] has done fits right into our problem of control."[21] Rockefeller's Max Mason thought that Westinghouse's new photoelectric cells could sense the dim light from stars and amplify them to drive the telescope to keep the stars in view. Mason read Hazen's 1934 papers on servomechanisms and recognized that the Palomar problem of controlling electric servos from a light-sensitive source resembled the automatic curve follower that Hazen and Brown had built for the differential analyzer. Mason suggested that McDowell talk to Bush about adapting the automatic curve follower to control the telescope, and he sent Warren Weaver to MIT "to see how much B. knows about F.'s mechanical computing devices."[22]

At MIT, Weaver had a detailed conversation with Bush about his analyzer and the

Ford machines. During this visit Bush made his successful pitch for Rockefeller to fund his next analyzer, though he admitted that Ford could make one equally well.[23] Soon Bush went to Pasadena to visit the Palomar project and consult on its control system.[24] Hannibal Ford also spent several weeks there on consultation. Max Mason was anxious to get the two men together. "It seems to me," he wrote to McDowell, "that they [Bush and Ford] have hold of the two handles of the differential analyzer and the servo-mechanism game, and their combined facilities and abilities ought to mean a lot."[25]

Bush did indeed invite Ford to MIT, but no records indicate that he visited.[26] Nevertheless, the navy now recognized the power of the two men's combined expertise, as well as the similarities between MIT's servo research and military control systems. McDowell reported back to Admiral Furlong, now chief of BuOrd, that Bush and Ford "make a very good combination I think and one of which the Bureau may want to take advantage sometime."[27]

After soliciting a proposal from Ford Instrument, the telescope's management elected, against McDowell's advice, to use a control system designed by an in-house engineer, Sinclair Smith. When Smith died tragically in 1938, Bush and Ford recommended as a replacement a former MIT student, Edward Poitras. The young engineer had graduated in 1929 from MIT, writing his master's thesis under Bush on photoelectric cells, which gave him useful experience for this electro-optical project.[28] He then worked for Ford Instrument and became chief engineer for the Lombard Governor Company in Boston. Soon Mason noted that Poitras "brings a complete knowledge of the techniques of Bush and Ford to the job and is working out very well indeed."[29]

Although the design was developed in-house, it did make use of navy technology, including 68 selsyns from Ford Instrument, which had to be procured by special permission.[30] McDowell's 1938 article on the telescope in *Scientific American* said almost nothing about the control system, probably owing to the navy's security concerns.[31]

As it turned out, the Palomar instrument did not need very powerful controls, just precise ones. The instrument itself proved to be so well balanced, and it moved so smoothly on its oil bearings, that aiming it required motors of only fractional horsepower. An array of selsyns, electric motors, feedback loops, and remote controls allowed users to aim the telescope from any of several control stations (Fig. 6.1). Two mechanical computers formed the core of the system for moving the telescope, track-

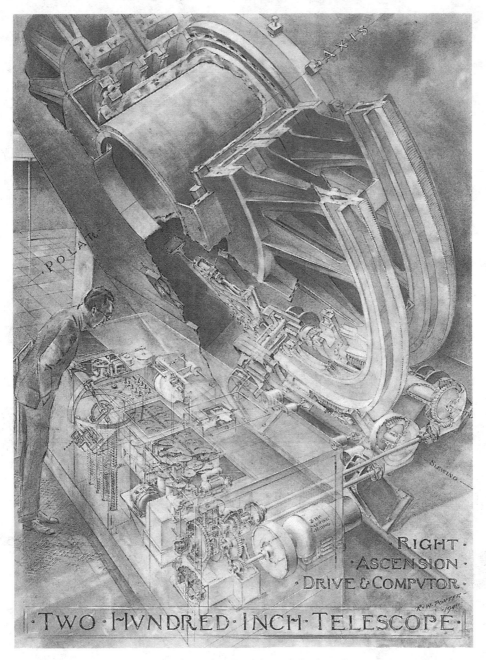

Fig. 6.1. The right ascension drive and computer control system for the Palomar telescope. The sel-syns driving dials on the operator's console are similar to those in the Ford Rangekeeper. Drawing by R. W. Porter, California Institute of Technology.

ing the earth's rotation, and correcting for atmospheric distortions and mechanical errors. An automatic recorder printed the position and movements on paper.

At the core of the system, sensors measured the actual position of the instrument and fed it back to the motors, where it was subtracted from the desired position to produce an error signal. The operator would set the position and push a button, and the control system then automatically moved the device to the proper place and ensured that it stayed there.[32] A small model of the telescope, called a phantom, mimicked the motions to provide feedback to the operator about the telescope's position. The simple, black dials both inherited the style of the Ford Rangekeepers and reflected the aesthetics of clean precision of the streamlined decade.[33] Palomar's control system design was largely complete before the war, but the telescope itself would not be finished until 1948. It was laid aside as engineers and machinists at Cal Tech turned their attention, and their newly acquired skills in optics and precision machinery, to wartime problems of fire control.[34]

The four horsemen at MIT and the Palomar project represented more than the training of a few officers and the construction of a scientific instrument. In these men and their projects we see the foundations for the military, industrial, and scientific networks of World War II. These relationships were not always smooth; Mason fired McDowell in 1938, for example, because of the friction between his naval/industrial management style and the scientifically oriented engineering staff at Cal Tech.[35] Still, the interactions brought several of the key players into contact and began to work out the subtle relationships required to bring the intellectual world of universities into contact with the rough practicality of the military. When Bush took over the Carnegie Institution in 1939, one of his first moves was to endow the operation of the telescope that Rockefeller had built. Soon he would tap both Warren Weaver and Edward Poitras for a much larger project.

# Organizing for War

## The Fire Control Divisions of the NDRC

### The Engineering Fabric of Wartime Research

World War II transformed American science and technology. The nature and impact of that change has been a dominant theme in the history of twentieth-century science. Scientists and their institutions participated in military projects as never before, on an entirely new scale. The visibility of the Manhattan Project and the atomic bomb propelled scientists into major roles in formulating American policy. The federal government began to fund research in ways that previously belonged primarily to private foundations, and it continued to do so long after the war ended. In the postwar world, "big science" and bureaucracy came to the fore, overshadowing the smaller-scale, privately funded world of the 1920s and '30s.[1]

Recently some historians have begun to open the black box of wartime technology, placing the NDRC within a longer history of science and the federal government. Hunter Dupree called "the great instauration of 1940" the period, eighteen months before Pearl Harbor, when the scattered military research of the thirties began to coalesce around Bush's new agency. New types of research contracts shaped the universities' role, and the NDRC division chiefs and their staffs personally connected scientific ideas to the military.[2] Scholars have also argued that the 1940 watershed was as conservative as it was revolutionary. Bush's committees resembled the industrial associations envisioned by Herbert Hoover in the 1920s as much as anything from the New Deal. Measured in terms of its dollar value, the research conducted by the NDRC represented only a fraction of wartime research. Bush had to fight numerous boundary battles to maintain the autonomy of his agency. And the NDRC may ultimately have been counterproductive for Bush's vision of postwar research.[3]

Most important, we are gradually becoming aware of the technology component of the OSRD and of the role of engineers in addition to scientists. Bush himself created some of the bias, as he used the mantle of science to distinguish his men from those from companies like Sperry Gyroscope and Ford Instrument: "When I came to work closely with the army and the navy I found it essential to introduce all of my people as scientists, for the word engineer to them meant too often the sales engineer coming from one of their contractors. This finally went to the point where every man in my organization got called a scientist, although it was fortunately well permeated with engineers."[4] In the NDRC, as historian Larry Owens put it, "engineering was often more important than science, practice more important than theory, and the ability to mediate, to move comfortably among university, government, military, and industry . . . most important of all." Laboratory studies confirm the view, further opening the internal workings of wartime technology beyond the famous projects like the Radiation Laboratory and the Manhattan Project.[5]

An examination of the NDRC's work on control systems reveals how wartime research acquired the expertise of its predecessors, how it related to the services and to industry, and how it could innovate where others had failed. The story is not one of unmitigated success, and the failures of the NDRC are as instructive as its triumphs. Most important, the story of control systems elucidates the social fabric and organizational culture of these interactions and clarifies their influence in the postwar world.

## Organizing Research and the Antiaircraft Problem

The original impetus for the NDRC arose, at least in part, from what Vannevar Bush called "the anti-aircraft problem." In 1938 he left MIT for Washington to assume one of the central positions in American science, that of head of the Carnegie Institution. From this lofty vantage point, directing one of the major sources of scientific patronage, he could survey the landscape of scientific research from an unusually broad perspective. An additional position as chairman of the National Advisory Committee on Aeronautics (NACA) focused Bush's attention on the dramatic strides in military aircraft and their fearsome implications. In the spring of 1939, while Europe was still at peace, he grew alarmed about "the anti-aircraft problem."

Bush wrote to his hero, the retired engineer-president Herbert Hoover, for help. As NACA chairman, Bush wrote, he saw the rapid progress aircraft were making toward higher speeds and greater altitudes. He also understood that such progress made airplanes difficult, if not impossible, to hit with traditional gunnery. High-frequency-radio research at MIT and Stanford, he continued, held promise as a way to detect and locate aircraft, but "the precise and rapid control of guns" required connecting such equipment into systems.[6] Hoover had no advice for Bush, but he found support from other colleagues closer to home. He wrote to Frank Jewett, then president of Bell Labs, that his interest in national defense arose from both the NACA and "a private conviction that antiaircraft is not receiving the attention it should have."[7]

With the outbreak of war in Europe in September 1939, German Stukas screamed across the skies as the blitzkrieg stormed across Europe and dramatically demonstrated the airplane's central importance in modern warfare. In 1940 Bush proposed to President Roosevelt that he form a council to coordinate defense research. While the NACA coordinated aeronautics research, Bush wrote, "no similar agency exists for other important fields, notably anti-aircraft devices."[8] On 27 June 1940, just as the four horsemen were completing their theses at MIT, President Roosevelt approved an order establishing the NDRC, directing it to fund scientific research into military problems. The notation "OK FDR" on Bush's letter gave him the authority he sought. Bush assembled a committee of leaders in American science and engineering: Jewett, now also president of the National Academy of Sciences; James Conant, president of Harvard; Karl Taylor Compton, president of MIT; Conway P. Coe, commissioner of patents; Richard C. Tolman, of Cal Tech; one liaison each from the War Department and the Department of the Navy departments; and Bush himself.

The group tilted toward academia (even Jewett's Bell Labs had a decidedly academic flavor), and overall, the NDRC would heavily favor MIT when it came to letting contracts. This bias would be simultaneously the NDRC's strength and its weakness. Ph.D. scientists and engineers brought fresh ideas and a rigorous quantitative approach to military problems. Many professors and researchers, however, were novices in fields in which others had already built careers and knew little of military problems, techniques, and cultures. The NDRC's eagerness could shade into arrogance, both intellectual and organizational: army and navy laboratories, industrial contractors, and any number of government agencies would seek to restrict its influence.[9] "There were those who protested that the action of setting up NDRC was an end run, a grab by which a small company of scientists and engineers, acting outside established channels, got hold of the authority and money for the program of developing new weapons," Bush wrote in his memoirs. "That, in fact, is exactly what it was."[10]

### D-2 and Division 7

To structure his organization, Bush surveyed the armed services for pressing problems. He set up four divisions within the NDRC: Division A, armor and ordnance, under Tolman; Division B, bombs, fuels, gases, and chemistry, under Conant; Division C, communications and transportation, under Jewett; and Division D, radar, fire control, and instruments, under Compton.[11] Bush also included a Uranium Committee, which would later transfer to the army and become the Manhattan Project. Division D divided into four sections: D-1, detection and radar; D-2, fire control; D-3, instruments; and D-4, heat radiation. Section D-1 oversaw the development of microwave radar at MIT's Radiation Lab. D-2 had responsibility for control systems.

In the five years that followed, D-2 and its successor, Division 7, let 80 research contracts and in the process fused prewar approaches to control systems.[12] Serving as a kind of central technology bureau, the NDRC transferred information between groups, set standards, and charted new directions for investigation. The members of D-2 and Division 7 had to craft this role carefully, however, employing a combination of research contracts, technical authority, and political skill. To lead this effort, Bush chose one of the country's most skilled men negotiating the lines between money, power, and scientific knowledge: the Rockefeller Foundation's Warren Weaver.

When Weaver heard that Bush was beginning to mobilize science for military purposes, he offered to come and help, even if that meant resigning his job at Rockefeller. The foundation did not want him to resign but rather supported him, continuing to pay his salary throughout the war.[13] Bush responded quickly and asked Weaver to head Section D-2, on fire control.

### Warren Weaver's Private Patronage

Bush's selection of Weaver underscores the NDRC's continuity with the polite foundation patronage of the prewar world. Bush selected men of his own status or higher. In 1940, Bush was relatively new to his national role at the Carnegie Institution, whereas Weaver had been supervising major research programs for eight years. Weaver was intellectually equipped to deal with technical matters of fire control, but he had little experience with either the military aspects of the problem or the previous years' work in feedback controls and theory. He had had some exposure to gyroscopic stabilization during World War I, and years earlier Bush had sent him Hazen's papers on servomechanisms, but that seems to have been the extent of Weaver's exposure to control problems. Nevertheless, Weaver had important prior experience that served him well in this time of crisis.

As a student at the University of Chicago, Weaver, like Frank Jewett, worked with physicist Robert Millikan. Weaver spent the 1920s and early '30s on the mathematics faculty at the University of Wisconsin. In 1932 he moved to New York to become director of the Natural Sciences division of the Rockefeller Foundation, a job of central importance in the interwar scientific community.[14] Weaver, a talented teacher and administrator, brought a midwestern pragmatism to the foundation's work of granting funds for scientific research. Despite the frustrations of the Depression, he crafted a system of science patronage characterized by interdisciplinary programs and project-oriented grants and revolving around his role as a "manager of science" at Rockefeller.[15] Though he never considered himself a creative mathematician (some disagreed with his assessment), Weaver did have a talent for cross-pollinating established disciplines. He brought these practices, as well as this talent, to the NDRC.

Weaver also had a philosophy of managing scientific research. "When seeking information," he wrote, "officers should, in the language of radiation theory, be good absorbers and very poor emitters."[16] Weaver believed that the foundation officers should be scientifically competent to judge the work they were funding and to evaluate the individual investigators and the atmosphere of their laboratories. Still, this was not hands-off grant making; Weaver believed his program officers should take an active role in shaping research agendas. During the 1930s, for example, Weaver brought together mathematicians, physicists, and biologists in a coordinated research program on "vital processes," a program for which he coined the term *molecular biology*.[17] He would shape the NDRC around this activist, interdisciplinary approach as well.

Curiously, one arena that did not interest Weaver while he was at the Rockefeller Foundation was engineering, which he thought was close enough to profit-making industries to support itself.[18] To this rule, he wrote in 1952, "there have been no exceptions in twenty years." Weaver seemed not to include under the category of engineering the tens of thousands of dollars Rockefeller spent on Bush's differential analyzer, a measure of Bush's success at framing it as a scientific instrument. Bush met Weaver in 1935, and the two became friends. When Rockefeller began sponsoring the new differential analyzer, the relationship grew closer during Weaver's many visits to Cambridge. As Bush contemplated his move to the Carnegie Institution in 1938, he conferred in confidence with Weaver. When Bush did take over Carnegie, the men became peers. Over the next few years they corresponded frequently to coordinate their foundations' support of various projects, including the Palomar telescope.

Weaver also brought administrative practices to the NDRC. At Rockefeller, in "an old, proven, and treasured tradition," Weaver and his program officers wrote up a "diary" entry every time they had an interview with a scientist, visited a lab, or spoke on the telephone. The diaries circulated among the officers, reflecting and facilitating the intensely personal nature of foundation support, based on personal relationships "as opposed to buying out of a catalog."[19] The abundant resources of the foundation ensured that these diaries were annually bound and indexed (a luxury the NDRC could not afford). For the most part, Weaver shut down the Natural Sciences division during the war, and the diaries measure the activity: during his career at Rockefeller, Weaver's diary entries usually filled about one bound volume per year. In 1944 Weaver made only seven entries in his Rockefeller diary, compared with hundreds in an average peacetime year. The five years 1941–46 together produced only a single volume.[20]

The NDRC records fill this gap. They paint a detailed picture of the technology and its politics as they unfolded. When Weaver or his committee members visited a facility, attended a meeting, had an important phone call, or even made a relevant observation, they wrote up a diary entry and distributed it to the rest of the group. These "entirely highly classified internal documents," recalled Harold Hazen, "gave free scope to uninhibited expression by creative individualists. They were often brilliant, salty, and very flavorable."[21] The secret diaries, together with the NDRC's correspondence and technical reports, provide remarkable insight into the technical and operational nature of the organization as it evolved and insight into the influence of Weaver and the Rockefeller Foundation.

Weaver's presence at the NDRC confirms Nathan Reingold's assessment of Bush's "instauration" as "the triumph of the old order,"[22] although the old order itself had a new, federal patron. In the person of Warren Weaver, Bush imported more to the NDRC than a prestigious colleague. He brought a style of science patronage, an approach to managing the creation of knowledge, and administrative practices designed to support them. The nature of the organization would change as the war evolved, but the government's agents started out as gentleman-scientists, skilled in the arts of private patronage, distributing federal dollars. They would not always get along with engineers, who were used to the scrappy pragmatism of the industrial world.

## Setting Up

In early July 1940 Weaver assumed the chairmanship of the NDRC's D-2, devoted to fire control, and immediately began assembling experts. He quickly made an important connection: "The problem of training a gun against a moving target in the sky is in principle exactly the same as having a telescope follow a star, except that the angular velocities are of a different order of magnitude."[23] Weaver turned to the Palomar project and called in Edward J. Poitras to be his chief technical aide. Poitras, the former student of Bush and Hazen (and former Ford Instrument employee), was working in Pasadena on the telescope but immediately left for Washington. Weaver also invited Thornton C. Fry and Samuel H. Caldwell to join as members of Section D-2. Fry, director of mathematics at Bell Labs, had been a colleague of Weaver's in the math department at Wisconsin. Caldwell, head of MIT's Center for Analysis, had been Bush's graduate student and had collaborated with him on the differential analyzer. By his choice of personnel Weaver began to fuse the prewar threads of control systems.

Security clearances for these men took some time, so Weaver spent the summer on his own and with the main NDRC committee gathering information. On 9 July, Weaver met with Bush, who briefed him on the history of fire control. The first gun directors had been built by Hannibal Ford for heavy naval guns before World War I, Bush told him, and the first antiaircraft fire control had been built in the mid-twenties. For setting up committees under his auspices, Weaver divided the field into electronics, optics, mechanical design, and mathematics. This early plan included no separate research for servomechanisms, feedback, or theory.[24]

The following week, Weaver met with the head of the army's Anti-Aircraft Artillery Board (AAAB), who described the Sperry gun directors they were using. He expressed disappointment with Sperry's "human servomechanisms." Wherever possible, he told Weaver, the army desired to eliminate the pointer matching endemic to the Sperry system, but the power controls for moving the guns were weak and unreliable.[25] From the first, Weaver was faced with automating the human role in antiaircraft systems.

To explore the academic side of things, Weaver visited MIT and met with Gordon

Brown and Sam Caldwell. Caldwell discussed the differential analyzer and the Center for Analysis, which included MIT's computing facilities. Brown briefed Weaver about MIT's program in servomechanisms and mentioned the four officers sent by BuOrd.[26]

Returning from his MIT visit, Weaver came up with an agenda for his research program. The topics he listed as "pressing problems" show how his thinking had progressed in four weeks:

1) Development of much more rapid, reasonably accurate, automatic controls for lighter AA guns

2) Possible improvement of existing fire control equipment for heavier guns

3) Increased accuracy by (a) simplification of equipment or procedure (b) by combining units (c) by substituting automatic for manual controls

4) Consideration of any special problems referred to us by the fire control groups of the Army or Navy (e.g. automatic fuze setting, improvement of rangefinders)

5) Theoretical analysis by (a) overall analysis of errors (b) analysis of complex systems of servomechanisms, particularly a determination of most effective and simplest type of intercoupled damping to secure stability when several servomechanisms are connected in series (c) analysis of function of computer, including higher order derivatives (d) probability analysis of risks involved in various dispositions of equipment

6) Basic program of development of servomechanisms (MIT group, Brown & Caldwell)[27]

Here Weaver showed that he already understood much of the prewar work and moved toward a conception of fire control as a system. The list includes the mechanism for D-2 to respond to problems raised by the services (4), and unlike his listing a few weeks earlier, it notes the need for special attention to theory and statistics (5). "Intercoupled damping" and connecting servos in series (5b) refers to the navy's stability problems with the Mark 37 antiaircraft director.

Bush had chosen the right man. Weaver quickly grasped the salient problems of this complex field, demonstrating the competence as a science manager he had developed at Rockefeller. As he had with molecular biology, Weaver easily gained knowledge of "the kind that insiders have of their own fields, a knowledge of personalities and interests as well as ideas and experimental practices."[28] This early agenda anticipated and shaped the work of the next five years.

D-2 held its first formal meeting in Hanover, New Hampshire, in September 1940. It met during a conference of the American Mathematical Society at Dartmouth. At this moment, the Battle of Britain raged, and the conference held a special session, "War Preparedness among Mathematicians," to connect academics with military research. The conference also earned a place in the history of computing, for here George Stibitz of Bell Labs demonstrated his "Complex Number Computer," or "Model I Relay Computer," over a telephone line to New York, cited as the first demonstration of remote computing. Stibitz had been experimenting with telephone relay calculators since 1937, and his research director (now D-2 member) Thornton Fry had urged him to build a calculating machine to aid electronic-filter design at Bell Labs. Those who attended Stibitz's demonstration included Norbert Wiener and John Mauchly (who later designed the ENIAC), among others, probably including the members of Section D-2.[29]

The following week, Weaver made one more visit without his committee, to the army's Aberdeen Proving Ground, in Maryland. He was accompanied by Bush, Compton, the NDRC's army and navy liaisons, and Alfred L. Loomis, a wealthy New York lawyer with an interest in microwave radio who now headed Section D-1, on radio detection. The group saw the Sperry M-4 director, which was at Aberdeen for testing with new electro-hydraulic gun servos. "There is no hunting," Weaver observed of the Sperry servos, "but the motion is frequently jerky and the rates are slow." The navy representative, noted Weaver, "who is, of course, familiar with the Navy automatic control, takes WW to one side and agrees that this is a pretty unsatisfactory device."[30] The day at Aberdeen demonstrated the limitations of Sperry's interwar antiaircraft program, and a naval officer was only too eager to point out weaknesses in army technology.

In later months Bush, Compton, and the members of the NDRC proper would not be so directly involved as on this day at Aberdeen. At this early stage, however, the organization remained small enough that a few men could attend to the whole thing, particularly as Bush was concerned about antiaircraft from the start. Also, Weaver would need unprecedented access to military and government facilities for his upcoming work. Bush, with Roosevelt's executive order behind him, paved the way.

### Learning the Field

Once organized, Weaver and D-2 set about surveying existing work in fire control and control systems. Because of the navy's clear advantage in fire control, they initially concentrated on the army's problems of land-based antiaircraft fire, especially for heavy artillery. Naval fire control had a well-earned reputation as a closed community. Until late in the war, in fact, BuOrd and naval fire control would remain largely outside of the NDRC's domain, costing the navy its lead in the technology.

The army, by contrast, was unhappy with its equipment and extended a welcoming hand to the NDRC. Colonel William S. Bowen, president of the army's Coastal Artillery Board (CAB), aggressively recruited NDRC help. On 3 October, Weaver and the committee (Fry, Caldwell, and Poitras) visited Bowen at Fort Monroe in Virginia. Bowen explained the dissatisfaction in the service with the current M-4 director, and especially with Sperry Gyroscope as the sole manufacturer of such devices (Sperry had not yet contracted with the Ford Motor Company for additional production). Directors incorporating electrical rather than mechanical techniques had been proposed by Bell Labs, Bowen added, but such equipment would have to be very rugged to be useful. Microwave detection techniques had also been proposed, but presently their accuracy was not suitable.

Bowen described the human limitations on the accuracy of existing fire control. Tracking of targets with handwheels, for example, was not smooth enough. Again, Bowen had a low opinion of the pointer matching in Sperry directors. "The use of personnel for the matching of gun dials is quite undesirable," Weaver noted, "large errors occurring under conditions of firing. . . . They [CAB] feel that mechanical loading of the guns is a step in the right direction to minimize and make more constant dead time." Weaver and D-2 saw that for all its strides, Sperry's replacement of human operators by servomechanisms remained inadequate. Bowen "emphasized that servos are to be used wherever possible in place of manual matching of dials. The saving of manpower in this way is not important but the accuracy is of great importance."[31]

To hear another view of the army's antiaircraft directors, D-2 members visited Sperry Gyroscope. They met with Sperry's chairman, Reginald Gillmor; president, Preston R. Bassett; director of research, Dr. Hugo Willis (now a member of NDRC Section D-1, on radar); and director of fire control development, Earl Chafee. Chafee explained the details of Sperry's fire control computers and the advantages of its plan prediction method. Sperry was currently modifying its M-4 director, Chafee added, to incorporate the suggestively named "aided laying," later called *rate control*, which partially automated tracking. The company was also increasing the M-4's range and including a provision for radar inputs.[32] Despite its flaws, Sperry's prewar program with the army served as a baseline against which to measure new approaches.

Weaver and D-2 did not consider naval fire control to be as pressing as the army's problem, but they did try to incorporate its experience. On a visit to the Naval Gun Factory, in Washington, D.C., they examined directors, computers, and rangekeepers and were given copies of the user manuals for these machines. They went aboard the USS *Quincy*, interviewed its gunnery officers, and examined its Ford Rangekeeper, its thyratron servos, and its antiaircraft directors. When the committee visited G.E., in Schenectady, they observed the company's switchboards, electronic computers, and a wide variety of servos. They also met Edwin Hooper, one of the four navy students who had recently graduated from MIT, now with the Naval Inspector's Office at the company. At Ford Instrument, in Long Island City, the section members saw the company's delicate rangekeepers and a machine for making ballistic cams. Sam Caldwell visited Edwin Land, at Polaroid, to discuss the optics of rangefinders.[33]

Finally, the D-2 members learned about electronics, feedback amplifiers, and communications engineering. They visited Bell Labs, in New Jersey, which had just begun building an electronic director using feedback amplifiers (see chapter 9). At nearby Fort Monmouth, the Army Signal Corps research lab was supporting the Bell Labs project and hoping to integrate it with its new microwave detection sets. The army retained a cultural bias against electronic equipment, which it believed too unreliable for field service. The prevalence of radio, however, and the anticipated importance of radar were forcing greater acceptance of electronics.[34] Weaver and Poitras also traveled to England, where they spent six weeks in the heart of the "blitz" and a night with an antiaircraft battery in combat to observe their technology and procedures. Despite a bad automobile accident, the two men returned full of confidence in their own program and full of ideas for improving on British methods.[35]

In this initial investigation, D-2 tapped the four prewar traditions of control systems. The busy first months demarcated the landscape, both geographic and technical, of fire control as it existed at the beginning of the war: industrial firms, military sites, and a university. The bulk of the NDRC's fire control work, and the bulk of its contracts overall, would occur within the confines of the industrial region bounded by Virginia in the South (Fort Monroe), Massachusetts in the North (MIT), concentrated in New York and New Jersey. Central technical problems were those Sperry had failed to solve, those they could take from the navy and apply to other areas, and those arising from new technology, especially electronics. D-2 absorbed this diverse technical knowledge into its own fledgling organization.

With key problem areas identified and a budding core of expertise, D-2 began to define its program. At an October meeting Weaver outlined critical areas and assigned responsibilities. Fry would coordinate systems, statistical analyses of errors, and research in servomechanisms. Caldwell and Poitras would investigate electrical analogs for mechanical computers and servos. The committee overall would support research on optical rangefinders, evaluate rangefinder operators, perform efficiency studies of manual procedures for loading guns, and improve instruction books for antiaircraft systems. They also sought a standardized graphical language for mechanical computers, similar to the one Bush created for the differential analyzer, to enable consistent notation across projects. They suggested Claude Shannon, now at Princeton, as the person to create that language, because he had created similar notation for MIT's computers and relay circuits. In a similar vein, the committee suggested adopting a standard nomenclature for the antiaircraft problem itself and compiling a table of equivalent symbols used by the army and the navy. Weaver soon circulated a memo by Thornton Fry with suggestions for this new language of control.[36]

In November 1940, after this short but intensive two months of study, Section D-2 began letting contracts for research. Before examining these projects, however, and D-2 and Division 7's management of the research, it is worth considering the remarkable novelty of these arrangements. How could a group of university professors and industrial researchers usurp one of the military's most secret and most complex technologies? What resistance did they meet?

### D-2's Fresh Approach

For at least ten years the nation's leading manufacturers of control systems had tremendous difficulty making progress in antiaircraft. Now D-2, a civilian group, acquired proficiency in a highly technical, highly specialized, and highly classified military technology. In less than three months' time they identified the pressing problems and began directing research toward their solution. The members of D-2 had at least three advantages that enabled them to tap new sources of innovation. First, they were either Ph.D.'s or academically trained engineers. Weaver and Fry were among the country's top minds in applying mathematics to practical problems, and Caldwell was among the most experienced with calculating instruments. Even Poitras, an engineer with a master's degree who worked at an industrial firm, had worked closely with scientists on the Palomar project. This academic background enabled D-2 to use mathematics and theory that had been largely absent from previous approaches to the problem.

The section's second advantage stemmed from the highly secret and compartmentalized nature of fire control. Backed by a presidential order and holding the strings to a large purse, these men had unprecedented access. In their first two months, D-2 members achieved an overall view of fire control previously enjoyed by no one. They were shocked to find almost no communication, indeed outright hostility, between army and navy fire control designers (Section D-1, on radar, had a similar experience, noting the army and the navy were unaware of each other's programs).[37]

Third and finally, Weaver, Fry, Poitras, and Caldwell were new to the problem, as yet unencumbered by institutions or traditions. They brought not only fresh perspectives but also "a range and breadth of experience over a variety of fields," Harold Hazen later recalled, "that could see relations between fire control and many varied fields of endeavor that, superficially viewed, are unrelated to it."[38] Just as Hazen had seen the fundamental features of servomechanisms because he employed them in calculating instruments, now these men began to perceive the fundamentals of fire control because of their unique vantage point as outsiders, apart from each service's culture. The novel organizational conditions of the NDRC allowed them to see fire control as a particular case of a general problem of control: a feedback problem, a stability problem, and a problem of representing the world in machines using electrical signals.

## The NDRC Reorganizes

Within a year of its founding, the NDRC spent more than $6 million (it would spend more than $500 million between 1941 and 1946) and grew to such a size and complexity that reorganization became necessary.[39] In June 1941 an executive order created the OSRD, which incorporated the NDRC along with a number of other committees, including the Committee on Medical Research. Once the NDRC became part of the OSRD, its responsibilities expanded to include more design, pilot production in some cases, and less fundamental research.

In December 1942 the NDRC itself reorganized, taking a more bureaucratic form.[40] The original 4 divisions now became 17, numbered instead of lettered. D-2 became Division 7, still responsible for fire control. Other divisions were responsible for ballistics, missiles, subsurface warfare, and electrical communication. Possibly because of friction at D-2 (see chapter 8), Warren Weaver moved to head the newly created Applied Mathematics Panel (AMP), which collected mathematicians to provide analysis services to the divisions. Weaver also joined Division 14, which oversaw radar development, and remained a member of Division 7 as an adviser and as a liaison with AMP. In Weaver's absence, the servomechanism theoretician Harold Hazen headed Division 7.

D-2's projects were transferred to the new Division 7, except for several "of an essentially mathematical character," which went to the Applied Mathematics Panel.[41] Weaver co-directed the panel with Thornton Fry, and it resembled Fry's vision for the Mathematics Department at Bell Labs, where the industrial mathematician was "a consultant, not a project man."[42] Weaver's new position was no doubt more suited than the messy industrial world of military contracting to his penchant for fundamental research and his talents as a science manager, although he had difficulty with the mathematicians as well.[43]

As a mature organization, Division 7 became more established, more bureaucratic, and more procedural than D-2. Hazen the engineer, administrator, and department head replaced Weaver the science manager. Division 7 meetings became more budgetary and contractual than the mix of administration and engineering that characterized D-2. The division funded no "fundamental" research that did not show immediate promise of contributing to the war effort and indeed canceled several significant projects.[44]

The character of the contracts reflected this shift away from fundamental work. Division 7's research became more industrial than D-2's and also more focused on offensive, as opposed to defensive, projects. Of 52 contracts let before 1943, 44 percent went to academic institutions, the remainder to companies (including industrial labs). Under Division 7 only 18 percent went to universities. Committee logistics, and hence committee culture, also echoed the trend toward industry. Under Weaver, the committee met 55 floors above New York City in the plush Rockefeller Center headquarters of the Rockefeller Foundation. Under Hazen, about half the meetings remained in New York, but the other half rotated among industrial and research labs, including Sperry Gyroscope, G.E., the Franklin Institute, the army's Aberdeen Proving Ground, and MIT's Radiation Lab. Usually such gatherings lasted two days, with one day devoted to laboratory tours and equipment demonstrations. Despite its different style, Division 7 carried on D-2's work, retaining its contracts and letting new ones. The easygoing Hazen recalled Division 7 meetings as "'family affairs' . . . among friends in which the discussion was often brutally frank and in which no punches were pulled."[45]

Division 7 divided into a number of subsections (Fig. 7.1). Their organization in

Fig. 7.1. Division 7 of the NDRC, 1943. *Top, left to right:* George Philbrick, Foxboro; Karl Wildes, MIT; Duncan Stewart, Barber Coleman; Warren Weaver, Rockefeller Foundation; Preston Bassett, Sperry Gyroscope; Harold Hazen, MIT. *Bottom, left to right:* Al Ruiz, G.E.; George Stibitz, Bell Labs; Ivan Getting, MIT Radiation Lab; Lawson McKenzie, BuOrd; Thornton Fry, Bell Labs; Samuel Caldwell, MIT. Reprinted from Getting, *All in a Lifetime,* 200.

1943 indicates the growth in complexity and variety of fire control problems in the two years since Weaver's initial assignments:

*7.1. Ground-based antiaircraft fire control*

Chief: Duncan Stewart, president of the Barber Coleman Company

*7.2. Airborne fire control systems*

Chief: Samuel H. Caldwell, MIT

*7.3. Servomechanisms and data transmission*

Chief: Edward J. Poitras, Ford Instrument

*7.4. Optical rangefinders*

Chief: Thornton Fry (replaced by Preston C. Bassett, president of Sperry Gyroscope)

*7.5. Fire control analysis* (administrative connection to the Applied Mathematics Panel)

Chief: Warren Weaver

*7.6. Navy fire control with radar* (added in 1944 as liaison with the Radiation Lab)

Chief: Ivan A. Getting, MIT Radiation Lab

Throughout the course of the war, Division 7 members and technical aides also included J. R. Ragazzini of Columbia, George Valley, Karl Wildes, and Charles Stark Draper of MIT, George Stibitz of Bell Labs, George Philbrick of the Foxboro Company, Walter MacNair of Bell Labs, John Taplin (who had identified the similarity of feedback amplifiers and servomechanisms), and John D. Tear, director of research at Ford Instrument.[46]

## Management Style

For five years D-2 and Division 7 supervised the research and development of control systems applied to wartime problems. They let 80 contracts (see appendix B), totaling just over $10 million.[47] These contracts formed the core of control systems work in the United States during World War II and synthesized prewar engineering cultures of feedback, control, computing, and communications. Projects originated in several different ways. Sometimes the services requested work on a difficult problem. Sometimes the army or the navy turned over existing research projects for the NDRC to administer. Others arose from committee discussions that pointed to a promising or neglected path of inquiry. Sometimes contractors made proposals of their own. Often ideas came up informally, from preliminary arrangements made through members' personal contacts.[48]

The research contract itself represented a significant innovation. Traditional governmental procurement practice usually dictated the delivery of a physical piece of equipment. Sperry Gyroscope, for example, financed its development of antiaircraft devices in the 1930s by selling pilot production lots to the government. By contrast, NDRC contracts allowed the government to buy research itself, freeing wartime engineers and scientists from the strictures of procurement and assuring a free and flexible control of money. To safeguard this separate sphere, Bush consistently resisted military requests for the NDRC to manufacture the machines its research programs designed (except for small, urgent pilot runs). Preferably, NDRC contractors (companies or universities) would turn production blueprints over to manufacturers when the research contracts ended. These arrangements also allowed scientists and engineers to remain in the employ of universities or companies rather than become military personnel.[49] More important, the government would pay the *full cost of research,* which included not only equipment and salaries but also indirect costs, the now famous factor of *overhead*—the entire administrative and institutional cost of supporting the laboratories. Government-funded programs could now grow arbitrarily large without draining the resources of their host institutions.

D-2 and Division 7 also developed their own methods of operating, distinct from those of other divisions. Several other NDRC groups created central laboratories for their work. To develop a proximity fuze, for example, Division T (named for its leader, Richard Tolman) set up a lab at Johns Hopkins, which later became that university's Applied Physics Laboratory. Similarly, Division 14 (radar) concentrated its resources in a single institution, the MIT Radiation Lab, which was the one most expensive NDRC project, with expenditures 40 times greater then those on fire control.[50]

In the tradition of the Rockefeller Foundation, D-2 and Division 7 did not set up laboratories of their own but relied instead on existing organizations. Still, following Weaver's lead, they took a hands-on approach, acting, in Hazen's words, as "a closely knit group of experts . . . studying, analyzing, and formulating service needs in terms of possible projects, then obtaining and directing contractors in the carrying out of such projects."[51] Hazen ran the division from a special office at MIT; Poitras managed the main office in Washington. Members took to the road supervising contracts, observing demonstrations, and meeting with military services. Every month or so the division would meet to discuss projects, report progress, solve problems, and discuss technical direction. This arrangement embodied a more multiple and flexible approach than the other divisions' efforts, but the lack of a centralized laboratory also had disadvantages. Outside of its small group of members, D-2 and Division 7 could not build up an institutional culture, and they sometimes had to resist the local cultures that arose at the laboratories they sponsored.

### Like Bees Pollinating Flowers: Diffusion and Standards

Despite the lack of a central laboratory, D-2 and Division 7 served as more than simply a source of funding or technical consulting. "Like bees pollinating flowers," to borrow Merritt Roe Smith's phrase, the members transferred information, techniques, and equipment between the contractors, the services, and research groups who had not previously been in contact.[52] D-2 clearly articulated this role at a meeting to define its relationship to industrial contractors. Ed Poitras and Warren Weaver met with Hannibal Ford and R. E. Crooke, of Ford Instrument; William L. Maxson, owner of another military contractor; Preston Bassett, president of Sperry Gyroscope; and Al Ruiz, of G.E. As Poitras told the consultants, "NDRC can make contracts which are decidedly long shots which the [military] services can hardly do." He offered to help the companies make university contacts for mathematical studies and to help them plan and finance test programs. To Bassett, the most valuable thing the NDRC could do was standardize testing procedures for gun directors. Hannibal Ford emphasized the need for "coordinated designs of directors with microwave and/or optical range-finders." Industry envisioned the agency as an information bureau, providing intellectual infrastructure. "Those present," Weaver noted, "had in mind a working arrangement comparable to the Bureau of Standards."[53]

To build this infrastructure, D-2 and Division 7 standardized symbols and vocabulary and created a common language of fire control (though not one uniformly adopted by contractors). More important, the NDRC developed a means of testing fire control devices, creating a standard measure for new machines. The NDRC's broad view of the secret activity in a number of laboratories—industrial, academic, and military—allowed it to act as a clearinghouse, a potent source of technology transfer, innovation, and synthesis.

Yet the wartime climate constantly opposed such knowledge diffusion. D-2 and Division 7 confronted military secrecy, proprietary industrial information, and antagonism between the army and the navy. Despite their large budgets and frenetic activity, Division 7 and the NDRC controlled only a portion of wartime research funds. Government laboratories and industrial firms carried on their own relationships with the military services. Companies and universities worked together as well; turf battles often ensued. Companies had much to lose from the new agency, for without it they had held a monopoly in fire control. Sperry, for example, entered into no contracts with Division 7. The company already had a relationship with the army and was funding research at MIT (by Charles Stark Draper and Gordon Brown) under a navy project. Similarly, Ford Instrument had no NDRC contracts, continuing instead its decades-old relationship with BuOrd.

Bush preferred that the NDRC steer clear of conflict with industrial organizations by emphasizing the fundamental aspect of the research and eschewing production. Still, despite the academic nature of the contracts, NDRC-sponsored research had a decidedly practical bent. Even "fundamental" research was expected to lead to military applications. Division 7's standard questionnaire for reviewing the status of projects conveys a sense of the NDRC's goals:

1) Date of completion of First Phase:
2) Date of transition to development:

3) Date at which Mr. Gordon's office [the engineering transition office, charged with "few quick" pilot production] becomes involved:

4) Date of first field trials:

5) Date of first effect on military or naval action (few quick in action):

6) Date of extensive use (effect of mass production):

7) What is the status of procurement of devices or equipment the new device will supplant?

8) If you had more money and men, what time schedule would result?

9) What is the section estimate of the military significance of the work?

10) Is there a shortage of personnel, equipment, or materials in the research or contemplated program?[54]

As we shall see, tensions between fundamental work, engineering development, and industrial production would characterize the NDRC's choice of research contracts throughout the war, as well as its relationships to contractors and investigators.

## The Contracts

When D-2 began letting contracts in November 1940, it was still more than a year before Pearl Harbor, but the country's scientists and engineers were mobilizing. The shock of 7 December 1941 is barely evident in the NDRC's working documents: by that point its members had been on a wartime footing for many months. Still, Pearl Harbor surely strengthened the case for advanced control systems. Before the Japanese strike, few questioned the need for antiaircraft defenses, but few also had found it urgent. Afterward, the fear of air attack was etched into the American consciousness.

The character and distribution of D-2 and Division 7's 80 contracts map the world of control systems. Twenty-nine contracts went to academic institutions, the remaining 51 to industrial firms or laboratories. The largest contract cost about $1.5 million, for the Bell Labs gun director; the smallest $2,000, for Norbert Wiener and his assistant to study the theory of prediction. The average cost was about $145,000. The longest contract lasted nearly five years, the shortest four months, and the average about two years. More than half of Division 7's contracts went to institutions along the East Coast, with the remainder concentrated mostly in the Midwest and California. Most of the contracting organizations remain familiar today: Western Electric/ Bell Labs, MIT, Cal Tech, Princeton, the Franklin Institute, Eastman Kodak, Polaroid, Foxboro, RCA, Bausch and Lomb, Bristol, and Leeds and Northrup, to name but a few.

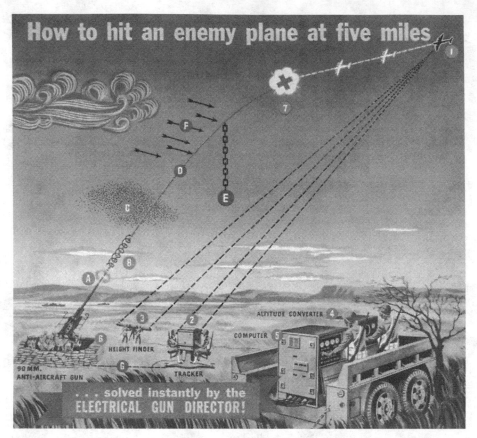

Fig. 7.2. Vannevar Bush's "antiaircraft problem": schematic of antiaircraft-system elements from a promotion for Western Electric's M-9 electrical gun director ca. 1944.

Enemy plane (1) is spotted and followed by tracker (2) and height finder (3), which feed information into altitude converter (4) and computer (5). The computer plots the aircraft's range, course, and speed and sends aiming directions to the gun (6), as well as a time for the shell to burst (7). The computer takes into account muzzle velocity of gun (A), shell drift due to spin (B), air density (C), time of shell's flight (D), gravity (E), wind (F), and parallax between tracker and gun (G).

"One must always remember that a fire-control system is more than the sum of component parts," wrote Harold Hazen at the end of the war. "It is an integrated whole with interrelated functioning of all its parts and one is safe in considering the parts separately only if one always keeps in mind their relation to the whole."[55] Before the war an engineering vision of control as a general principle had just begun to take shape, and the NDRC completed the formulation by concentrating on each element in the system and then on their integration.

When the NDRC began operations, Sperry Gyroscope and others had already defined the components of antiaircraft fire control (Fig. 7.2). Instruments of perception, in the form of optical rangefinders and tracking telescopes, provided range, bearing, and elevation of the target. As the war progressed, radar took over these functions, at first only for rangefinding and later for tracking. A central computer or gun director integrated these data with settings for wind, terrain, and ballistics, which depended on the particular gun and shell. The director predicted the future location of the target based on its speed and direction and calculated azimuth, elevation, and fuze setting. These data were articulated to the guns, which pointed automatically, with hydraulic or electric power controls, or manually, based on follow-the-pointer indicators. Gunner's "cut" the fuze time for shell by rotating a dial before they loaded it into the gun.

Of the 80 projects D-2 and Division 7 funded, more than 60 addressed one of these components of the land-based antiaircraft system. Some built individual elements, some worked on interconnection, some studied the human operator, and some worked out theory. Most projects addressed the army version of the problem; some added speed, pitch, and roll sensors for the navy. The remaining projects concerned gun controls for airplanes, torpedo and rocket directors, regulators and governors, guided bombs and bombsights.

## Beyond the Contracts

Appendix B describes and summarizes the D-2 and Division 7 contracts, tracing the scope of the research program and the problems it attacked. As units of analysis, however, the contracts can be misleading, for they were far from equal. Many contracts were small, short in duration, and insignificant. Some encompassed numerous smaller projects. A number produced significant advances and were consistently extended. Others showed no promise and were unceremoniously terminated. Many successfully completed their initial assignments and ended. A few created important machines that went into production and into combat. Typically, contracts produced prototypes, pilot studies, and reports. Some projects were initiated redundantly as "insurance" against the failure of larger, more central efforts. When the primary approaches succeeded, the backup designs were not needed.

To understand more completely the wartime work in control systems, we must examine some D-2 and Division 7 projects in detail. Institutional cultures, individual engineers, and international events all shaped the technologies of control. Chapter 8 narrates the NDRC's evolving relationship with MIT and Sperry Gyroscope and the friction generated over ownership of new technologies and appropriate definitions of research. D-2 and Division 7 oversaw a number of projects to combine new radar tracking devices with gun directors to make automatic "blind firing" systems. Chapter 9 examines the most prominent of these efforts, the M-9, produced by Bell Labs for the army. This device, when combined with a servo-controlled radar designed at MIT's Radiation Lab, became an automated antiaircraft system and achieved success against the V-1 "buzz bombs." Chapter 10 shows how the next-generation Mark 56 Gun Fire Control System, also designed at MIT, and produced by G.E. for BuOrd, carried the notion of system even further by designing the radar, computer, and actuators as a single unit and defining a new organizational role, the "system integrator."

To improve the performance and accuracy of these fire control systems, engineers and mathematicians at MIT and Bell Labs began to study the flow of data throughout the system as a problem of communications. Chapter 11 shows how this approach, that a true "signal" could somehow be separated from extraneous "noise," even when the noise was generated by the human operator, began to unify problems of machine control, communications electronics, and the manipulation of information. This approach influenced cybernetics and digital computing, although neither was fully accepted by the NDRC. Division 7 endeavored, with varying success, to understand fire control as a complete system, but under the constraints of wartime urgency and immediate field applications. Examining the NDRC's work in fire control, then, allows us to gain a detailed picture of the management of wartime research, with which to contemplate its influence on the technologies of the postwar world.

# The Servomechanisms Laboratory
# and Fire Control for the Masses

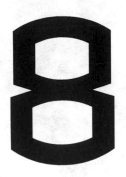

The NDRC's research program in control systems, though extensive and wide-ranging, did not cover the entire field. D-2 and Division 7, in fact, supplemented more than appropriated the existing landscape, which did not wither or evaporate. In fact, BuOrd radically reorganized its research and development, partly seizing control of the technology from its clique of contractors, who were themselves scrambling to meet the production demands of a wartime boom. With the vast resources of wartime ordnance procurement to back it up, BuOrd's new research division posed a formidable rival to the immature NDRC.

BuOrd would only rival the NDRC, however, if they competed for the same resources. Money was not the issue; there was plenty, if not too much, for all. Nor were materials; research, even with its need for special tools and machinery, made modest demands compared with production. Manpower, however, especially in engineering research, proved a resource over which competing wartime research agencies clashed. In the area of control systems, the engineers in one laboratory, MIT's Servo Lab, led by Gordon Brown, stretched in more directions, pulled by more actors, than any other. The Servo Lab was founded by D-2's very first contract, developed close relations with Sperry Gyroscope, and joined servomechanisms with telephone engineering. It has a prominent place in postwar histories of computing because of its Whirlwind computer, numerically controlled machine tools, and the SAGE air defense system. Less has been said about its earlier, wartime work. Engineers at the lab merged communications and control, so its founding, its relationship to the NDRC and BuOrd, and its intellectual identity bridge the prewar engineering culture at MIT and the postwar projects.

### Defining the Outside: Industrial Process Control

At D-2's second meeting, in October 1940, the committee reviewed the recent work at MIT on servomechanisms. They requested a copy of Mustin and Rivero's thesis on fire control against attacking aircraft.[1] The next day, Sam Caldwell wrote to Gordon Brown, who was just completing his paper "Behavior and Design of Servomechanisms." Caldwell asked that the paper not be published for "open circulation" but rather that the NDRC "undertake limited publication at its own expense."[2] With this request the curtain of military secrecy descended on servomechanisms, or at least on MIT's version of it, just as Brown was about to present it to his professional peers.

What was the state of control engineering at the time, and how did the NDRC reconfigure it? The main professional group in America, to which Brown belonged, was the ASME's Committee on Industrial Instruments and Regulators (the American Institute of Electrical Engineers [AIEE] dealt with automatic control through its Committee on Automatic Substations). As the title suggests, the group was primarily concerned with industrial controls, factory instrumentation, and the various pressure, temperature, and flow regulators those arenas employed. Despite, and perhaps because of, Brown's withdrawal from their 1940 meeting, the ASME committee became aware of the NDRC's activities. Its president, Edward S. Smith, who founded the group in 1936, wrote to the NDRC in early 1941 and offered the organization's services for the wartime effort. He sent a membership list and an evaluation of the members, rating each one with an A, B, or C. Of the 89 members, 41 were associated with companies, 17 with universities, and 3 with government agencies. Twenty-eight listed no affiliation, being either independent consultants or maintaining membership as individuals and not through their employers.

Smith included no fire control companies in his list. The member companies tended to be either large, process-oriented firms like chemical producers (Dow Chemical, Gulf, Monsanto, Standard Oil) or smaller firms that made instruments for those processes (Bristol, Taylor Instrument, Foxboro, Leeds and Northrup). Although Elmer Sperry Jr. belonged to the group, neither he nor any other member listed as an affiliation Sperry Gyroscope, Ford Instrument, Arma, or any of the Sperry Company subsidiaries (one member was from G.E.). The university members included Gordon

Brown, Harold Hazen, Charles Stark Draper, and Professor Trinks, from the Carnegie Institute of Technology. Government representatives came from the National Bureau of Standards, the U.S. Patent Office, and the U.S. Navy Postgraduate School at Annapolis (the only member with a potential fire control connection). Smith also supplied the NDRC with a bibliography of relevant literature that reflected a similar bent. Nearly all 26 entries related to process controls, except for two papers by Draper on aviation instruments. The bibliography did not include papers by Black, Nyquist, or Bode or either of Hazen's papers on the theory of servos.[3] In 1940 the professional group of control engineers was dominated by industrial process control.

If Brown had been allowed to present his paper, it would have brought to this community the high-performance, transient-analysis approach that originated in MIT's power systems studies, which Hazen had applied to servomechanisms and which Brown and his navy students had developed. It would have brought *control* to a community still defined by *regulation*. But it was not to be; D-2 classified the paper and kept it from public view. The ASME group, along with other professional associations, would play almost no role in the NDRC's control systems work over the next five years. In 1943, for example, Weaver and Hazen killed a proposal to hold a session on servomechanisms at an AIEE meeting, declaring it "undesirable to use the words servomechanisms or even automatic control" in the public announcement.[4] D-2's action defined the community of control negatively: it was still not clear who would take up feedback as a wartime cause, but it would not be this established, industrial, civilian group.

### Defining the Inside: Brown's Initiating Text

With the outside defined, the inside began to take shape behind a curtain of secrecy. Brown's paper became the initiating text for a new community of military control systems. Upon first becoming associated with Section D-2, an administrator, a researcher, a military officer, or a company received a copy of Brown's paper. Because of the controlled distribution, this process carefully documented itself. A list in the NDRC archives of who signed out Brown's paper tracks in detail the growth of this new community of control from 1940 to 1945.[5] Of course, all classified papers were similarly tracked, but no trail in D-2's files displays the breadth of this paper's path. It was the foundation, the initiation.

Consider the first few entries, made in the latter half of 1940, when D-2 was organizing and building its network. The first person to sign out the paper was "Dr. Weaver," followed by "Mr. Poitras." The next six copies went to "G.B. Davis, Bureau of Ordnance," demonstrating the early interest of naval fire control in Brown's work. Eight copies then went to Brown's students from MIT, two each to Hooper, Ward, Mustin, and Rivero. Soon thereafter telephone engineering applied its methodology to fire control; copy 17 went to Donald Parkinson and Thornton Fry, of Bell Labs. Copies 19–43 went to Gordon Brown himself. Before the end of 1940, recipients of the paper included Brigadier General Somers, of Army Ordnance; Ray Stearns, of G.E.; Thomas Doe, president of Ford Instrument; Arthur Davis, founder and president of Arma; Harry Vickers, of the Sperry subsidiary Waterbury Tool; Theodore von Karman, of Cal Tech; Sam Caldwell; Carroll Wilson; and "Dr. Bush." A later entry signs the paper out to "The Manhattan Engineering District." The distribution of Brown's paper traces the diffusion of control theory as it covered and defined the secret wartime landscape.

### The Servomechanisms Laboratory

This canonization of Brown's paper helped spread his ideas and enhance his reputation, but it also stripped him of his control of the process. Brown no longer possessed the academic's primary tool for selling an intellectual program: publication. Yet selling he needed, for in early 1940 Brown and his navy students had begun to set up a laboratory, with equipment donated by Sperry Gyroscope. Brown soon grew frustrated over a lack of manpower and hardware; without government-sponsored research he would be unable to get either. Warren Weaver visited MIT on his initial tour in July, and Brown seized the opportunity, proposing that the NDRC fund projects in servomechanisms.

Brown's initial proposal made an important leap: he proposed to study the overlap of network and circuit theory with feedback principles, "a broad exploration of the properties of bridges, tuned circuits, non-linear tubes, reactors, materials, frequency modulation, frequency proportional to signal systems, television principles, from the viewpoint of their possible usefulness in establishing error, error-derivative, and error-integral signals for actuating control devices in servomechanisms."[6] His language reflected the scientific bent of MIT electrical engineers from the 1930s, but it also demonstrated a keen new interest in communications techniques ("frequency modulation, frequency proportional to signal systems, television principles").

Brown was making a bold move, but he was not communicating it clearly. The proposal was academic to the point of pedantry, discussing highly technical aspects of control and circuit theory. It was hardly comprehensible to Weaver at the time, and it was unreadable by anyone in the military. That summer, before D-2 classified his paper, Brown remained a civil scientist, but one who envisioned bridges between telephony and servomechanisms.

Weaver nonetheless saw the worth of MIT's expertise, and Brown learned quickly. At Weaver's request, Brown rewrote the proposal, this time displaying considerably more acumen in attracting government support: "There now exists at the institute [MIT] a background of experience which has come first, as a result of the work conducted here during the past decade on calculating machines and associated mechanisms, and second, as a result of a formal program of teaching and graduate research on servomechanisms inaugurated a year ago in connection with a program of graduate training for U.S. Naval fire control officers." Now Brown emphasized facilities, personnel, experience, and direct military application, "the foundation on which we could build a program of research on fundamental problems in fire control."[7] Brown was cannily defining his boundaries, reaching outward and inward, connecting his work to military problems and addressing it to the basics of feedback.

The second proposal succeeded, and Weaver recommended an NDRC appropriation for Brown's laboratory. Under the contract, MIT would pursue five projects in fundamental studies of servomechanisms. These included "the problem of the control of an hydraulic gear used as a follow-up system for military purposes" to extend Hazen's theory to hydraulic controls, building on Hooper and Ward's thesis on gun turrets. Also, the MIT lab would investigate high-power continuous control servos of up to 500 watts.[8] On 1 November 1940 the NDRC allocated $24,500 for this work; it was D-2's first contract, Project 1. The following week, Poitras visited Brown at MIT and picked up his paper on servomechanisms, taking it back to Washington for restricted publication.[9] Brown's original idea, theoretical studies of communications and servos, seemed forgotten.

A servo is not a computer or a system or an idea, but a thing, akin to a motor. Brown's program did not address distributed or integrated control systems nor even mechanical computing, but the servos themselves: electrical or hydraulic motors for controlling the position of heavy machinery from precise, low-power signals. Brown's program inherited MIT's approach to rotating machinery, transient phenomena, and engineering science. At this point things should have carried on smoothly. Brown had achieved official recognition, backed up by funding, of his previous work. If D-2 was looking for a central laboratory, it found the ideal candidate. The Servo Lab in effect was set up as the nation's primary facility for fundamental studies of servos, control systems, and control theory. But Brown was a practical man, and he loved the immediacy of machinery. Fundamental studies by themselves would make a dry contribution to the war effort, and he had other suitors besides the government.

## Sperry Borrows Brown

Much to the irritation of the NDRC, Brown did not drop everything for fundamental research. While appealing to Weaver for support, Brown had also pursued an industrial avenue: Sperry Gyroscope. Sperry had become interested in Brown's work through Charles Stark Draper, and the company aggressively courted Brown. He consulted for Sperry and Ford Instrument during the summer of 1940 and returned to MIT with the trunk of his car filled with hydraulics and servo equipment. "I was simply taken out to their junk pile," he recalled, "and told I could take any of those hydraulic transmissions I wanted." Before long, Brown got not junk equipment but the latest high-performance servos for his laboratory.[10]

Industry gave Brown equipment, pressing problems, and relevance. It steered his attention to hydraulics. Brown's original proposal to the NDRC, in fact, mirrored Sperry Gyroscope's own research agenda at the time. Brown's proposed study of a "means for measuring and indicating time rates of change of error," for example, had much in common with Draper's work on a gyroscope to detect rates of change of motion, already in process at MIT under Sperry direction. In fact, Brown probably prepared his original proposal not for Weaver but for Sperry Gyroscope. Brown credited Sperry's research director, Hugo Willis, with sparking his interest and giving him the phrase *fresh, fundamental approach* to describe his research on servomechanisms.[11]

In the summer of 1940 Britain strained under air attack and France fell to the Nazis. Sperry began to work with the British merchant marine to help defend their ships from aircraft.[12] This was the project that Mustin and Rivero stumbled on when they asked Draper for help; he was converting an aircraft instrument into a lead-computing sight for light, naval antiaircraft guns, what Draper called "a disturbed line of sight" device. When calculating the lead, the device would offset the reticule, so that all the operator had to do was keep the target in the cross hairs and the lead would be applied automatically. The sight mounted on a "dummy" platform, and a selsyn transmitted the gunner's motions to a gun (or guns) located some distance away. Powerful servomechanisms then "slaved" the gun to the motions of the gunsight. By the fall of 1940 Draper had completed the theoretical work on this idea, and Sperry had undertaken a full development program.[13]

Sperry wanted Brown to help with the servos for moving the remote gun. Brown's servomechanisms paper, which the NDRC sent to Sperry in November, made it clear that his analytical skills would be useful to the company.[14] In December, Sperry and MIT's Division of Industrial Cooperation signed a contract, just like the NDRC's, for Brown's lab to undertake "fresh, fundamental research" in servomechanisms.[15] Brown was already obligated to the NDRC, so Willis and Draper proposed that Sperry "borrow" him to help with the Draper sight. Brown was having good luck assembling engineers for his Servo Lab, they argued, but the work was proceeding slowly due to lack of equipment.[16] Indeed, by this time Brown had recruited a respectable staff: Albert C. Hall, a graduate student from MIT's Measurements Laboratory, Donald P. Campbell, a new graduate student from Union College, and George Newton, an undergraduate.[17] Soon Brown also added Jay Forrester, Robert Everett, William Pease, and Stephen Dodd to the list. But Brown was having difficulty making the military and industrial connections necessary to acquire the unusual equipment required for his work. Weaver agreed to release Brown from NDRC contracts, but only if he continued to teach navy students and returned to NDRC work after three months. In Brown's absence, Harold Hazen would run the Servo Lab, which would focus on building an automatic fuze-setting machine.[18]

In the fall of 1940 the army's Watertown Arsenal provided a gun mount to Brown's laboratory for experiments. Brown installed it in a confined basement room at MIT.[19] Designing a fast, powerful, yet stable servo to drive a large gun posed a considerable challenge. Existing devices suffered from a number of performance defects, particularly *velocity lag*. This error occurred when the gun tracked a target with continuous motion, causing the gun position to fall a fixed amount behind the commanded position, in a *steady-state error*. Sperry's practical feedback artists could not solve this problem, so it fell to the more theoretical MIT engineers. Here was a chance to apply Hazen's theory and its fruits to a practical military and industrial problem.

Brown and his student Jay Forrester solved the problem of velocity lag with a special correction mechanism. The intense, cerebral Forrester was raised on a cattle ranch in Nebraska and came to the Servo Lab in January 1941, after a year at MIT's High Voltage Laboratory. Forrester and Brown patented their design and assigned the rights to

Sperry Gyroscope.[20] In the spring of 1941 they tested the new mechanism with a powerful servo on a gun mount at Fort Heath in Massachusetts. A Draper sight connected to a selsyn data transmission drove a gun that fired live ammunition over Massachusetts Bay. As an operator directed the sight by hand, the gun mount remotely followed the sight.[21] The tests were successful. Sperry's arrangement with Draper and Brown paid off: Draper's instrument of perception drove Brown's articulated gun through a human operator who integrated the system.

## Smug Attitudes and Practical Experience

As Sperry's three-month "borrow" of Brown neared its end early in 1941, Weaver grew concerned. Brown did not profit financially from his work for Sperry, earning only a token consulting fee of one dollar, but his commercial work did overlap his contract for the NDRC. Sperry treated Brown's results as proprietary information and would not allow him to release them to Weaver's group. Such secrecy threatened D-2's status as a clearinghouse. Weaver wanted to use any and all products of work at the Servo Lab for NDRC purposes, which naturally alarmed Sperry Gyroscope. Just because the company supported some research at MIT, it argued, should not mean that its technology belonged to the government. Weaver appealed again to Karl Compton, who declared that all members of Section D-2 should have complete access to all research and development work done by Draper and Brown at MIT for Sperry.[22]

Compton's dictum upset Sperry Gyroscope, which tried to protect itself from the "capture" of its technology by the NDRC.[23] After significant wrangling, Caldwell, Compton, and Brown finally agreed to cancel those parts of Brown's NDRC contract that overlapped with work done by Brown for Sperry. Brown consulted with his mentor, Harold Hazen, who was temporarily in charge of the Servo Lab, and agreed that the more applied work should go wholly to Sperry but that the "fundamental" aspects of servo control and "reducing to practice" ideas already developed should remain under MIT auspices.[24] Still, Brown liked the immediacy of the industrial work and dived deeper into Sperry projects, while continuing to avoid his NDRC contracts. Weaver terminated Project 1 in September 1941, as scheduled, rewriting the contract to cover only the work already done. Brown had spent only about a quarter of the budgeted funds.[25]

This friction between Weaver, Brown, and Sperry illustrates the difficulty the NDRC fire control committee faced in beginning its development program. The new organization could work as a clearinghouse for organizations with common interests, but established military contractors had little to gain from the NDRC. To Sperry, information exchange meant loss of ownership. From its point of view, the NDRC knew little about the company's technology and would only appropriate it and give it to others. Sperry made its reputation developing advanced and proprietary technology. The company placed heavy emphasis on trade secrets, local knowledge, and patents, whether classified or not. It was not about to let the technology out of its control just because the government had a new agency for research.

Sperry Gyroscope must have been particularly sensitive in the area of fire control, in which it had put so much effort and had so little to show. From mid-1940 to mid-1941 Sperry's dominance, indeed, monopoly, in land-based antiaircraft fire control steadily slipped away. In the company's eyes, this erosion stemmed largely from the NDRC and its fire control committee. Studies under D-2 contracts showed Sperry's solutions to be inadequate, if not plain wrong, and D-2 let contracts to other companies to correct flaws in Sperry's antiaircraft directors. The committee also funded an electronic version of the director at Bell Labs, which would soon end Sperry's business in the area altogether. And Sperry Gyroscope was transferring plans for its M-4 antiaircraft directors, the pride of its prewar development program, to Ford Motor Company for quantity production. "Sperry gets nothing out of this deal," lamented Sperry's president, Preston Bassett.[26] The company had begun some promising research in radar in the late 1930s, but the NDRC, with its new Radiation Lab, threatened to usurp that technology as well. Hence Sperry's relationship to D-2 and Division 7, while not strictly one of competition, began in tension. Despite Sperry's leading position in control systems before the war, it entered into no contracts with the NDRC's section D-2.[27]

Experienced industry hands saw NDRC members as novices in the complex (and not entirely rational) world of fire control contracting. Academic scientists and engineers, for their part, saw this world as bureaucratic and inefficient. Weaver had the impression that Brown's project for Sperry Gyroscope was replicating work done several years before by Ford Instrument. When describing the situation to Karl Compton, Weaver articulated the NDRC's tense relationship to the secret politics of military contracting during 1940−41:

It is somewhat peculiar that the relations between the Sperry Company and the Ford Instrument Company are such that the Sperry Company needs to go to an outside man to get a job done which could have been done (and very probably better done) by engineers in the Ford Instrument Company. . . . To the best of my knowledge, two factors have brought about this situation. First, and most important, the Ford Company is a Navy Contractor and the Sperry Company an Army contractor, and they have always been instructed that information was to be kept secret between the two companies. But it is also true, I am let to believe, that the Sperry Company has frequently taken a somewhat smug attitude that they had a great deal of practical experience and that there was very little necessity for them to learn from other sources. This later point, I believe, has some bearing on their opinion of Section D-2.[28]

Nearly ten years after their integration within a single company, Sperry Gyroscope and Ford Instrument remained separate universes, mirroring the tension between the navy and the army. The NDRC, acting as a clearinghouse, was trying to bridge these worlds. To the NDRC, efficiency dictated that they should mix, but tradition and commercial interests conspired to keep them separate.

Weaver's perception of Sperry's "smug attitude" was probably accurate, but the company must have been aware of the problems with its controls. On a typical visit, in May 1941, Sperry's Bassett, Chafee, and Willis told Ed Poitras of their problems, admitting that they "avoid the cascading of servos for power drives." To work around these stability problems, they still used the old approach of putting a human in the loop.[29] The company, especially Willis, recognized that the academics could help Sperry solve such stability problems and that Draper and Brown could help Sperry regain its dominance in fire control.[30]

Ultimately, both the NDRC and Sperry wanted Brown to solve the same problem: the inadequacy of earlier mechanical gun directors. If the NDRC owned the solution, it could go to any contractor for production. If Sperry owned it, it might regain the favor of its old patron, BuOrd, grown rich with emergency.

## BuOrd's Antiaircraft Revolution

While the NDRC had to confront existing interests in fire control, those interests themselves were hardly static. In fact, in the year before Pearl Harbor, BuOrd changed radically. William Furlong, who headed BuOrd's Fire Control Section after World War I, took charge of BuOrd in 1937. Despite his earlier innovative role introducing G.E. and synchronous systems, Furlong now represented the conservatism of an established technology and entrenched contractors. BuOrd could boast of its fine systems for main battery control and heavy antiaircraft directors, but it had no machine gun directors, no fire control radar, no antiaircraft directors for small ships. Both cause and symptom of these problems was BuOrd's reliance entirely on its captive contractors; it had no development or test facilities of its own. It was not lost on the navy that the Japanese attack on Pearl Harbor pitted aircraft against battleships, with many of the latter ending up on the bottom. A few days later the British lost the battleship *Prince of Wales* and the heavy cruiser *Repulse* off Malaysia to Japanese aircraft, driving the point home. Aircraft threatened not only the navy's ships but its social order as well; the elite gun club could not survive against aviators if it could not shoot back.

BuOrd defended its ships with guns, technology, and administration. In 1940 a new chief radically altered BuOrd's policies for antiaircraft. William H. P. "Spike" Blandy, a 1913 Naval Academy graduate, had excellent gun club credentials: he had done postgraduate work in ordnance and had served as gunnery officer on the battleship *New Mexico*, which had one of the original Ford Rangekeepers, and also aboard the *West Virginia*, which had a new G.E. system. He had even spent time observing production at the Midvale Steel Company, where Frederick Winslow Taylor did his pioneering work in scientific management. Blandy pushed computers as replacements for manual plotting, argued for innovations in training, and won his ships numerous gunnery trophies.[31]

Ironically, in 1938 Blandy saw the future of naval warfare while serving as commander of one of the oldest battleships in the fleet, the USS *Utah*. The *Utah* had been converted into a floating antiaircraft gunnery school and a target for aerial bombing practice. Sitting on the bridge as the passive recipient of simulated air attacks, water-filled bombs, and dummy torpedoes, Blandy developed a passion for new defenses against dive bombers and torpedo planes. He came to BuOrd in 1940 to coordinate antiaircraft work and to expedite antiaircraft gun production.

Blandy's personal mission became a top priority for the Navy. In December 1940 an Antiaircraft Defense Board reported that antiaircraft defense "constitutes the most serious weakness in the readiness of the Navy for war."[32] In February 1941 Blandy, the navy's antiaircraft expert, was promoted over a hundred senior officers to head BuOrd. At age 50 he was the youngest line admiral in the navy. Antiaircraft was to define BuOrd's mission during the war, as the U.S. Navy underwent a veritable antiaircraft revolution. BuOrd spent $4 billion on antiaircraft defenses during World War II, its largest single expenditure, as ships began to bristle with antiaircraft guns (Fig. 8.1).

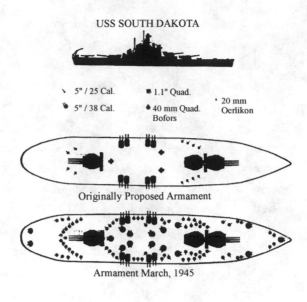

USS SOUTH DAKOTA

\ 5" / 25 Cal.          ■ 1.1" Quad.
♦ 5" / 38 Cal.          ◆ 40 mm Quad.          ⸱ 20 mm
                            Bofors                    Oerlikon

Originally Proposed Armament

Armament March, 1945

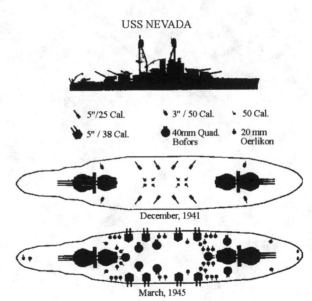

USS NEVADA

\ 5"/25 Cal.           ♦ 3" / 50 Cal.          ⸱ 50 Cal.
✋ 5" / 38 Cal.        ● 40mm Quad.           ♦ 20 mm
                            Bofors                    Oerlikon

December, 1941

March, 1945

Fig. 8.1. The Blandy antiaircraft program: growth in anti-aircraft weapons on typical battleships during World War II. Reprinted from W. H. P. Blandy Papers, Library of Congress.

At the center of this revolution were light, short-range guns that could fend off dive bombers and torpedo planes.

Blandy soon became frustrated with the conservatism of the fire control clique. He found the companies to be disconnected from practical problems. Ford Instrument, G.E., and Arma all had machine gun director projects under way, but they produced ponderous, impracticable solutions under expensive contracts. The main antiaircraft director in the fleet, the Mark 37, was having bad problems with stability in its servos.[33] Decades of secrecy, isolation, and peace had divorced the contractors from the changing tactical threat, and the pressures of production had frozen complex designs in an obsolete or inadequate state. (BuOrd would soon face similar problems with the infamous exploder mechanisms on its torpedoes, a tragic scandal that embarrassed Blandy and the bureau.)[34] "When I arrived here," Blandy wrote to a colleague, "I found the Fire Control Section and all of the civilian engineers of the commercial companies could think only in terms of the complete solution, namely: to make the 1.1 [the 1.1-inch gun] capable of bringing down any plane on any bearing and any position angle. . . . Well, as you can imagine, such a director involves enough gyros, cams, potentiometers, etc. to make your head swim, plus a great deal of weight, cost, and time to deliver." Blandy agreed on the need for an ultimate solution, but he also wanted something that compromised completeness in favor of expediency. He pressed for control systems that were small, lightweight, and amenable to mass production. As though anticipating Pearl Harbor, in the fall of 1940 Blandy clamored, "I want something in a hurry."[35]

To develop these machines, Blandy reorganized BuOrd. He established a new research and development division, designated "Re," to conduct fundamental studies, design systems, and engineer production runs.[36] "Re" comprised a number of groups, with Re14 responsible for fire control design and Re4 responsible for fire control in general. BuOrd now had its own version of the NDRC's D-2. Like the NDRC, BuOrd contracted for private research, spending about $34 million on industrial research during the war and almost $700,000 at educational and research institutions, much of it at MIT.[37]

In addition, BuOrd transferred more than $40 million to the NDRC/OSRD, though the bureau viewed Bush's innovative agency not as an independent sponsor of research but as a device "to put the laboratories and scientific agencies of the country at the disposal of the Army and Navy." Indeed, the NDRC's section T, which developed the proximity fuze, became a virtual extension of BuOrd. The new research group also allowed BuOrd to erode the dominance of its fire control clique, ending "the condition where we are totally dependent on a few fire control companies, such as Ford, Arma, Sperry, and General Electric, for fire control development."[38]

When Commander M. Emerson Murphy took charge of Re4, he immediately received five copies of Gordon Brown's paper on servomechanisms to initiate him into the new world of control and into BuOrd's crisis. "There appeared to be a crying need," he wrote, "for a small, simple director capable of being produced by the thousands."[39]

## Fire Control for the Masses

BuOrd's antiaircraft revolution depended on new guns and new controls. Blandy pushed the procurement of two new antiaircraft guns that would cover American warships for the duration of the war: the Swedish 40 mm Bofors and the Swiss 20 mm Oerlikon. At the start of the war, the navy's guns used 1.1-inch and 30 and 50 caliber guns to defend against close-in aircraft. The first was simply not a good gun, and the latter two were too weak. They were aimed by tracer bullets, which made the gunners feel good, but the bullets' seeming accuracy proved illusory. In comparison, the 40 mm Bofors gun was powerful and fast, firing 2-pound projectiles at 160 rounds per minute, but it needed a director to be accurate against moving targets. Mounted in single, double, and quadruple mounts, it became known for its ruggedness and reliability. The Oerlikon, more like a heavy machine gun, fired 450 rounds per minute and was light, was easily maintained, and required no external power; it could be bolted down anywhere on a ship. A man could freely swing the gun in all directions with muscle power.

The Oerlikon entered service beginning in late 1941; nearly 150,000 were produced during the war. The Bofors faced difficult production problems but entered the fleet in mid-1942, with nearly 40,000 produced during the war. (Both guns were used by Britain and Germany as well.)[40] These guns put antiaircraft defense in the hands of the common sailor. Now Murphy needed "fire control for the masses."[41]

Here BuOrd's investment in an MIT connection began to pay off. Blandy assigned the gunnery officer and former MIT student Lloyd Mustin to help with production.[42] Blandy set up a special antiaircraft section, under the direction of the ballistics expert Captain E. E. Herman, who reported directly to Blandy. Blandy also created a new Radar Desk; at its head he placed Mustin's master's degree partner, Horacio Rivero.

Rivero quickly seized the opportunity to bring his MIT connections to BuOrd. He mentioned to Captain Herman that Professor Draper, at MIT, had been working on a gyroscopic sight for an army gun, for which Gordon Brown had designed a servo and which might be small enough to direct the navy's new Oerlikon gun. Murphy and Rivero went to see the sight in May 1941 and made another visit soon thereafter with Herman and other navy officials.[43] Favorably impressed, they supported Sperry and Draper's continued development of the device, both as a gunsight and as a small director for a remotely controlled gun.

In June the navy ordered 12 pilot models of the sight.[44] Soon BuOrd ordered 2,500 of the sights for its 20 mm Oerlikon guns, officially designating it the Mark 14 sight, also known as the Sperry-Draper gunsight. The Mark 14 succeeded not because of the quality or precision of its computations but rather because of its compromises. Estimating range provided the most significant shortcut. Rather than using a bulky and slow rangefinder, the operator merely estimated range by eye and then dialed it in by hand—a rough approximation, but the range to an attacking airplane was likely changing rapidly anyway. Moreover, because the device employed polar coordinates, such errors diminished in significance as the target got closer (in contrast, the prewar Cartesian directors exacerbated errors at close range).

Like the Ford Rangekeeper decades earlier, the Mark 14 sight had the right combination of precision, ease of use, and simplicity in the tactical situation for which it was designed (though it was easily damaged by gun blast, and smoke from the barrel frequently obscured the view). By the end of the war more than 85,000 had been produced. A variant of the device that replaced the operator's muscle power with a remote servo became the Mark 51 director, of which about 14,000 were produced (Figs. 8.2, 8.3, 8.4, 8.5). Handling the contract with Draper was James E. Webb, Sperry vice president; two decades later, as director of NASA, he would again contract with Draper, for the Apollo navigation and control system.[45]

The Mark 14 sight represented the return of Sperry Gyroscope to naval fire control and a triumph of the company's simple, tight coupling of operator and machine over the complex, integrated systems produced by the fire control clique. "The new sight broadens the mental powers of the gunner," wrote Sperry president Tom Morgan, "frees him from tasks requiring judgement, and enables him to devote his entire attention to accurate 'tracking' of enemy aircraft."[46]

The Mark 14 gunsight also expressed in solid form a combination of industrial, university, and military technology. As "fire control for the masses," it embodied the relationships between Sperry, MIT, and BuOrd, relationships that existed entirely outside of the auspices of the NDRC and its fire control committee. BuOrd, with its new

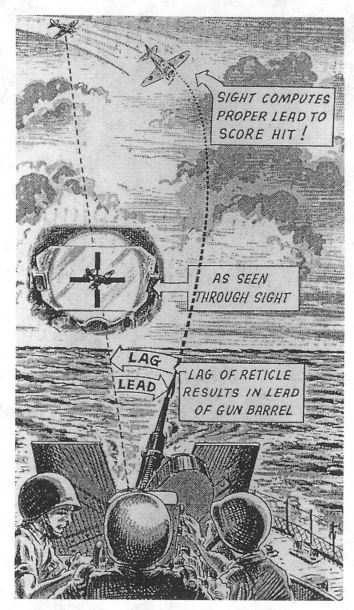

Fig. 8.2. Theory of operation of Sperry-Draper Mark 14 lead-computing gunsight. The gunner was no longer responsible for leading the target but rather kept the reticule directly on it, while the gyros in the gunsight disturbed the reticule, causing the gunner to lead with the gun. Reprinted from "Gun Sight Mark 14, Gunner's Operating Bulletin," United States Fleet, Headquarters of the Commander in Chief, SGC Papers, box 20.

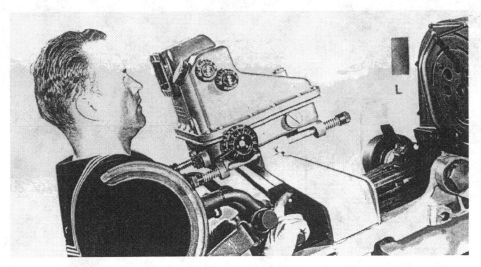

Fig. 8.3. The Mark 14 lead-computing gunsight connected to Oerlikon 20 mm antiaircraft gun. Courtesy of Hagley Museum and Library.

research and development organization, its recent MIT connections, and a private contractor capable of large-scale manufacturing, remained on the cutting edge of naval fire control. You could see it in the gunners' hands.

## The Servo Lab's Continuing Work

While BuOrd's ties with university researchers blossomed, Brown's relationship with Warren Weaver soured. As Sperry's "borrow" ended, Brown still wished to work for the NDRC and may even have been tiring of his relationship with Sperry.[47] But Weaver distrusted Brown's interest in fundamental research, believing "that B[rown] will never be satisfied, having once tasted blood [i.e., industrial work], to deal exclusively with a patient long-time academic general program."[48] Hazen proposed extending the Servo Lab contract to fund graduate students in Brown's absence, but Weaver blocked the move.[49]

Despite this personal friction, Brown's work was going well, and he was learning a great deal from his Sperry experience. In addition to the gunsights, Sperry was producing small, high-power electro-hydraulic motors for aircraft turrets, and Brown recognized that such devices could be deployed in a wide array of war machinery.[50]

Fig. 8.4. The Mark 52 director was a Mark 14 sight modified for longer range and to remotely drive a 40 mm or 5-inch gun mount. Photograph by the author aboard the USS *Massachusetts*, Fall River, Mass.

Rather than propose a project to Weaver, however, Brown brought it up with MIT President Karl Compton. Displaying a notably nonacademic interest in manufacturing, Brown predicted that the army and navy air corps might need as many as a quarter- to a half-million units. Compton read between the lines: Brown wished to do more work for the NDRC, but he could not propose it directly. Two days later Compton passed the message to Weaver, as well as to Bush and the army.[51]

Indeed, Brown's proposal did match an army interest. In early 1941 the army adopted the British antiaircraft director known as the Kerrison predictor, which had been used in the Battle of Britain. It employed a servo to drive a 40 mm Bofors gun (the same one the navy had adopted), and the army standardized the director as the M-5. Sperry, already under contract to do pilot production, was assisting Singer Sewing Machine and Delco to go into full production.[52]

But the army was concerned about manufacturing the British hydraulic pump and motor that drove the guns. Firestone Tire and Rubber had a model in production, but it barely worked. Brown and the Servo Lab were working with a servo designed by the Oilgear Company of Milwaukee, which might replace the troubled British design. To study the problem, Division 7 let contract 35 to the Servo Lab, and soon Brown's group redesigned the Oilgear servo to eliminate velocity lag and make it interchangeable with previous systems.[53] As a concrete design for a production device, this project was a long way from the "fresh, fundamental research" of the previous Servo Lab contract.

The NDRC then contracted with Westinghouse to do a production design, but the

Fig. 8.5. A Mark 51 director *(right)* operating quad-mount 40 mm guns with servos designed at the MIT Servo Lab, 1943. Courtesy of Hagley Museum and Library.

company ran into trouble with the hydraulics. Brown wanted to help smooth out the problems, but the Army Ordnance Department and Weaver himself were adamant that the NDRC should stay out of manufacturing.[54] The tension between the two men, which had been building for nearly two years, came to a head in the summer of 1942. Weaver, by his own account, "unfortunately loses his temper and tells Gordon Brown several things which should have been made clear for him by his mother long ago." Weaver blamed "war nerves" but recommended that the project be discontinued, and Poitras then refused Brown's request for further funds.[55]

Still, Brown's lab continued to help Westinghouse under the direction of the Army Ordnance Department. The company, it turned out, had built the pistons in the pumps to too close a tolerance, which made the servos unstable. One of Brown's students showed that oil leaking around the piston was equivalent to a damping term in the servomechanism, which actually improved stability, so making the pistons to a wider tolerance improved the performance. "If there hadn't been any MIT [people] leaning over their shoulders," Brown recalled, "if there hadn't been people looking at these equations long enough and [who] had some insight into all of the factors that contribute to the instability," the device never would have worked.[56] For Brown, finding the problem with the Oilgear servo was the ultimate contribution of an academic lab to the war effort, and of theoretical study to industrial problems.

Weaver and Brown simply had different notions of how the NDRC and the research it sponsored should contribute to the war effort. Brown saw little distinction between his role as a consultant and his role as a professor. "I see nothing wrong with a university engineering school participating in these [prototyping] programs," Brown later said of the Servo Lab, "taking conceptual ideas, bringing them down to at least a first manifestation."[57] In wartime, he would do whatever he could to make automatic control useful in the field, even if it meant working on the factory floor. Weaver, in contrast, held a more traditional view, consistent with his background as a foundation science manager. However effective Brown's work might be, his casual crossing of institutional boundaries was unacceptable under NDRC policy. With institutional threats coming from every corner, not least from BuOrd's new research division, Weaver would have to carefully define D-2's boundaries to avoid becoming dissipated as yet another wartime bureaucracy.

To help resolve these differing philosophies, larger forces intervened. The NDRC reorganized in December 1942, and Weaver left to head the Applied Mathematics Panel. Section D-2 became Division 7, headed by a man who could not have been more friendly to Brown and his program: his mentor and department head, Harold Hazen. Weaver remained a consultant to Division 7 and continued to voice his opinions of Brown. No documents confirm a link between Weaver's departure from D-2 and his disagreements with Brown, but it remains a remarkable coincidence that in a time of intense conflict, Brown's harshest critic was replaced by his closest friend.

## What Was the Servo Lab?

When D-2 stabilized as Division 7 in late 1942, the Servo Lab's identity stabilized as well. Brown had spent two years creating an institutional and conceptual home for servomechanisms, servo engineers, and the Servo Lab. "We were seeking an identity," Brown recalled of the period.[58] The identity he found did not encompass broad notions of systems or computers, but only the servos themselves: powerful motors harnessed by control systems to move with precise elegance.

Despite the capital letters in its name, the Servo Lab had no formal status within MIT. "There was no formal announcement," recalled Brown of founding the lab, "it just grew, because people had a kind of competence that fitted the bounds of a particular application." The Servo Lab was a place, at first in a basement lab and then in a much larger building on the MIT campus. "Servo Lab" was also a label for about a hundred people, including engineers, students, machinists, administrators, and a handful of professors. Under Brown's direction the lab maintained an educational mission; he and his staff continued to teach MIT students and officers from BuOrd. The lab also wrote technical manuals for servos to instruct military users in their new art. And the Servo Lab consisted of equipment: gun mounts, electric and hydraulic servos, measuring instruments, and a rolling platform to simulate a ship at sea.

No company related to the Servo Lab like Sperry Gyroscope, and the two remained intimate throughout the war. Sperry personnel worked full time in Brown's laboratory, "as part of the family." Brown made use of Sperry facilities as well, enjoying "the chances and privileges of going down to Sperry [on Long Island]."[59] With Sperry Gyroscope as a catalyst, the Servo Lab also retained a close tie to Draper's Confidential Instruments Laboratory. Partly as a result of the Mark 14 and Mark 51 projects, Brown's group was seen as the servo arm of Draper's gunsight work, building the power drives that responded to Draper's delicate mechanisms. Brown gladly accepted the position and explicitly modeled the Servo Lab on Draper's laboratory.[60] Draper already had reputation and prestige, whereas Brown was building them, and the association could only benefit his new group.

### Servos as Feedback Amplifiers

The Servo Lab's identity also hinged on an intellectual transformation, from transient analysis to the frequency domain. Brown included communications theory in his original lab proposal, and frequency response techniques became "very important to the growth of the Servo Lab and the development of servo theory."[61] Initially, Brown and his engineers used transient analysis, derived from his and Hazen's work of the 1930s. Still using the operational calculus, they worked directly from the differential equations that specified a system, adjusting design parameters to get the desired transient response. Robert Wieser, for example, found that in the Servo Lab "the drilling I'd gotten as a power engineer in transient behavior of systems was immediately applicable to servomechanisms."[62] Brown and Forrester's early work for Sperry on power drives utilized transient techniques exclusively, as did nearly all Servo Lab work until the end of 1942.[63] Typically, engineers would receive an existing servo as a sample from a company or military service and test it for transient response to a step input or a constant-velocity input. It was quite difficult, however, to translate these responses into meaningful design criteria. The Servo Lab's practice of closing feedback loops around existing actuators highlighted the inadequacies of the transient approach.

Radar stretched the transient approach to its limits. One of the original Servo Lab projects was to make a radar drive a servo to move an antenna and automatically track a target. But radars produced unruly electrical signals, corrupted by noise from a number of different sources. A servo that tried to translate radar echoes directly into motion could produce grinding gears, jerky motions, and instability. Lab member Albert C. Hall remembered that in 1941, "while we had designed a fine experimental system for test, we had missed completely the importance of noise, with the result that the system's performance was characterized by large amounts of jitter and was entirely unsatisfactory." Control engineers needed a new way to conceptualize their loops. "The advent of radar," noted Hall, "required the controls engineer to design equipment to operate well in the presence of signals that he could not even describe in terms then in general use."[64] Radar, with its affinities with communications, pushed Servo Lab designers to understand control systems as processors of signals, according to their frequency content.

Herbert Harris, another member of the Servo Lab staff, explicitly drew the analogy between servomechanisms and feedback amplifiers. In a restricted report published by the NDRC, Harris applied the notion of frequency response, "used in the radio and telephone arts," to a servomechanism.[65] Building on the work of Taplin and Brown, Harris brought together MIT's servomechanisms with Bell Labs' feedback amplifiers. "The recognition of the similarity between servomechanisms and feed-back amplifiers," he wrote, "makes available to the automatic control engineer many valuable analytical tools developed by communications engineers." Harris proposed a general vision of control, based on a system's abstract building blocks and not on its physical structure. In servo design, as with feedback amplifiers, frequency response provided powerful tools for analyzing stability. Harris employed Black's characteristic equation, Nyquist's stability criterion, and Bode's magnitude-phase relationship. He discussed a mechanical system using the terms and methods of the communications engineer, probably the first such discussion in print. When he left MIT in the spring of 1942, Harris went to work for Sperry Gyroscope.

Following Harris, Albert C. Hall explored the implications of a frequency response approach to servos. In his classified 1943 dissertation, he formulated "a servomechanism design procedure based primarily on an analysis of the system response to sinusoidal inputs of various frequencies."[66] Unlike Harris, Hall did not build his analysis entirely on the analogy between servos and feedback amplifiers. He acknowledged the similarities, but also some important differences: servo designers are concerned with precision, amplifier designers are not, and servos work in a much lower frequency range than feedback amplifiers, so the electronics are easier. Hall formulated a set of trade-offs between transient and sinusoidal representations of the control system. Frequency domain data were easier to measure, but servo performance was ultimately specified in the time domain, as transient response to a varying input. Hall did not offer a design procedure; rather, he offered a set of guiding principles for the servo engineer. He developed a number of graphical and analytical techniques that the designer could use as tools, including a "transfer locus" plot, compensating networks, lead controllers, and integral controllers. Reflecting the Servo Lab's abiding interest in reducing the velocity lag of tracking servos, Hall devoted two significant sections to "minimum velocity error servos."[67] Hall's analysis reflected the realities of working in an academic lab tied to industrial concerns.

Harris and Hall's work articulated and accelerated the changing practice of control engineering at the Servo Lab. Using frequency response, Servo Lab engineers injected sine waves of varying frequencies into servos under study and plotted the magnitude and phase of the response—the technique Harry Nyquist had developed for feedback amplifiers. Until 1943, Servo Lab reports presented only transient analysis. Reports written after that year all include frequency response plots and Nyquist diagrams.[68] Ironically, control engineers at MIT used a graphical technique borrowed from another industry to define their own technical practice.

## Conclusion

The Servo Lab was born through exclusion, as the NDRC appropriated the theory of feedback controls from civilian research. During the war the NDRC distributed 294 copies of Brown's paper to its secret colleagues, defining a new landscape of control.

Beginning with this paper, the Servo Lab built its identity and its practice between the research world of the NDRC and the industrial world of Sperry Gyroscope. While beginning with a "fresh, fundamental approach," the new control engineering came to be characterized by collaboration of academic engineers and industrial concerns on high-performance, fast-acting mechanisms. No longer were fire control computers large, centralized machines, no longer was servo behavior studied as a purely transient phenomenon, and no longer did the navy's fire control clique have a monopoly on the field's technical secrets. Now BuOrd aimed small, light computers at fast, proximate aircraft. Engineers designed in both the time and frequency domains. And now academic experts could provide designs and advice to military organizations, circumventing the established fire control contractors.

Initially, the Servo Lab understood servos as manifestations of classical governors. It built servos that articulated the output of a larger control system, including instruments of perception and integration, designed by the lab's collaborators, Sperry Gyroscope and Draper's Instrument Lab. Gordon Brown's original laboratory vision remained tied to the powerful, precise servo motors. Harris and Hall began to expand this local vision with their frequency domain approach. Now the mechanisms themselves became processors of signals, just like any other component in a larger system. This vision drew heavily on telephone engineering, and it was telephone engineers who began using feedback techniques for overall systems, rather than just for individual servos or amplifiers.

# Analog's Finest Hour

> The purpose of a servo is to reproduce a signal at a place, or at a power level, or in a form that is different from the original signal, but is under its control. It is, therefore, a signal-transmitting system. It uses negative feedback to minimize noise and distortion. All this can be said of a telephone repeater, a public address amplifier, or an intermediate-frequency amplifier . . . the basic problems of stability, bandwidth, and linearity are just the same.
>
> Enoch Ferrell, "The Servo Problem as a Transmission Problem"

### Robots versus Radar

In June 1944, one week after D-day the *Luftwaffe* replaced its pilots with automatic controls; the V-1 buzz bombs began to cross the English Channel. The odd craft were exceedingly difficult for people to fly (several died trying), but feedback loops kept the contraptions stable and on course. Launched from shallow ramps on the Continent pointing at London, the rocket-powered V-1s flew with a racket and seeming mindlessness that would have made them comical were it not for their 1-ton warheads.

Few in the first wave of automated attackers made it across the Channel, and fewer still hit the barn door called London. Their poor performance stemmed from the inexperience of the launch crews, trained as antiaircraft artillerymen. But they learned quickly. Within a week the V-1s flew from more than 50 sites, crossed the English Channel by the hundreds, and likely as not reached the city. From then on, for nearly three months, an average of 100 missiles per day ascended from the railroadlike launchers on the Continent.[1] More than 7,000 V-1s were fired against London, and many more against targets in Belgium. Of those crossing the Channel, about half were destroyed. The rest made their way to the city, cut off their engines, and glided silently to the ground before exploding. They killed thousands.

Not that the flights were easy or unopposed. Over the English coast the V-1s faced an array of aircraft, balloons, guns, and even tanks massed and coordinated by tens of thousands of people to shoot them down. Defenders called the attackers "PACs," for "pilotless aircraft," or "divers" for how they ended their trips. They were fast, small, and low-flying, which made them exceedingly hard to hit. But in their automatism lurked a fatal flaw: the buzz bombs flew absolutely straight and level, which no human would do under a barrage of fire. The robot bombs were predictable.

The buzz bombs also faced an ethereal enemy, locking onto them with the singular purpose of a stable feedback loop. As they approached the coast, regular pulses of microwaves began to bounce silently off the machine's metal skin, nearly 2,000 times per second. A tiny fraction of that energy reflected back to the source of the waves, an MIT-designed radar with a parabolic antenna. Once in the system, the reflection became a signal and flowed through a series of filters, cables, and vacuum tubes. Circuits, motors, and mechanisms filtered out noise and amplified the signal before it entered the servos that moved the antenna. The sign, of course, was reversed, so whenever the signal weakened, the antenna moved to increase it; the diver's motion literally drove the antenna. As it locked on, the feedback loop followed the airplane, continuously measuring its range and bearing with great precision. Inside the machine, an analog of the target's motion modulated the rotating beam on a cathode-ray tube (CRT), producing a glowing picture on an oscilloscope of the target flying through the sky above.

Human operators distinguished the pips of the arrow-straight flight from background noise, identified the peak, and matched the pip to send it on to represent range, bearing, and elevation. The three variables then traveled through an electrical cable and entered a computer designed by telephone engineers at Bell Labs, where a series of servos performed the classic fire control prediction: calculating the target's course and speed and extending them out into the future. The results of the prediction then passed through custom-shaped wire cards that calculated ballistics and traveled through still more cables until reaching the gun, where Sperry servos amplified them into the motions for precise aim. The gun fired. As the shell flew, a miniature radio fuze designed at Johns Hopkins University's Applied Physics Laboratory transmitted a signal and listened for proximate echoes. When the signal reflected back, indicating that an object was near, the radio set off the shell, closing an overall feedback loop and, ideally, destroying the target.

This was how things were supposed to work, anyway. At first, when the mocking bombs flew over, these subtle loops did not close and the robots looked down on technicians scrambling to learn their equipment. The gunners gallantly fired while the men with the electronics and computers flipped through instruction manuals. Before long, however, a few confident civilians showed up, men in their twenties, technical experts who had helped design the machines. They instructed the operators and helped them align, calibrate, and connect the diverse apparatus into a coherent system. Radar operators learned to read the glowing phosphors on their displays, to sense the response of the tracking servos, to quickly find dead vacuum tubes and faulty cables. The people and their interconnected machines gradually came into their own, and the statistics showed rapid improvement. Within a few weeks more than a third of the robot attackers were downed by gunfire.

This tense confrontation pitted computers against autopilots, servos against servos. Through these and related episodes, the late years of World War II put feedback, controls, and human operators into new combinations, generating new ideas about systems and how to design, organize, and produce them.

### The Bell Labs Gun Director

"At first thought it may seem curious," wrote Warren Weaver in 1945, "that it was a Bell Telephone Laboratories group which came forward with new ideas and techniques to apply to the AA problems." What would a group of experts in transmission lines, modulators, and switching have to add to the Newtonian problem of fire control? "First, this group not only had long and highly expert experience with a wide variety of electrical techniques . . . Second, there are surprisingly close and valid analogies between the fire control prediction problem and certain basic problems in communications engineering."[2] These analogies allowed Bell Labs engineers not only to advance the art of fire control but also to blur the boundaries between communications, servomechanisms, and computing.

Like the feedback amplifier, the Bell Labs project has its own mythology of origin, starting with the dream of a staff member, the physicist Donald B. Parkinson. In the spring of 1940 Parkinson was working on a device to record the logarithm of an applied voltage on a strip of paper. Parkinson's circuit included a logarithmic potentiometer connected to a pen, and "to all intents and purposes this small potentiometer could be said to control the motion of the pen."

"I had been working on the level recorder for several weeks," Parkinson recalled,

when one night I had the most vivid and peculiar dream. . . . I found myself in a gun pit or revetment with an anti-aircraft gun crew. . . . There was a gun there . . . it was firing occasionally, and the impressive thing was that *every shot brought down an airplane!* After three or four shots one of the men in the crew smiled at me and beckoned me to come closer to the gun. When I drew near he pointed to the exposed end of the left trunnion. Mounted there was the control potentiometer of my level recorder! There was no mistaking it—it was the identical item. . . . It didn't take long to make the necessary translation—if the potentiometer could control the high-speed motion of a recording pen with great accuracy, why couldn't a suitably engineered device do the same thing for an anti-aircraft gun![3]

Parkinson realized that a well-designed amplifier could bridge the gap between controlling a pen and controlling a gun, between articulating a laboratory instrument and articulating a fire control computer.

About 1 June 1940 Parkinson proposed the idea to his superior, Clarence A. Lovell. Parkinson outlined three Bell Labs technologies that could contribute to an "electrical predictor":

1) A coil winding machine which can wind potentiometers on any shaped card thus giving a rotation which was a rational function of the voltage applied.

2) An electrical differentiator proposed and tested for another job . . . capable of measurement of extremely small angular velocities.

3) We have designed extremely high-acceleration electrical servos, [based] on electrical feedback circuits which operate at high speeds and are critically damped. It should be possible to extrapolate them to larger size and make them swing the gun around automatically.[4]

With no prior experience in fire control, Parkinson grasped the essence of the problem. It required a means of solving equations electrically (1), a means of deriving rates for prediction (2), and a means of moving the guns in response to firing solutions (3). In the coil-winding machine, he also recognized the importance of manufacturing.

Lovell liked Parkinson's idea and proposed it to his boss, Mervin J. Kelley, then director of research at Bell Labs, who in turn presented the proposal to Frank Jewett, who took it to the Army Signal Corps. Later in June, Parkinson, Lovell, Kelley, and several other Bell Labs engineers met with representatives from the Signal Corps at Fort Monmouth, New Jersey. There the Bell Labs group inspected a Sperry M-4 director and other fire control equipment and received manuals and books on antiaircraft and fire control. The Bell Labs engineers also presented their ideas to the navy, which had no use for the project, being content with its own directors. The signal corps, however, expressed great interest.[5]

During their initial period of exploration Parkinson and Lovell put together a group of Bell Labs engineers for preliminary analysis. Lovell himself had visited Bush's lab and seen the differential analyzers during the '30s, and both he and Parkinson had previously designed servos for measurement work in acoustics.[6] They produced a study, "Electrical Mathematics," that examined the functions required for fire control equations: addition, subtraction, multiplication, division, integration, differentiation, and looking up tabulated data. Lovell recorded in his notebook an idea for a machine based on electrical feedback mechanisms. He described how "servomechanisms may be used directly in making transformation from one coordinate system to another without the necessity for setting up [mechanical] scale models having to be considered."[7]

Lovell had picked up a general knowledge of the Sperry directors at Fort Monmouth a few weeks earlier. He recognized that Sperry systems incorporated servos in their calculating units (replacing the human servomechanisms of earlier models), but only to transmit information between stages. Lovell's servos actually calculated, placing a mathematical element directly in the feedback loop. By tending to reduce the difference between their inputs and their outputs—the error—to zero, the servos could solve equations. This new application echoed Hendrik Bode's use of telephone feedback amplifiers as equalizers to invert the distortion of a transmission line and Hazen's use of "back coupling" to solve equations with the differential analyzer. Bell engineers commonly referred to the technique as "electronic" but acknowledged that it was really "electro-mechanical": the servomotor turned a mechanical potentiometer whose output voltage was a function of angular position.

Lovell noted that his innovation, modeling mathematics with servomechanisms, could make not only a gun director but also a calculator. He saw his computing elements as analogous to the mechanical ones used in earlier computers: "The availability of accurate differentiators and servo-mechanisms make possible the solution of differential equations. . . . machines of the same character as the Differential Analyzer of Bush and Caldwell can be made to operate electrically by the use of the means at our disposal and . . . a machine can be built to solve systems of simultaneous differential equations."[8] In his notebook, Lovell sketched a differential analyzer made entirely out of servomechanical computing devices.

Like Lovell, Parkinson displayed growing understanding of fire control and computing. He suggested a variation on Sperry's ballistic cam, a "space potentiometer," which would solve functions of two variables, rather than the single variable of his logarithmic potentiometer. Such a device, he noted, could look up data in ballistics tables and replace the cumbersome ballistic cams.[9] The wire-wound potentiometers could be wound by relatively unskilled, female workers. "This offered the possibility," Lovell recalled, "of getting a new and relatively trainable type of labor into the manufacture of these things instead of the very high precision mechanics that were necessary by using the prior method [i.e., ballistic cams]."[10] Parkinson recognized, as Sperry Gyroscope had several years earlier, that electrical computers offered manufacturing advantages over mechanical ones. Once again, a change in the mode of representation entailed a change in labor.

### D-2 Funds the Bell Labs Director

While the Bell Labs group sketched their ideas for a new director, Weaver was assembling NDRC Section D-2. After learning of the Bell Labs project from the army, D-2 visited Bell Labs and met with Kelley, Lovell, Parkinson, Harvey Fletcher, and other Bell engineers, who explained their new ideas for electrical gun directors and computers to the visitors.[11]

Weaver and D-2 were enthusiastic. An electronic machine would provide a needed alternative to Sperry's directors, whose shortcomings were becoming clearer every day. Bell engineers argued that electronics worked with greater accuracy and speed and at lower cost than mechanical computing. The NDRC, however, had other reasons for its interest. It thought that an electronic fire control computer would be easier to reconfigure and correct in case of errors, by simply rewiring the components. By contrast, in a mechanical computer the algorithm was tightly connected to the physical structure of the machine and was hence difficult to change. Furthermore, an electronic director could be built by the vast manufacturing capacity of Western Electric, which was at the time underutilized for war production. Sperry's resources, as well as those of many precision mechanical manufacturers, were already stretched thin.[12] Bell Labs was also a successful laboratory with a good reputation and an organization familiar to the members of D-2. Its founder and former president, Frank Jewett, was a leading member of the NDRC.

So far, Bell Labs had funded the project internally. The army was willing to provide funds but proposed that the NDRC take it over "during the development stage, when flexibility of contract is important." Weaver agreed, and Section D-2 let its second contract, Project 2, effective 6 November 1940.[13] Under this contract, Bell Labs would construct an electrical gun director, designated T-10, to drive a 90 mm antiaircraft gun via Sperry hydraulic power controls. An optical rangefinder would provide altitude input (range and elevation angle), but the machine would include the provision for radar inputs. It would also keep the constant-altitude assumption of previous directors. The Frankford Arsenal, which had directed Sperry's work in the 1930s, would act as liaison.

Over the next few months Bell Labs engineers continued gathering information. They studied materials from army training courses, operating manuals for gun directors, and ballistics tables. Lovell visited the army's antiaircraft gunnery schools and the arsenals responsible for development. He requested samples of telescopes, synchronous transmitters, receivers, and other equipment. Frankford Arsenal sent him blueprints for Sperry directors.[14] Ed Poitras, of D-2, sent Parkinson copies of Gordon Brown's paper (numbers 17 and 18), thus admitting Bell Labs to the secret world of wartime control engineering.[15] In less than six months, the Bell Labs electronic director had gone from being an individual's dream to being one of the leading control systems projects in the country.

During most of 1941 Lovell, Parkinson, and their colleagues designed and built the T-10 director, with help from the Bell Labs mathematical research group, under Hendrik Bode. Throughout, they conceptualized the servos in the language of communications. As one engineer put it in his notebook, using terms from radio and telephony, "A servo, in general, involves a carrier, and a means for modulating that carrier according to some function."[16] Since the mathematical quantities in the T-10 were all represented by DC voltages, the amplifiers and servos needed to maintain precision and stability in the face of variations in temperature, age, moisture, and other factors—the very requirements Harold Black had addressed for telephone repeaters nearly fifteen years earlier.

### "The Computer as a Servo"

Appendix C contains the detailed block diagram for the T-10 computer. Because an electrical director was itself a new idea, Bell Labs engineers chose to exactly replicate the algorithm from Sperry's earlier directors "in order to achieve a direct and easily

Fig. 9.1. M-9 gun director, tracking head with operators. One follows the target in elevation, the other in azimuth. The unit and the operators rotate while tracking. Courtesy of AT&T Archives.

interpretable comparison" between the mechanical and electrical techniques.[17] The T-10 had four servo motors and three selsyn transmitters for sending firing data to the gun. The unit weighed 1,600 pounds, nearly twice as much as the Sperry directors, and nearly 3 tons when combined with its supporting units. The human operators sat on a small, rotating "tracking head" that mounted the telescopes and transmitted data to the "computer," a rack of electronics on a separate trailer (Fig. 9.1). Overall, the system was "ballistically complete," that is, it included all known factors in the ballistic calculation. Unlike in the Sperry machines, with their follow-the-pointer operations, the T-10's only manual inputs were the two tracking telescopes and a rangefinder.[18]

Bell engineers envisioned the T-10 director as a feedback system at every level, from amplifiers to servos to the computer as a whole. Where the Servo Lab engineers used frequency response methods to study their servos, those at Bell Labs also eroded the distinction between servos and feedback amplifiers. "Servo performance is readily studied by the highly developed method of feedback analysis," they wrote. "That a servo is a feedback system becomes apparent from a comparison of its action and that of the feedback summing amplifier." Indeed, Bruce Weber, another Bell Labs engineer, modeled the director's computations using amplifier stability theory.[19]

A section in the T-10 final report, "The Computer as a Servo," explains the feedback in the prediction loop. Were it not for the many corrections and firing data within that loop, that section notes, the entire prediction could be performed by a single servo. Overall, "the system has a structural resemblance to a feedback amplifier with multiple loop feedback, and may be analyzed by the usual feedback methods. . . . the whole system is stable whenever there is a physical solution, provided the individual servo loops are stable."[20] Like their counterparts at MIT, the Bell Labs engineers now realized that Black, Nyquist, and Bode's stabilized feedback amplifiers had applications far beyond telephone repeaters and could themselves perform computations.

At Columbia, another group of Division 7 researchers picked up these applications of feedback amplifiers from the Bell Labs project and used them for modeling dynamic systems. John Ragazzini, Robert H. Randall, and Frederick A. Russel used amplifiers, based on those developed for the T-10, to model missiles and guidance systems in the laboratory. They called their devices "operational amplifiers," alluding to the operators in Heaviside's calculus.[21] Operational amplifiers, high-gain devices that amplify the difference between two inputs for use in feedback loops, are still among the most common building blocks in electronics.

### "A Rather Devastating Device": The Dynamic Tester

For all its innovation, the T-10 posed a problem for the members of D-2: they had no way to judge it. Did it work better than the Sperry machines? Did it work at all? What did *better* mean? In traditional live firing tests, a sleeve or sock was towed from an airplane as a target. Using the director, gunners fired at the target while theodolites and cameras observed the shell explosions. The scheme was chaotic and uncontrolled; numerous parameters changed from test to test and even from moment to moment. Different operators produced different test results, as did the same operator on different days. Furthermore, towed sleeves could mimic only the most basic attacks, not dive bombing or low-level runs. "It was literally impossible," Sam Caldwell complained, "to make a decision regarding any fire-control equipment from an appreciation of realistic, quantitative data."[22] Before the NDRC could evaluate the prototype, it needed a way to compare, and hence to define, performance.

Standardized testing became the ideal contribution for the new NDRC group. The function satisfied industry's desire for a kind of military National Bureau of Standards, and measurement fit the self-image of the academically oriented members of D-2. A solution came in 1941, when Duncan Stewart, president of the Barber Coleman Company, of Rockford, Illinois, visited the NDRC seeking work on fire control. Barber Coleman, a medium-sized manufacturer (1,800 employees) of machine tools, textile machinery, and air-conditioning accessories, had experience with thermostats, follow-ups, and other types of control devices. Stewart, whom Harold Hazen described as "a very shrewd savvy Scot," designed an antiaircraft director and presented it to the army. Unimpressed, the army put him in touch with D-2 and Warren Weaver. Weaver, it turned out, had been Stewart's teacher at the University of Wisconsin, but he was equally unmoved by Stewart's director design. Still, Weaver thought Stewart and the company had promise.[23] With the T-10 director nearing completion, the NDRC needed a means for evaluation. In August 1941 the NDRC let a contract for a fire control testing machine, known as the Barber Coleman dynamic tester (Project 25). Stewart himself would also become an active member of the NDRC's fire control committees.

The dynamic tester simulated the inputs to an antiaircraft director by mimicking the hand motions of a perfect human operator. It turned the handwheels on a director, using servos that followed a set of machined cams based on flight profiles for an imaginary target airplane. Three cams determined a particular flight profile in azimuth, elevation, and range. Dive bombing, level flight, or close-in attack could be set by changing the cams in the tester, which also had a "perturb" switch to add noise to the data. The output of the director then fed back into the dynamic tester. The tester compared the actual solution with an ideal solution (programmed using another set of cams) and recorded the difference between the two, the error, on a chart; errors resulting from mechanisms were class A errors, while those resulting from theory or mathematics were class B errors.

Most important, as its name indicates, the dynamic tester evaluated not only absolute accuracy but also dynamic performance. It measured the director's response to transient inputs or sine waves of different frequencies, reflecting the analytical, signals-based approach of the NDRC. Thus, researchers could characterize the director in both the time and frequency domains to determine its mathematical behavior as a control system. Here the Sperry and other mechanical directors came up short, for even when they settled on accurate solutions, the inertia of their mechanisms made them lag behind rapidly changing signals. They also resonated at certain frequencies, greatly exacerbating errors.

The dynamic tester became a fixture at Fort Monroe, and groups involved in fire control brought their machines to be evaluated by the device. D-2 and its contractors could now quantitatively compare control systems under laboratory conditions, allowing them to make fine distinctions between new techniques under a mantle of objectivity. "The Dynamic Tester was a rather devastating device," Caldwell recalled, because "it had no respect for the opinions of experts, including those within section D-2, and it gave no credit for lucky hits."[24] Barber Coleman built several copies of the machine and distributed them to contractors, including Sperry Gyroscope and Ford Instrument.

With the dynamic tester the NDRC brought fire control into the laboratory and under the control of its scientist-engineers. It redefined a gun director as a black box with inputs that could be simulated and outputs that could be measured. Success no longer meant hitting practice targets but rather achieving measurable transient and dynamic performance, manifested as traces on paper. The machine also provided data on the performance of operators by separating errors due to the machine from those due to human performance. With this machine D-2 acquired the authority to compare new technologies and determine their veracity; the NDRC literally built its expertise in fire control. The dynamic tester's paper recordings did not, however, persuade the army.[25]

### Delivery and Testing of the T-10

After several months' delay, the T-10 prototype shipped to Fort Monroe the day before Pearl Harbor. When hooked up to the dynamic tester, the T-10 performed about as well as, or perhaps a bit worse than, the Sperry directors. Duncan Stewart, who joined D-2 as a member, oversaw the project with Bell Labs. He found the test data inconclusive.[26] George Stibitz, of Bell Labs, now a member of D-2 as well, shared Stewart's reservations. He believed that "the mechanical inaccuracies in T-10 are completely swamped by poor use of data" and that a "smoothing network" or other method of eliminating noise and jitters was required to improve its performance. Stibitz thought the army, overly impressed with Bell Labs and the new machine, was foolishly rushing into production and that the design ought to be improved first.[27]

The army believed the T-10 to be about equal in performance to the mechanical directors. D-2 agreed, and suggested that a pilot production lot be made quickly for field trials. The Army Ordnance Department, however, did not believe in the dynamic tester and accorded little authority to D-2's test results. For Army Ordnance, the advantages of production and procurement outweighed any possible deficiencies in performance. It did not wait for Bell Labs to complete its own tests, it did not care about D-2's approval, and it tested the unit only by traditional methods. "If a good supply of instruments [the T-10] were available which were not even as good as the Sperry M-4," Army Ordnance told Weaver, it "would still feel compelled to purchase this supply."[28] The department kept its word; in November 1941 it ordered 200 of the directors, and it soon standardized the T-10 as the M-9 director. Ordnance manuals honestly touted the T-10 only as "just as accurate" as the mechanical models.[29] In the tense weeks after Pearl Harbor the army needed quick action on new technology. The NDRC's instrumental approach to fire control could not counter the imperatives of mobilization.

The army's decision to buy without test results threatened the D-2 members' status as experts and their control of Bell Labs' research. "There is some unhappy evidence," Weaver noted, "that the higher BTL authorities . . . seem to be more interested in production than they are in improvement."[30] To remedy the situation, D-2 acted the only way it could, funding more research. It extended Bell Labs' contract (Project 2c) to improve smoothing and reduce errors in the T-10 before it entered production. This work, in the spring of 1942, achieved its intended results in two ways. It brought the T-10's performance to a level that satisfied D-2, and it allowed D-2 to sign off on the device without losing face. Duncan Stewart, for his part, remained concerned that the M-9 would have "prohibitive" field troubles.[31]

The first M-9 directors began rolling off Western Electric production lines in October 1942 and included components from subcontractors Ford Instrument and International Harvester.[32] During the war, Western Electric produced more than 1,500 of the M-9 director and its derivative models (modified for different ballistics).[33] Ironically, the T-10/M-9 director, the primary NDRC fire control technology to become operational during the war, originated not in a university but independently, in an industrial lab.

## The T-15 Director

Production numbers could not overcome the machine's shortcomings. The M-9 remained the result of a rush project to design an electrical director and get it quickly into production. It introduced no innovations in computation, merely using the same algorithm as the Sperry M-7, implemented with electrical components. The Sperry algorithm, however, had basic problems. Its plan prediction method derived the target's velocity directly from its position, by differentiating. Observed position data unavoidably contained roughness due either to the jerky nature of human tracking or to electrical noise in a radar signal. Thus, the instantaneous rate or velocity derived from this signal fluctuated wildly. Smoothing could average out the errors but only by introducing lags and delays, sending stale data to the predictor.

To overcome these problems, in February 1941, only months after the T-10 project began, D-2 suggested an electrical director based on new algorithms. This became Project 11, "Fundamental Director Studies," with Bell Labs. Bell Labs designed a new machine, the T-15, under the direction of Walter MacNair. Hendrik Bode, as part of MacNair's team, applied his experience with electrical networks and feedback amplifiers to smoothing networks for the T-15, which used AC and not DC electronics. Instead of using Sperry's plan prediction method, the T-15 worked entirely in polar coordinates. Using a "memory point method" for deriving rates, the director stored an initial position for the target in a mechanical "memory." It derived the target's velocity by subtracting this initial position from a current position and then dividing the difference by time. Because it used subtraction and not differentiation (D-2 came to call this the "one plus" method), it dealt with quantities of relatively small magnitude, required comparatively less accurate computing mechanisms, and inherently smoothed-out perturbations.

When the T-15 design was completed in November 1941, D-2 gave Bell Labs a contract to build the device (Project 30). When it was completed, about a year later, tests showed the T-15 to be more accurate than the M-9 by a factor of about 2 and to settle on a solution two to three times more quickly. The M-9 was already in production, however, so the army never adopted the more accurate T-15. Still, the project produced useful results, advancing electrical computing and analytical understanding of the fire control problem. Moreover, though it used the same assumption of constant course and altitude as the Sperry and the M-9 director, T-15 engineers began to consider the possibilities of predicting the position of airplanes taking evasive action, or "curved flight prediction." D-2 let further contracts to Bell Labs, Norbert Wiener, and others to study this problem. When Wiener came up with his mathematically optimal predictor, the T-15 served as the baseline for comparison and proved about as accurate.[34]

### Fire Control at the Radiation Lab

Another advantage of the M-9 director was that it could accept electrical inputs and hence readily adapt to radar, which was rapidly rising on the army's technical horizons. During the 1930s the Signal Corps had tried to incorporate new "radio ranging" devices into existing mechanical gun directors. This work led in 1937 to the SCR-268 radar (which Western Electric began producing in 1940), designed to supply fire control data to Sperry's M-4 director.[35] The two machines were poorly matched, however, as the M-4 was designed for its own optical inputs. Early radar sets performed similarly to the old sound-ranging equipment they replaced: They were useful for detecting incoming aircraft and providing an idea where they were but not as precision inputs to fire control systems. The SCR-268, however, worked much better than acoustic devices and could direct searchlights to track a target. It could adequately measure target range but had problems detecting elevation, especially at low angles.[36]

The SCR-268's poor accuracy stemmed in part from its relatively long wavelength, of 1.5 meters. Existing vacuum tubes could not generate higher-frequency (shorter-wavelength) signals at high enough power for aircraft detection. Thus, when Bush looked for solutions to the antiaircraft problem in 1940, shorter wavelengths, or *microwaves,* became part of his program. The NDRC included microwave research under Section D-1, the Microwave Committee. During the summer of 1940, when Weaver and D-2 toured the field and learned about fire control, the Microwave Committee did the same with respect to radar. Like the fire control group, the radar group realized that the army and the navy were unaware of each other's work. They found very little research on tubes capable of producing waves below 1 meter and none for microwaves with wavelengths of 10 cm and below.[37]

American radar radically changed in September 1940, when a British technical mission led by the physicist Sir Henry Tizard came to the United States and met with the NDRC. In a remarkable act of technology transfer, the "Tizard mission" revealed the *cavity magnetron* to the Microwave Committee. This device could produce high-power microwaves at wavelengths of 10 cm and below. Not only did high frequencies produce more accurate beams, but their small antennas could be carried aboard aircraft. The cavity magnetron jump-started American microwave research.

While best known for the magnetron, the Tizard mission also established a technical exchange on fire control. A team of fire control experts, fresh from the Battle of Britain, accompanied Tizard. They brought with them the 40 mm Bofors gun, soon to be a key part of the arsenal, and the Kerrison predictor, which Sperry manufactured as the M-5 gun director. The British mission visited BuOrd and witnessed demonstrations of Sperry directors. Soon, engineers from Sperry and other companies visited Britain to better incorporate recent combat experience into their development programs.[38]

The British mission intended for Bell Labs and Western Electric to begin research and production of the prized magnetrons, which they did. Bush and the NDRC, however, continuing their "end run," set up a central laboratory for microwave research, not at Bell Labs but at MIT. The Radiation Laboratory, often referred to as the "Rad Lab," became the NDRC's single largest endeavor and one of the best-known laboratories of the war. The lab began with three projects, reflecting British priorities from early in the war: Project I sought airborne intercept radar for aircraft, Project II sought automatic fire control, and Project III aimed at long-range navigation (it eventually produced the LORAN radion navigation system).

Project II was headed by the Harvard physicist Kenneth T. Bainbridge, who brought with him a young physicist from his laboratory, Ivan Getting. Getting, the son of Czechoslovakian diplomats, had grown up in Europe and Washington, D.C. He attended MIT on a scholarship and did an undergraduate thesis in physics under Karl Compton in 1934.[39] After completing graduate work as a Rhodes Scholar at Oxford, he returned to the United States as a member of the Harvard Society of Fellows. In November 1940 Getting joined Project II, "to demonstrate automatic tracking of aircraft by microwave radar of accuracy sufficient to provide data input to gunnery computers for effective fire control of ninety-millimeter guns."[40] Getting was put in charge of the "synchronizer," the master timing device "which tied the system's operation together."[41] In February 1941 the University of Pennsylvania physicist Louis Ridenour arrived to head the group, which also included Lieutenant Colonel Arthur Warner, the electrical engineers Henry Abajian and George Harris, and the physicists Lee Davenport and Leo Sullivan.

## Automating Tracking: The XT-1 and the SCR-584

In 1940 tracking with radar remained a manual activity. Operators performed *pip matching*, the electronic equivalent of follow-the-pointer operations. Viewing radar data on an oscilloscope screen, the human operator used an electronic pointer, controlled by a handwheel, to select the radar echo that was indeed the target. Then the blip, or "pip," and not the actual radar signal, went on to indicate the valid range or bearing. Training manuals included courses in "pipology," the skills for detecting and interpreting the electronic blips and correlating them to targets in the physical world.[42] The operators worked like the "human servomechanisms" in Sperry's directors: distinguishing signals from noise, smoothing, and renewing the signals before sending them to the next stage.

The Radiation Lab leadership, aware of MIT's strength in automatic control, suggested that Project II automate this tracking and eliminate pip matching. If the radar signal itself could drive servos to move the antenna, it could automatically follow the target as it moved. This problem carried Harold Hazen's conception of servomechanisms as amplifiers to an extreme, for it amounted to amplifying the faint radar echoes by many orders of magnitude to drive large antennas, all without becoming unstable or corrupted by noise.

To replace pip matching, Getting and his team developed *conical scanning*. This technique involved rotating an off-center radar beam around the axis of the antenna 30 times per second. The overlap formed a narrow, precise beam, in a 3° cone, for tracking. The spinning beam could detect when the target was off its centerline and direct a feedback loop to move the antenna to return the target to the center. If the target was moving, like an airplane, the antenna would thus track its motion.

Conical scanning required fast and precise servos to move the antenna, and they came not from academia but from industry. The Radiation Lab obtained a machine gun mount from G.E. to move the antenna. Along with the mount came G.E. engineer Sidney Godet, who had designed the amplidyne servos for tracking. Getting, Godet, and their team first tested conical scanning at the end of May 1941 on the roof the Walker Memorial Building at MIT.[43] It worked, and the antenna tracked aircraft around Boston. "It was very impressive," Getting recalled. "You could look through the telescope mounted on the radar mount, and the airplane would go behind a cloud, and you wouldn't see anything but a cloud. When the airplane emerged from behind the cloud, there was the airplane right on the cross hair. It was just like magic."[44] Soon they had copied this "roof system" and built a prototype, designated XT-1. Getting and Davenport bought a truck and modified the radar to fit inside. An elevator raised the parabolic antenna from its storage position in the truck through the roof for tracking.

Fig. 9.2. Army SCR-268 fire control radar, the army's only fire control radar before the SCR-584. The operators are unprotected and require hoods to see the oscilloscope signals in daylight. Courtesy of MIT Museum.

The truck added not only mobility but enclosure. Earlier army radar sets (like the SCR-268) mounted displays and operators directly on the rotating antenna platform (Fig. 9.2). This arrangement reflected the army's conception of the radar operators: like soldiers on the battlefield, operating a piece of equipment like a telescope or radio. To Getting and the Radiation Lab this seemed foolish. The operators' eyes could not adjust to see the cathode-ray displays in bright sunlight. They were exposed to rain and snow. Their hands got too cold to precisely tune the equipment. The truck protected the operators from these elements and from the chaos of the battlefield, "which might otherwise react on their nervous condition."[45]

Getting and his engineers saw the operators not as soldiers but as technicians reading and manipulating representations of the world. The XT-1 truck brought the operators inside a darkened, insulated space—a control room, a laboratory. Enclosure allowed the operators' eyes to adjust to the delicate blips on the CRT, freed their hands from the cold, and isolated their ears from the sounds of battle. It also gave them a plan position indicator (*PPI*), a round tube displaying a rotating beam tracing out a virtual map of the area being scanned (deriving its name from the plan position

Fig. 9.3. The SCR 584 automatic tracking radar, originally designed for fire control, was the most versatile Allied ground radar in World War II. Courtesy of MIT Museum.

method of calculation used in Sperry's directors). Now radar operators and their commanders could perceive and manipulate the field of battle as a map, not simply as electrical reflections. Glowing radar screens created an analog of the world, collecting data from a broad area and representing it in Cartesian form.

In late 1941 Getting's group drove the XT-1 truck, painted with MIT colors, to Fort Monmouth, New Jersey, for testing. The first tests failed because of a mismatched gear ratio, but the problem was easily corrected. A second series of tests proved successful: the army could compare the radar's output with that of an optical tracker and feed the data into a Sperry M-7 gun director, where operators matched pointers to enter them into the machine. The tracking was a bit jerky, and the device still needed a PPI display to search a broad area, but the army liked what it saw, finding it "superior to any radio direction finding equipment yet tested . . . for the purpose of furnishing present position data to an anti-aircraft director." It provided output signals for azimuth, elevation, and range. It had synchro outputs that could feed into the Sperry M-4 or M-7 directors and custom potentiometers to drive the new Bell Labs electrical director. Compared with the SCR-268, its tracking errors were 10–20 times less, and the data outputs were in a more suitable format (Fig. 9.3).[46] Even though the signal corps had a significant commitment to its longer-wave 268, it was easily converted to the virtues of microwave tracking. While they were celebrating their success in Manhattan, the MIT engineers heard the news that the Japanese had attacked Pearl Harbor.

## Connecting Machines and Laboratories

While the XT-1 had convenient interfaces to gun directors, organizational connections within the NDRC proved more delicate. The Radiation Lab was working on fire control under Section D-1, so it had to be careful not to tread on Section D-2's terrain. Early on, Weaver recognized the potential for conflict. He wrote to Alfred Loomis, who headed D-1, of his desire for "a reasonably definite understanding of the location of the fence between our two regions of activity . . . a wire fence, through which both sides can look and a fence with convenient and frequent gates." Weaver proposed that the relationship between the organizations mirror that of radar to a computer, "saying that your output (three parameters obtained from microwave equipment) was our input (input to a computer or predictor)."[47] Karl Compton, in charge of Division D, agreed with Weaver's proposal and set up a special committee, called "D-1.5," as a liaison between D-1 and D-2. It included Edward Bowles of D-1, Ridenour and Getting of the Radiation Lab, and Caldwell and Fry of D-2. This group only existed for about a year, but during that time it conducted a comprehensive survey of all radar development in the United States and Canada.

Industrial relationships proved just as sensitive. Sperry Gyroscope, with its background in fire control, should have been the obvious choice to build these new systems. The company was working with the Varian brothers on klystrons, which gave it some experience with radar as well. Yet the army, aided by the dynamic tester, was just beginning to appreciate the shortcomings of the Sperry mechanical directors and doubted the company's ability to develop a new system. It requested only that Sperry connect its M-4 director to the SCR-268 radar, which the army already possessed in large numbers.[48]

Yet the army and the NDRC drew on Sperry corporate knowledge in another way as well. Sperry's fire control director, Earl Chafee, joined the Army Ordnance Department and was assigned to survey existing technology and to propose a gun director to make use of the latest research. Chafee was to work with D-2, and not only was he to examine individual components but "emphasis is to be placed on the over-all aspects of the *system* . . . and on the role which radar should play in such a unified system."[49] When completed, the Chafee report placed the existing programs within the historical context of other work at Sperry and elsewhere. Most important, Chafee argued that the NDRC needed to see radar as more than simply a replacement for optical equipment: radar altered the entire system of fire control; all of its components needed to be engineered together rather than as separate units.[50]

Such new approaches were already under way. During the D-1.5 surveys, Ivan Getting, at the Radiation Lab, learned of the new Bell Labs director. He immediately began working with Bell Labs to connect his XT-1 tracking radar to the new T-10 director. The designs of the two devices proceeded together, and Bell Labs stayed in touch with the MIT group throughout. Ridenour and Getting, of the Radiation Lab, and Stibitz and Lovell, of Bell Labs, visited back and forth, exchanged information, and discussed interfaces between the machines. Getting was particularly interested in *time constants,* measures of how quickly the T-10 could respond to inputs. When designing his antenna and tracking unit, he had to know how fast the T-10 could keep up with incoming data, that is, its response to different frequencies.[51] "Close liaison should be maintained between director designers and designers of radars and other tracking equipment," ran the T-10 final report. "The specifications on each unit should be written with full consideration of the features and capabilities of the other."[52] During this project the idea emerged that a system might be more than the sum of its parts; the added element was noise.

Just as Albert Hall found in his work at the Servo Lab, noise posed the biggest problem in trying to connect the MIT radar and the Bell Labs director. Servos worked fine as calculators when input data were smooth and ideal. Errors in tracking, however, "would produce prediction errors of dominating proportions" because differentiating the tracking signal tended to emphasize high-frequency perturbations.[53] Radar signals had several sources of noise, making the problem especially bad. For example, as a radar beam reflected off an airplane, it would shift from one part of the plane to another, as though the airplane were twinkling in the sun (known as "glint"). Both the Bell Labs and MIT groups dealt with the critical issue of noise in the signal spectrum and incorporated it into servo design with smoothing networks. A data smoother could eliminate short, high-frequency perturbations from the input data, but smoothed data were slightly delayed when they went to the predictor.

A number of theoretical questions arose around these smoothing networks, resembling those telephone engineers had been asking for some time. What was the optimal trade-off between smoothing and time lag for a network? How did the smoother distinguish proper tracking data from erroneous inputs? What effect did the time lag of a smoother have on the stability of a feedback loop? As Nyquist and Bode had shown, and as Harris and Hall were applying to servos, the answers depended on the frequency response of the components. Weaver put it best when he observed that "if one applies the term *signal* to the variables which describe the actual true motion of the target; and the term *noise* to the inevitable tracking errors, then the purpose of a smoothing circuit (just as in communications engineering) is to minimize the noise and at the same time distort the signal as little as possible."[54] At Bell Labs and the Radiation Lab, just as at the Servo Lab, building control systems meant rethinking the nature of electronic controls. Using radar to close a feedback loop required paying attention to connections as well as to components. Radar and electronic computers forced control engineering to become a practice of transmission, of signals, and of communications.

Neither the Bell Labs director nor the Radiation Lab's radar had been designed from the first with such notions of signals and systems. Rather, two groups tried to connect two separate machines. Neither had formal responsibility for coordination, but their cooperation paid off. In the fall of 1942 the army held a competitive test of radar-controlled "blind firing" at Fort Monroe in Virginia.[55] The XT-1 was matched against two other radars, each connected to a T-10 director and Sperry power drives for a 90 mm gun. Although problems remained, particularly extraneous electrical noise in the cables, the XT-1 system performed best. It demonstrated that a radar-controlled director could track a target, figure a firing solution, and aim the guns (although it still required human input for target selection, pip matching in range, and a number of other tasks). The radar-controlled gun destroyed the target with four shots, and "by the time the fifth round hit, three or four more rounds were already in the air and they each exploded along the track where the banner would have been," a dramatic and convincing effect.[56] The competing programs were canceled.

### Making the '584

After its successful tests, the army standardized the XT-1 as the SCR-584 radar system and put it into production. It ordered more than a thousand units from manufacturers, who redesigned the prototype as a production unit for field service. Manufacturing the SCR-584 proved no simple matter. It required 140 tubes and a host of specialized electronic parts, weighed a total of 10 tons, and cost about $100,000. It took months to make the transition from laboratory prototype to industrial product. Chrysler built the tracking unit, adding rugged gear trains derived from automobiles and stamping out the parabolic dish antennas with fender presses (Fig. 9.4). Freuhauf built the trailer, and the Palmer-Bee Company made the range-tracking electronics. Both G.E. and Westinghouse served as prime contractors and did final assembly on two interchangeable versions.

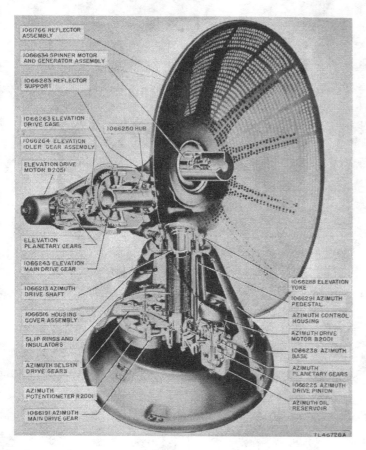

Fig. 9.4. The SCR-584 servo-driven antenna mount, manufactured by Chrysler. The spinner motor at the center of the parabola rotates the tracking beam. Included in the base are both selsyns, for transmitting data to Sperry gun directors, and potentiometers, for transmitting data to Bell Labs gun directors. Reprinted from *Radio Sets SCR-584-A and SCR-584-B Service Manual: Theory, Troubleshooting, and Repair,* U.S. War Department Technical Manual II-1524 (Washington, D.C., July 1946).

Radiation Lab personnel did not leave the project when the design was completed but acted as consultants through production and even in the field. Lee Davenport conducted much of the liaison with the companies, and as he recalled, "developed into a systems manager." Getting oversaw the work and learned to mediate between his engineer-scientists and the military. As he described it, "I was essentially the merchandiser, and the politician." The Radiation Lab's Edward Bowles, now an adviser to Secretary of War Henry Stimson, lubricated the process from the top.[57]

The Radiation Lab group, many of whom had trained as scientists, learned how much the factory and the battlefield differed from the laboratory and the proving ground. "We started this program as physicists," Davenport recalled, "looking at things with much more theoretical attention. We quickly found out we had to be practical engineers to make anything work." Young men who never thought they would work outside a university now found themselves dealing with multimillion-dollar con-

Fig. 9.5. Interior of control van for SCR-584 radar, showing radar operators at left. The manual plotting board was later replaced by a servo-driven plotter built by Bell Labs. Courtesy of MIT Museum.

tracts, negotiating with corporate executives, and training fresh draftees to use and repair their equipment. "Each step increased our awareness that reliability and repairability were virtues that could not be overlooked. . . . we did learn very early that nothing ever works in the field as well as it does in the laboratory."[58] It was mid-1943 before the machines began rolling off the assembly lines; by early 1944 more than 700 units had been delivered to the army.

When the SCR-584 finally made it to the field, it proved a remarkably able device. In search mode it could scan the skies out to 70,000 yards (38 miles) and display the results on a PPI display. Once a target was identified, the '584 could lock on and track an aircraft with an accuracy of one-twentieth of a degree from a range of 32,000 yards (18 miles). One operator managed the automatic tracking, and another matched pips in range. The '584 had selsyn outputs to transmit data to the Sperry M-4 or M-7 directors and shaped potentiometers (of the type Parkinson had invented at Bell Labs) to transmit signal voltages to the Bell Labs M-9 director. It also incorporated telephone circuits for communicating with the operators of the gun director (Fig. 9.5).[59]

One official history called the '584 "the answer to the antiaircraft artilleryman's prayer." Indeed, the '584 became the most successful ground radar of the war, with nearly 1,700 units eventually produced. By 1944 the combination of the '584 and the M-9 gun director had made its debut in the European theater as an automatic antiaircraft fire control system. At the beachhead in Anzio, Italy, in February 1944 several systems were deployed to cover the landing force. Together the SCR-584 and the M-9, combined with Sperry power drives to move the guns, shot down more than 60 enemy aircraft that harassed the troubled landings.[60] On D-day 39 systems landed at Normandy, floated ashore in waterproof boxes to protect the invasion force against air attack.

### A Moment of Automated War

For all its innovations, the new system still maintained the constant-altitude assumption of the prewar Sperry directors. Rushed into production in 1942, it did not incorporate the latest results on predicting curved flight from Bell Labs and MIT. In June 1944, however, a new threat emerged from Nazi engineers that fit the constant-altitude assumption exactly because it had no human operator: the robot bomb, or V-1. The V-1 relied on an automatic control system to fly and hence was the perfect target for the automatic antiaircraft gun.

Germany unleashed the V-1 blitz against London just after D-day, in June 1944, launching almost 7,500 buzz bombs against the English capital during the following summer. The English scientist Duncan Sandys estimated that they were eight times more difficult to destroy than an ordinary aircraft.[61] Smaller than a typical fighter, they flew faster than the bombers of the day (380 mph) and at low altitudes (2,000 to 5,000 feet); indeed, fast and low would become the archetypal radar-evading strategy. And they proved remarkably robust against shellfire, sometimes taking several hits before falling. The feedback loops in the well-designed German autopilot proved remarkably stable. One V-1 hit by a shell burst was observed to turn over completely, then right itself and keep on going.[62]

Here the M-9/SCR-584 combination had its "finest hour," to borrow Winston Churchill's famous phrase from the Battle of Britain. In the words of the head of the British Antiaircraft Command, "It seemed to us that the obvious answer to the robot target of the flying bomb . . . was a robot defense."[63] In response to a special request by Churchill, Radiation Lab engineers rushed hundreds of systems out of production, loaded them on ships, and accompanied them to England (Fig. 9.6).

Automatic tracking required not only servos but also training and persuasion. In early 1944 the Radiation Lab engineer Henry Abajian traveled to England and began teaching courses on the '584 at weapons depots. While the Radiation Lab had paid careful attention to training manuals and procedures for maintenance, it had neglected the critical task of training operators. Abajian found that the operators still relied on the old-style pointer matching to provide human input. Both the army and the navy still insisted that tracking required a human operator in the loop. The Radiation Lab engineers, however, believed that closed feedback loops could effectively filter out the noise. In fact, the question of pip matching arose repeatedly throughout the development program; Getting and his group continually resisted pressure to add electronics to the units to allow for pip matching. When Abajian followed the units into the field, he found that the operators, proud of their ability to detect signals from noise, did not trust the automatic loops. The '584 had no pip-matching capability (except for range), but the operators were actually interrupting the data transmission to the computer in order to improve the signals. "Everyone is convinced that the trackers can smoothen the data," he wrote to the signal corps; he had to convince the men that the feedback loops could track as well as human eyes and hands.[64]

Abajian had to convince the operators because even with the automatic tracking, the

Fig. 9.6. A glimpse of automated war. The SCR-584 radar, driving the M-9 gun director, and 90 mm guns with Sperry servo drives. The 584 itself is in the foreground, as well as buried into a revetment as part of the system in the background. This system proved successful against the V-1 buzz bombs in 1944. Courtesy of MIT Museum.

SCR-584 was still heavily dependent on skilled users. Initially, the accuracy of the gun batteries against the V-1s was relatively poor, as the gunners had never before fired at targets at such low altitude and high speed. The simple, straight flight lines of the V-1s made gunnery errors strikingly apparent. To help in the training, members of the original SCR-584 design group (Davenport, Harris, and Abajian) and other Radiation Lab staff members traveled along the English coast from battery to battery, calibrating equipment, training operators, and tuning the radars—conveying tacit laboratory knowledge to crews in the field.[65] Also critical were careful preparation of the firing sites and the discipline of the team in setting up the system. As one field report put it, "You cannot emphasize too much orientation, synchronization, and alignment."[66] The crews also learned from experience and held frequent conferences to share ob-

Fig. 9.7. Continuous and discrete signals: this remarkable photo shows the SCR-584/M-9 system shooting down a buzz bomb. The missile enters at the left, is hit by a shell, crashes, and explodes. The white dots are shell explosions, which continue to statistically track the predicted position of the target. It is clear that the buzz bomb did not descend of its own accord, for the continuous glow indicates that the engine did not cut off. Reprinted from Massachusetts Institute of Technology, *Five Years at the Radiation Laboratory*, 31.

servations and improve their setups. The engineers added a number of improvements based on this field experience: an electronic gate to screen out noise and jamming, a plotting board to track engagements, and fixes to components and cables.

As the operators gained experience, the number of hits increased.[67] Between 18 June and 17 July 1944 the automated guns shot down 343 V-1s, or 10 percent of the total attack force, and 22 percent of those shot down (the others were hit by aircraft, barrage balloons, and ships) (Fig. 9.7). Geography improved the results. For the first six weeks of the V-1 attack, the antiaircraft batteries were deployed in a ring south of London, and their ability to fire was limited to avoid hitting the fighters that also pursued the buzz bombs. The guns could fire only on positive identification of the target and if no fighter was in pursuit, giving aircraft the first chance to shoot down the missiles. In mid-July, in a massive and rapid mobilization, the antiaircraft batteries moved to the coast, where they could fire without limit over the channel. Firing over water also increased the low-angle performance of the radar because the clutter from the sea surface was less than that from land.

One other technology completed the system: the proximity fuze, developed by Division T, named for its leader, the Carnegie Institution physicist Merle Tuve. Though part of the NDRC, it operated under BuOrd. Traditional antiaircraft fuzes exploded the shell a fixed time after the shell was fired, after being set or "cut" by a loading crew. By contrast, the VT (for "variable-time") fuze placed in each shell a miniature radio that sensed when the shell neared the target and set off the explosion.[68] Before the VT fuze, antiaircraft, with all its feedbacks and controls, remained an open-loop system once the shell left the gun. The proximity fuze closed the loop, making each shell a guided missile with one degree of freedom, capable of sensing its environment and reacting accordingly. The navy would not allow the fuzes to be fired over land, lest a dud shell fall into the hands of the enemy. When the automated batteries began firing over water, however, they could use the proximity fuze, which nearly doubled their hit rate.

As the summer of 1944 wore on, and as the operators acquired experience and learned from the Radiation Lab engineers, automated antiaircraft fire began to show impressive results. From 17 July to 31 August the automated guns accounted for 1,286 V-1 kills, or 34 percent of the attack force and 55 percent of those shot down. That autumn, the M-9/SCR-584/VT-fuze combination defended Antwerp from the V-1 with similar success, downing 57 percent of the missiles engaged, or 52 percent of the total attack force (though it still required an average of 285 rounds of 90 mm ammunition to make a kill).[69]

For the NDRC, this success validated the hard work of bringing electronics, radar, and communications engineering to the fire control problem. The academic engineers had shown the military men and met an unforeseen, automated enemy. General Pile, head of British air defense, thanked the Americans for sending the equipment and commended the Radiation Lab engineers as "a grand lot of chaps."[70]

In principle, the '584 radically changed the technology for predicting trajectories; in practice, the machine proved remarkably versatile. Military users quickly devised new applications, many of them variants on the original intent of machine control, with additional human intervention. SCR-584 operators directed fighter pilots over enemy territory for close-support bombing. The radar could track mortar shells back to their source and could even track individual trucks and soldiers across enemy lines at night. One application even improved traditional fire control. During testing at Fort Monroe, the radar tracked shells fired from the army's 90 mm guns and produced plots that revealed a significant error in the 90 mm firing tables. The tables had been calculated on a Bush differential analyzer, but it turned out that the machine had some incorrect gearing in its setup, an error that all the Sperry M-7 directors included in their mechanisms.[71] Getting and the Radiation Lab even used the '584 to control the buzz bombs themselves, using American copies of the V-1, modified to include radar receivers, to receive remote-control inputs.

The '584 also adapted to an early form of ballistic missile defense. It could track the V-2 rockets from 70,000 yards away, plot their trajectories, and identify their launch sites. Radiation Lab engineers even looked into using the M-9/SCR-584 combination to shoot down the ballistic missiles, but the V-2 presented too small a target from immediately below. One solution was to place the radars in the suburbs and connect them through the telephone network to gun directors in London. The idea was simulated on the Rockefeller Differential Analyzer but never made it to the field.[72] The SCR-584 survived for many years in various military and tracking applications and some '584 systems, as well as '584 components, are still in use today.

## Conclusion

In the course of the Bell Labs project to build an electrical director, communications engineers applied their techniques to control systems. As the Bell Labs engineer Enoch Ferrell put it, "Normally, as communications engineers, they had dealt with current and inductance and band width and distortion. Suddenly they found themselves worrying about velocity and mass and lag and error. Instead of the problems of speech transmission, they had the problems of gun-pointing and bomb-sighting. Different quantities, different units, different equations, different methods of analysis and investigation. Or are they?"[73] To Bell Labs engineers, the T-10 director resembled a telephone network as much as it resembled the Sperry machines whose algorithms it shared. Before the war, Bell Labs engineers had begun to see the transmission of sound, text, and images through the common lens of the signal. Now fire control came under that purview as well. Indeed, their gunnery computer could not simply communicate inside itself, for it had to connect to other machines, especially the SCR-584 radar that would be its companion. That machine too emerged from an engineering group learning to think about signals, this time because unless they were properly filtered, the noisy radar echoes made a bad match for mechanical gearing and servos.

Despite its great success and its automated features, the M-9/SCR-584 system had limitations. It depended heavily on the skill of the operators to set up and operate. The SCR-584 could not search and track simultaneously, nor could it simultaneously use optical and radar data. As the users realized, manual optical tracking remained necessary because the radar circuits often could often not pull target signals out of noise from ground echoes, closely spaced targets, or jamming. (Optics also proved particularly accurate for tracking the buzz bombs at night, because of the fiery exhausts.) More important, the radar and optical sights were mounted on different equipment. Radar trackers sat inside the '584's trailer, while optical trackers and rangefinders sat outside on the M-9's tracking head. During combat, operators would constantly have to select between these varied sources of data, which proved difficult, error-prone, and fatiguing.[74]

Such troubles arose not simply from the operators' inability to keep up with the data flow but also from the relationships between design organizations, Warren Weaver's cordial "fence" between tracking and computing. Indeed, the M-9/SCR-584 was a combination of two separate units, designed by discrete groups with different philosophies. By the time the buzz bombs came across the Channel, however, Ivan Getting and his team at the Radiation Lab were already working to remedy this situation. Their collaborator was neither the army nor Bell Labs but the traditional fire control patron, BuOrd. At a meeting in 1943, just as the '584 was going into production, Getting emphasized to BuOrd that "it is not satisfactory to combine separate components" in a fire control system, and he urged "that this not be overlooked in any future program."[75] In describing what he considered the right way to build a system, Getting began to use the term *integrated*.

# Radar and System Integration at the Radiation Laboratory

> The wartime anti-aircraft systems were made up of independently developed components, each one designed to carry out its own function as well as possible and only secondarily was it made to work with the other units of the system. . . . in spite of the improvements made in wartime anti-aircraft, there was room for many more by properly integrating the components of the system or, one might say, by putting the responsibility for the system design in the hands of a single group of engineers instead of several groups each responsible for a component.
>
> Preston Bassett, President, Sperry Gyroscope, 1948

## New Ideas of Systems

In Europe the army countered a threat well matched to the strengths of its antiaircraft system: a missile in straight and level flight. In the Pacific the navy faced a different combination of men and machines that exploited the weaknesses of automatic controls. Kamikaze suicide attacks were fast, low, and unpredictable. They also shattered the assumption, built into the computers, that the attacking aircraft would try to drop a bomb or torpedo and then attempt to safely exit the area. Innovations like the Sperry-Draper gunsights for light guns improved defenses against these targets, but battle experiences highlighted their limitations. Merely damaging a suicide plane often meant that it would hit its target anyway, so it needed to be physically stopped as much as a thousand yards out from its target, and that required heavier guns at longer ranges. Captains complained that existing technology was inadequate, pointing especially to weaknesses in training, data communications, and "the integration of the radar with the fire control systems" (Fig. 10.1).[1]

These combat experiences had engineering implications, for a system could no longer be a set of interconnected components, no matter how well they worked together. Fast, dynamic targets called for fire control systems that incorporated perception, integration, and articulation into a single, high-performance unit. In 1943 BuOrd went to Division 7 and the Radiation Lab for help designing such a system. Ivan Getting and his group responded with the new Mark 56 Gun Fire Control System, which

Fig. 10.1. Another battle of cultures and controls: kamikaze attack on the USS *Missouri* off Okinawa in 1945. Note the gun directors 9 and 11 in the foreground and the gun mounts they control. Courtesy of Naval Historical Center, Washington, D.C.

seemed the ideal counter to the human-guided missiles. Getting and his group also developed a new organizational role, that of system integrator, to ensure its coherence.

### The Difficult Stepchild: Radar and Fire Control in the Navy

Before examining this project, however, and Getting's role as system integrator, we must understand the importance of radar for fire control in the navy and BuOrd's efforts in the area.

Early in World War II, radar radically altered warfare at sea, from navigation to night fighting. The Naval Research Laboratory (NRL), near Washington, D.C., had done some of the earliest work with "radio vision" in the 1930s and had installed sets in the fleet in 1940. But NRL came under the cognizance of the Bureau of Engineering, whose systems used long wavelengths that were not accurate enough for fire control. BuOrd had no interest in radar during the 1930s, believing it to be inferior to optical equipment (which was correct for the sets of the time). Only in the summer of 1941 did NRL set up a research group for fire control radar and did BuOrd achieve complete and official cognizance over fire control radar.[2] By that time BuOrd already had experts working to integrate radar with fire control systems: BuOrd's "four horseman" from MIT, now on their next assignments.

Horatio Rivero, after graduating from MIT, went to BuOrd in the fall of 1940 and, as head of Blandy's radar desk, was assigned to examine the navy's radio-ranging research to determine whether it had utility for fire control. He immediately recognized the device's potential. No one in BuOrd understood the technology, so Blandy gave the young lieutenant a free hand, putting him in charge of radar for the bureau. Along with Samuel Tucker, the head of BuOrd's Antiaircraft Section credited with coining the term *radar,* and an engineer on loan from Bell Labs, Rivero initiated a major program in fire control radar.

In these early years of the war, engineering responsibility for microwave radar roughly divided along service lines: army radar came out of the Radiation Lab, navy sets from Bell Labs and Western Electric. Bell Labs had had a small program in fire control radar for several years and had set up a field station for testing.[3] In the fall of 1940, as the NDRC was organizing, the Tizard mission brought the cavity magnetron to the United States, and Rivero directed Bell Labs to build microwave radars for fire control.[4] Before Bell Labs even had a prototype, Rivero ordered production to begin, much as the army had ordered the T-10 director without regard to test results.[5] Rivero personally assigned the sets to the fleet and had them urgently shipped for installation.

The first fire control radars for large, main-battery guns entered the fleet in the summer of 1941. In the hands of skilled operators they fundamentally changed fire control. Suddenly naval gunnery became a truly closed loop; the new instruments of perception could track targets, follow shells along their trajectories, and display shell splashes for spotting. Surface fire now became a matter of matching the target pips to the pips from the shell splashes. Spotting aircraft were soon removed from battleships.[6]

Edwin Hooper, another of MIT's four horsemen, pioneered the use of main-battery fire control radar in combat. In 1940 Hooper joined the gunnery staff of the battleship *Washington,* which soon received the first two main-battery fire control radars and four of the first antiaircraft sets. He cleverly adapted the new technology to its environment.[7] The radar display was designed to go into the director tower, but Hooper moved it down into the plotting room. The antenna was to be cranked by hand, but Hooper designed a servo to allow the rangekeeper to automatically drive the antenna as it tracked a target. He remembered an exciting, innovative time, despite the challenge of drilling holes for electrical cables in the thick armored deck of the *Washington.* "After every shoot," recalled a colleague, "Hooper would work on graphs, formulae, and functions far into the night."[8]

For twenty-five years the gun club had developed fire control in peacetime. Now Hooper tested his system in combat. Off of Guadalcanal in November 1942 the *Washington,* with Hooper at the gun controls, sank the Japanese battleship *Kirishima* from a range of 18,000 yards—the first surface victim of a U.S. battleship's guns since 1898. Throughout the encounter, Hooper recalled, his understanding of the behavior of the feedback loop of fire control corrections came from "my studies at MIT in servomechanisms and in dynamics."[9]

While radar easily transformed the comparatively slow (and mature) main-battery fire control, antiaircraft exposed the technology's limits, requiring much higher performance and speed. The first American fire control radars, dubbed FC and FD (later Mark 3 and Mark 4), came off of the production line at Western Electric in June 1941. They were mounted atop the Mark 37 director, the most common antiaircraft system in the fleet. The combination was first tested aboard the USS *Roe* in September 1941 before an audience of navy brass and scientists, including Rivero and the NDRC's Poitras and Caldwell, but the results were less than spectacular. "The firing was entirely wild," Caldwell recalled.[10] Their troubles notwithstanding, these devices, "the guinea pigs of fire control radar," went into production, began delivery in October 1941, and served the navy through much of the war. Many hundreds were installed, and nearly every ship in the fleet of destroyer size or larger had at least one Western Electric set. They played a part in nearly every American naval action of World War II.[11]

As in the army, these early navy systems were separate units not originally designed to work together, and prewar directors were ill suited to radar. The Mark 37 gunfire control system still used the Ford Mark 1 computer, an adaptation of the original Ford Rangekeeper. It was slow to converge on a solution, assumed that its target was flying straight, and had the old problem of increasing errors with decreasing range. The Mark 37's director also suffered from cramped space and restricted vision for the crew.

Line officers reported that the Mark 37 worked adequately well, but only because late in the war the proximity fuze provided "a shot in the arm to the basic system that enabled it to stagger through the war, partially concealing its inherent weakness." In general, the navy expressed great dissatisfaction with the existing fire control directors. Even the small, decentralized systems, like the Sperry Mark 14, appeared successful only because of the failings of the devices they replaced. The navy wanted integrated systems of the "ultimate" type.[12]

Antiaircraft fire control radar also had problems and proved "a stepchild slow to win affection." During the war, BuOrd spawned 27 different designs for fire control radar. Only 10 of these entered production, 7 actually saw action, and 3 became widely available. Yet none of these were microwave, 3 cm systems.[13] They had problems with reliability, maintenance, target discrimination, and especially automatic tracking. Jitter and noise in the radar echoes, such as glint and reflections from the plane's propeller, adversely affected the tracking servos. Only intensive human mediation—the old human servomechanisms—could produce high-quality electronic inputs for rangekeepers. Operators needed to pip match to eliminate noise, and they needed to manually follow targets with the antenna, much as they did with traditional optical rangefinders and telescopes.

As with the army system, operators routinely switched between optical and radar tracking, and the combination threatened to overload their attention. Yet optical tracking remained necessary because tracking radars could not discriminate between two aircraft attacking in formation, and the signals would frequently jitter between them. They also had particular trouble locking onto airplanes attacking low across the water, a weakness Japanese pilots used to tactical advantage. A special antenna was added to these systems solely to improve low-angle tracking.[14] Radar underscored the problem with antiaircraft fire control in general: it worked fairly well against high, straight targets but broke down when confronting fast, maneuverable, close-in attacks.

Well into the war the navy had no automatic tracking like the SCR-584 and no system for *blind firing*, where radar could direct the guns to fire automatically at night or through overcast. The new Combat Information Centers served to organize information and direct fighters from a central location, but they did not address the gunners facing attackers they could not see. Several projects tried to adapt existing control systems for blind firing. In 1941 BuOrd supported the Radiation Lab's development of a radar (Mark 9) to work with a director (Mark 45) then under development at Ford Instrument. The Mark 9 became the first Radiation Lab set to go into production, but

BuOrd soon dropped the inadequate and overweight Mark 45 and canceled the program. Similar fates befell other projects at Ford Instrument, G.E., and Arma. The Radiation Lab, working with NDRC Section T, added radars to the Sperry-Draper gyroscopic sight and its derivatives, but these only tracked in range (not in elevation and train), and they required human operators to move them.[15] BuOrd, even with the NDRC's help, simply could not design an automatic tracking radar and radar at the same time. Blind firing remained an elusive goal, fire control radar the frustrating stepchild.

## Getting and Systems

A solution began to emerge in 1942, when the Radiation Lab reorganized and created a Systems Division. Ivan Getting took charge of this unit, Division 8, devoted to army ground radar and naval fire control. The group included the physicist George Valley, who would later help transform the Whirlwind computer into the SAGE air defense system, as associate chairman. Nathaniel Nichols headed a servo group. Ralph Phillips headed a special subsection for mathematics and theory, which included the mathematician Walter Pitts and economist and future Nobelist Paul Samuelson, and R. P. Scott headed a subsection for systems.

With this organization Ivan Getting believed he could finally achieve true blind firing for BuOrd. Getting's vision entailed a new role for his laboratory, as Brown's had for his Servo Lab. Getting argued that BuOrd was failing at blind firing because it lacked a central, coordinating technical body that could oversee the integration of the system and "there was no attempt made to integrate the radar and the computer into a functioning whole." Getting believed that BuOrd, with its highly specified and compartmentalized contracting, still broke the fire control problem into component parts, technically and contractually. BuOrd did the "gross engineering" and parceled out "detailed engineering" to subcontractors who were unaware of the larger task. Getting wanted to start anew, with "a totally integrated effort starting from basic principles."[16] To do this, he needed to redefine the system. For him, an "integrated" system was one in which the radar, computer, and controller were designed simultaneously, considering signals, time constants, and feedback dynamics before physical equipment and mechanical components.

Getting found willing allies in the NDRC and BuOrd. Harold Hazen, as head of Division 7, recognized the value of coordinating radar and fire control design. Among

Division 7's priorities, Hazen announced in early 1943, would be "the overall design of fire control systems and the optimum use of radar on navy directors."[17] To smooth relations with the Radiation Lab, he invited Getting to join, making Weaver's wire fence a concrete bridge. Soon Hazen and Getting discussed a blind firing director with M. Emerson Murphy, head of fire control research at BuOrd, and proposed a new project. Murphy endorsed the idea, and BuOrd's chief, Blandy, concurred, creating a new project designated the Mark 56 Gun Fire Control System.[18]

### Organizational Challenges

Now Getting could start from scratch, defining both the machine and his position. His group would go one step beyond the NDRC's usual role of designing equipment, building prototypes, and preparing drawings. It would now oversee the selection and preparation of manufacturers, as well as a production run, not as consultants but as managers. This would allow the NDRC complete technical control of all phases of the project. But which part of the NDRC? A radar-driven fire control device fell within two domains: those of Division 14 (the Radiation Lab) and Division 7. Members of the latter argued that the Radiation Lab did not have sufficient experience with fire control and that the project should use M-9 director technology developed for the army (Bell Labs was then building on its M-9 director experience to make an electronic rangekeeper for BuOrd).[19] Getting's idea for the new system, however, had radar at its core.

To connect radar and fire control, Hazen created a special section of Division 7, dubbed 7.6, Navy Fire Control with Radar. Ivan Getting, a member of both Division 7 and the Radiation Lab's Systems Division, headed Section 7.6. He described the new section as "an attempt by Dr. H. L. Hazen to bring together the necessary elements which had been more or less artificially separated by organization, personality, and history."[20] Getting questioned the traditional division of labor between units: the NDRC's divisions dated from a time when fire control and radar were separate technologies. For earlier projects, such as the M-9/SCR-584 combination, the arrangement had worked well because of the high degree of communication between Bell Labs and the Radiation Lab. From that experience, however, Getting learned the value of coordination from the design stages all the way through production, and the value of controlling that coordination. The new Section 7.6 absorbed a few other Division 7 projects relating to navy fire control and undertook a number of small contracts, but the Mark 56 was its major work. Getting described the project as "the first fully-integrated radar fire control system that was not restricted by history or by prejudices."[21] History, prejudice, organization, and personality—Getting saw these as complicating factors to be overcome.

Yet he took advantage of history. For the new section, and for the Mark 56 project, Getting tapped members of BuOrd's fire control clique. He included vice presidents from Ford Instrument and Arma, Al Ruiz of G.E., Charles Stark Draper, and Robert M. Page, who had done the early long-wave radar work at NRL.[22] The committee did not actually meet until January 1944, by which time the Mark 56 project was well under way. Section 7.6's primary function then became as a forum for discussion, especially among BuOrd and its contractors.[23] But the contractors had other, secret projects with BuOrd and could not discuss their status or technical details. Nor did they wish to share such information with their competitors. The world of naval fire control, with its multilayered secrecy, frustrated Getting, who was used to the urgent excitement of microwave radar in its early days.[24]

And Getting himself faced competition. Despite his vision, nothing inherent in "coordinated design" dictated that a radar group should capture and hold the terrain. In fact, blind firing became the prestige project for BuOrd, and several groups vied for the technical spotlight. An argument could be made that Draper's Confidential Instruments Laboratory was best positioned for system integration, or Bell Labs, where research shared a corporate umbrella with Western Electric's manufacturing. Getting strongly opposed bringing in Western Electric even as a manufacturer. He disparaged his earlier work with the telephone company, writing to Karl Compton of "a considerable amount of bitter experience" and threatening to resign if production contracts for his design were given to Western Electric.[25] The contracts went instead to G.E., which had a longtime relationship with BuOrd and with whom Getting had worked so successfully on the SCR-584.

The most serious threat to Getting's vision came from within the NDRC itself. Section T, led by Merle Tuve, had developed the proximity fuze, which entered production in 1943. Tuve built the Johns Hopkins Applied Physics Laboratory in parallel with the development of the fuze, and he sought to capitalize on the fuze's great success. When Tuve and his staff looked for a new project, "fire control was the future."[26] Section T had little experience with control systems, but it did have an intimate and unique relationship with BuOrd: despite belonging to the NDRC, Tuve reported to the bureau and not to Vannevar Bush.

BuOrd chief Blandy requested that Tuve's group develop a blind firing director in collaboration with Draper and Sperry. Bush asked Division 7 to aid Tuve, raising the committee's ire. Division 7 resented Section T's relationship with BuOrd, an intimacy it had never enjoyed. "We recommend that the apparently anomalous relationship of Section T to OSRD be discontinued," Hazen's committee resolved after heated discussion, "and its status as a Naval agency be clearly recognized." Division 7 considered Tuve "extraordinarily able" but "wild and irresponsible" and refused to work with him. If Tuve's group were officially placed within the navy, however, Division 7 would offer assistance. The situation was, in Caldwell's words, "pretty sour," and it preoccupied Division 7 throughout 1943.[27]

Bush quieted the controversy by decreeing that Section T should undertake a short-term solution with Draper. Meanwhile, Bush directed, Section 7.6 would "undertake the development of a new fully integrated radar fire control system of an 'ultimate' type."[28] In this project, Getting's Radiation Lab "would act as central integrated clearing point."[29] This ambiguous division between long-term and short-term research blurred further as the war drew to a close. Still, Getting won the ideological victory: Tuve's group would combine existing components, while Section 7.6 would build the ultimate system from fundamentals. Getting's vision of radar designers at the center of systems design survived a serious challenge, but by a narrow margin.

### Building the Mark 56

Beginning in 1943 the Radiation Lab undertook the Mark 56 program, and it produced some of the last NDRC contracts in fire control (projects 71, 79, and 85). Its conical-scan, X-band (3 cm wavelength) radar could search broadly for targets and automatically track at the same time, even at low angles. Two sailors in the director on deck acquired and tracked targets optically and directed the radar antenna, for which processing occurred belowdecks (Figs. 10.2, 10.3, and 10.4). For the computers, the Radiation Lab, over Division 7 objections, did not defer to prior experience. Rather than the old plan position method of tracking, the Mark 56 did all its calculations in an inertial frame with a gyro that tracked the line of sight. In the Radiation Lab, the Czech exile and fire control expert Tony Svoboda designed a new mechanical computer using innovative four-bar linkages. To move the director unit, the Radiation Lab chose a G.E. amplidyne servo over an MIT Servo Lab design. The device was first tested on a specially constructed rolling platform at Fort Heath, north of Boston, in the spring

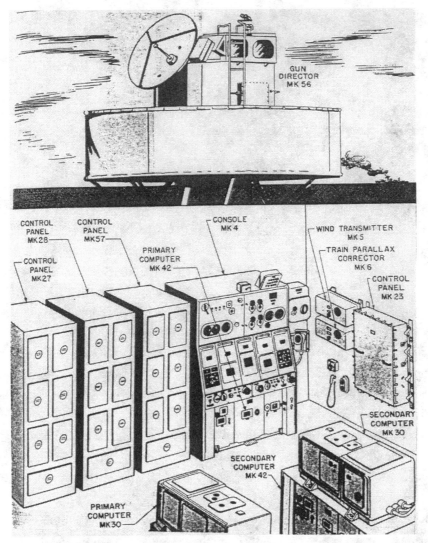

Fig. 10.2. Layout of the Mark 56 Gun Fire Control System. Two operators track optically from the deck positions, and two more work at the console in the control room belowdecks. Reprinted from U.S. Navy, Bureau of Navy Personnel, *Naval Ordnance and Gunnery*, vol. 2, *Fire Control*, 319.

of 1944. It first fired live ammunition, under automatic control, the following December.[30]

Getting was seeking a radical solution, but the project's ambitious goals adversely affected its timing. By 1945 BuOrd was tuned for wartime production and deployment and allocated priorities solely by anticipated delivery date. Because of its long-term

Fig. 10.3. Mark 56 tracking unit during testing. The target designator is on the left with binoculars, behind the radar operator. Courtesy of MIT Museum.

nature, the Mark 56 fell low on the list, and the schedule suffered. Getting, however, saw his "ultimate" system as a crash program to get blind firing to the fleet as soon as possible.

### The Radiation Lab as Systems Integrator

Throughout the Mark 56 project Getting continued to redefine the work of building control systems. This entailed two parallel moves: transforming the Radiation Lab from a radar group to a system integrator and transforming the human operator into a dynamic component. For the first, Getting elaborated the Radiation Lab's position, making it a coordinating technical body between the government and its contractors.

Earlier in the war, the urgency of the antiaircraft situation tended to smooth over political problems, and the NDRC's novelty provided a certain temporary authority. Furthermore, a new field like radar had no established expertise to resist the scientists' designs, so Getting had complete technical control. Late in the war, as things became more established, routine, and industrial, they also became more complicated. Getting was used to dealing with the army, a low-tech service awed by electronics; now he took on BuOrd, one of the most technically sophisticated—and entrenched—groups in the American military.

In a move similar to Gordon Brown's, Getting wanted to control not only engineering but production as well. Otherwise the role of the Radiation Lab would evaporate as the Mark 56 design neared completion. Toward this goal, Getting continued to cross established boundaries. He had joined Division 7, he had merged it with the Radiation Lab (7.6), and now he sought to place a man within BuOrd. Warren Weaver, by now experienced at compromises with the services, thought the plans were "discussed in over-pretentious terms" and suggested that "the way to work with the BuOrd is, so to speak, to work with the BuOrd."[31] Still, Getting got his way, and in March 1945 the Radiation Lab director, Alfred Loomis, ordered that Getting be assigned to BuOrd, "to devote your time and efforts to technical problems on fire control and their application to radar."[32] Getting made one of his men, Robert Patterson, of the Radiation Lab, the liaison he desired.

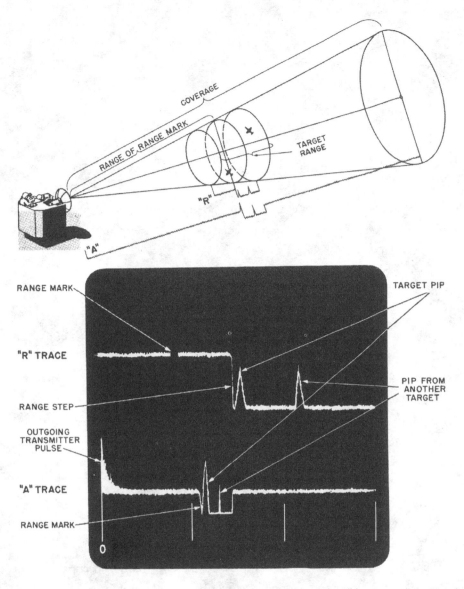

Fig. 10.4. Mark 56 radar tracking, demonstrating how targets in the real world map onto "pips" on the screen. Reprinted from U.S. Navy, Bureau of Navy Personnel, *Naval Ordnance and Gunnery*, vol. 2, *Fire Control*, 331.

Now Getting acquired the long-sought authority to delineate the role of the Radiation Lab. He formalized the lab's job of system integrator, which had previously been merely informal. Now it would

1) Make all technical information available to GE and the navy
2) Check and criticize designs at all stages of development

3) Send skilled representatives to participate in conferences

4) Report to the BuOrd on the progress of the project

5) Participate in testing of prototypes

6) Test pre-production models

7) Assist in establishing test and alignment procedures for manufacturing and acceptance tests

8) Assist in training programs[33]

Engineering, production, testing, alignment, and training were as much a part of Getting's vision as were time constants and signal spectra. To carry out these functions, the lab would have the following privileges:

1) To receive copies of correspondence between the navy and contractors

2) To receive copies of drawings and specifications prepared by contractors

3) To be notified when significant tests are carried out so representatives of the Laboratory may participate

4) To be notified of technical conferences and conferences where technical decisions are to be made so that representatives of the Laboratory may be present

5) To be given the opportunity to examine and criticize production designs or models before final design specifications are frozen

6) To have access to the establishments of the contractor and subcontractor by appointment, to confer with engineers or to inspect equipment

7) To receive one of the first production models for test and study if directed by the Navy[34]

Correspondence, drawings, specification, tests, conferences, and inspections embodied the relations between institutions. Getting needed to control them as much as he did the signal flows between components. These remarkable lists reflect the experience Getting had acquired in a few years of doing research and managing contracts for the NDRC. Each point seems to correspond to a particular episode in which he lacked necessary authority: being excluded from meetings, missing correspondence, being denied access to factory facilities. Getting defined system integration as an organizational as well as a technical task, and he argued that BuOrd by itself was not up to it. Because of its expertise in radar and electronics, Getting believed that the Radiation Lab had the best overall view of automatic control.

Where Getting appropriated authority from contractors, designers, and manufacturers, he also appropriated the work of the human operator. Unlike earlier system integrators, who organized and collated different types of data, Getting's operators were
to function purely mechanically, as dynamic components. In 1945, while fighting for
his project's priority, Getting wrote to Admiral Furer, the navy's coordinator of research and development, connecting his ideas for designing new integrated systems
with the principle of "automatic operation." Getting argued that wartime experience
had demonstrated the value of automation:

1) Human judgment introduced wrong guesses
2) Human operators succumbed to battle fever
3) The human mind reacts slowly compared to modern servo equipment
4) The intellectual processes were incapable of utilizing most efficiently all the ob-
    servable data.[35]

Radar, Getting continued, burdened rather than relieved the operator by increasing
the amount of information to sort through. Radar brought such complexity to military control that it strained human attention to hold the system together. Getting's
automation would rein in that human involvement, a strategy that resonated with
plans for demobilization, when men left the services but the machines remained.

To make his point, Getting invoked the success of the army's automated antiaircraft
fire control, the M-9/SCR-584 system. Getting used the authority he had gained by its
success in Europe to sharply criticize the navy's lack of automation, "In short the Navy
is an order of magnitude behind the army in heavy antiaircraft fire control and radar."
The solution, of course, was to grant highest priority to Getting's Mark 56, "a wholly
integrated operational system."[36]

Despite his efforts, the Mark 56 never made it into World War II. Getting lobbied
Chief of Naval Operations Admiral King to accelerate the schedule, but King added
requirements that slowed the project down. When Japan surrendered, five prototypes
were on order from G.E., two of them near completion. At the end of the war the
NDRC transferred its contract for the Mark 56 to BuOrd, which then ordered 100 systems. Further problems, delays, and changes by the bureau prevented Mark 56 production models from reaching the fleet until 1947. They did, however, proliferate
widely and remained standard equipment on U.S. Navy vessels into the 1970s.

### Dynamic Systems and Military Contracting

Radar's new subtlety accompanied new expertise: the Radiation Lab staked out a role as a system integrator. Organizational relationships solidified as technical systems, first the partially integrated but combat-tested SCR-584 radar and then the integrated Mark 56 Gun Fire Control System. The Radiation Lab also embodied its experience as textbooks, among its most lasting contributions. After the war the laboratory, with OSRD funding, published a 27-volume series on radar to distribute the results of its wartime work. Three of these volumes emerged from the work of Getting and his associates. Louis Ridenour's *Radar System Engineering* was the first published volume with "system engineering" in its title. Tony Svoboda's *Computing Mechanisms and Linkages* describes the innovative mechanical computers designed for the Mark 56. *Theory of Servomechanisms,* by the physicist Hubert M. James, Radiation Lab Division 8 servo engineer Nathaniel B. Nichols, and Division 8 mathematician Ralph S. Phillips, became a canonical postwar text of control engineering (and introduced the "Nichols chart," a graphical design technique for microwave systems). Along with other books from Bell Labs and the Servo Lab, the Radiation Lab volume initiated a generation of engineers into a newly constituted discipline.[37] For the Radiation Lab scientists and engineers, the boundaries of this knowledge derived from the boundaries of radar-driven fire control.

Getting wrote the introduction to the book by Nichols, James and Philips, reviewing the basic definitions of servomechanisms and the history of design techniques. "The work on servomechanisms in the Radiation Laboratory," he opens, "grew out of its need for automatic-tracking radar systems." Noting the field's lack of stable epistemology, Getting observes, "It is nearly as hard for practitioners in the servo art to agree on the definition of a servo as it is for a group of theologians to agree on sin." Getting and his co-authors acknowledged their predecessors; the 20-page introduction cites Hazen, Bush, Minorsky, Nyquist, Harris, Brown, Hall, Wiener, and Bode. Yet the book reflects Radiation Lab culture: design examples include the SCR-584 radar, numerous automatic and manual tracking schemes, filters for radar signals, and methods for dealing with noisy echoes. It makes no mention of Gordon Brown's Servo Lab. The Radiation Lab volume, while stabilizing control systems as a coherent body of knowledge, defined that stability by the work of radar scientists.

Before World War II, Harold Hazen defined the modular blocks of the differential analyzer so that he could manipulate and recombine them ad infinitum. By 1945

Hazen himself recognized the need for a new approach: "One must always remember that a fire-control system is more than the sum of component parts. It is an integrated whole with interrelated functioning of all its parts and one is safe in considering parts separately only if one always keeps in mind their relation to the whole."[38] Computer design, for example, depended on the bandwidth of the radar, the spectrum of its noise, and the capabilities of the human operator. In an integrated control system, each component affected all the others; the contracts could only be divided up and managed by experts in signals, dynamics, and control.

BuOrd's political economy, however, was still built on an earlier model. The navy itself divided up the problems, assigned pieces to separate contractors, and then assembled the pieces into systems. That approach only worked if a system really was the sum of component parts, but radar signals proved that it was more. For his project, Ivan Getting reconfigured the structure of contracting to suit a dynamic, noisy, error-prone model of a system. To embody the model in working systems, however, would require a set of engineering techniques and theoretical foundations to complement the institutional relationships. Those techniques also began to emerge during the war, driven by similar problems of noise, prediction, and the stubborn unpredictability of human operators.

# 11

## Cybernetics and Ideas of the Digital

Six months before Pearl Harbor, in May 1941, Harold Hazen wrote a memo to Warren Weaver which he called "The human being as a fundamental link in automatic control systems." As part of the NDRC research program, Hazen had observed the "human follow-ups" in a variety of control systems, including the manual servomechanisms in Sperry's gun directors. "The idea struck me more and more forcefully," Hazen wrote, "that we should know as much as possible of the dynamic characteristics of the human being as a servo and therefore his effect on the dynamic performance of the entire control system." Servomechanisms would surely replace pointer-matching tasks in the near term, he added, but people would still be part of the system for higher-level tasks for the foreseeable future. Studies of human operators were under way, Hazen noted, but they focused on adapting people to existing equipment rather than on designing equipment to match human capabilities. Hazen suggested that the frequency-response methods of communications engineering might shed some light on human behavior in control systems.[1]

How did human operators affect the stability of a control system? How should one design a machine to make it easiest to learn and use? How should operators be selected and trained for certain tasks? To address these questions, in response to Hazen's memo the NDRC's fire control section initiated a research program in human-machine interaction, focused on the problems of antiaircraft fire control. Not only did this work take an approach that would later be called "cybernetic" but it also influenced Wiener's formulation of his own ideas.

Human-machine interaction immediately raised the problem of how to represent the workings of the machine to its operators, and in turn how they should enter data about the world into the control system. Indeed, the NDRC dealt simultaneously with questions of analog versus numerical or arithmetical methods for modeling the world inside a machine. The NDRC struggled with the subtleties that surrounded these matters, as well as with how to apply them productively to the war effort. The NDRC's record is not one of unmitigated foresight or success. It had an ambiguous relationship with Norbert Wiener and cancelled his promising work at a critical moment. It rejected a proposal to build an electronic computer that eventually became the ENIAC. Yet the committee enthusiastically supported George Stibitz at Bell Labs to create digital relay computers during the war for antiaircraft applications. A comparison of the NDRC's handling of Wiener, the ENIAC, and Stibitz places cybernetics and digital computing within a broad landscape of engineers, beset with the uncertainties that surrounded human operators and control systems. For these researchers, questions of how to represent the world in a machine had no obvious answers, especially in light of their psychological, institutional, and material implications.

## Wiener and Bigelow: Prediction and Stability

In the fall of 1940, during the initial survey of fire control research, Ed Poitras visited MIT and met with Norbert Wiener, who wished to begin applying communications and network theory to servo problems. "[Wiener] wants to tackle the problem of solving for the controller of servos in terms of the input as the frequency spectrum," Poitras wrote in his diary, because "he believes that considerable of the present network theory could be applied to the servo problem."[2]

Wiener was drawing on his longstanding interests in network theory, harmonic analysis, and Fourier theory. On and off for nearly ten years he had worked with a former Bush student, Yuk Wing Lee, at MIT and in China, reformulating network synthesis and even building an analog computer. This work lacked, as Wiener later observed, "a thorough understanding of the problems of designing an apparatus in which part of the output motion is fed back again to the beginning of the process as a new input."[3] It did not include feedback. In 1940, when Wiener proposed a project to Ed Poitras applying network theory to servo design, the mathematician unknowingly sought to replicate Bode's work, which was published that year, and the ideas Gordon Brown had recently proposed to the NDRC concerning the study of communications and servomechanisms.

At the time Wiener was not yet familiar with fire control, and he expressed no interest in it.[4] In late 1940, though, he applied his knowledge of networks to the most difficult mathematical problem in fire control: prediction. The current methods, which required differentiating the target's position to derive its velocity, were limited to straight lines and were also highly susceptible to noise, so Wiener sought another approach.

Working with Samuel Caldwell, Wiener simulated an electrical prediction network on MIT's differential analyzer, which gave encouraging results. Caldwell, who was then beginning as a member of D-2, submitted a proposal for Wiener to build an "anticipator" network. D-2 let a contract, Project 6, on 1 December 1940 for "General Mathematical Theory of Prediction and Applications." Wiener then hired a research assistant, the electrical engineer Julian Bigelow. Bigelow had graduated from MIT in 1936 and then worked for Sperry Gyroscope and IBM as an electronics engineer.[5] Wiener and Bigelow devised a network whose output could follow a curve that represented the path of an airplane and estimate the value of that curve at some time in the future. During early 1941 they built a machine to simulate their ideas for prediction.[6]

They quickly ran into a stability problem: "the pieces of apparatus designed for best following a smooth curve were oversensitive and were driven into violent oscillation by a corner." In other words, like the classic prediction methods, Wiener's network was highly sensitive, even unstable, in the presence of high-frequency noise. "It became obvious that in any curve not precisely of the shape of a simple sinusoid or straight line, any attempt to use this method of prediction would lead to a failure because of lack of stability."[7] Here was a cousin of the stability problem electric power had faced twenty years earlier: transient inputs caused high-frequency oscillations. Engineers at Sperry, and increasingly at Bell Labs as well, knew only too well that jerky tracking and rapid maneuvering of the target would introduce such perturbations. Wiener quickly realized that the problem was fundamental, "in the order of things"[8]—he compared it to Heisenberg's uncertainty principle—and would need a new approach.

Wiener and Bigelow now turned to statistics and designed a new predictor based on "a statistical analysis of the correlation between the past performance of a function of time and its present and future performance." The network calculated a future position of the target based on the statistical characteristics of its past performance (its autocorrelation). It then continually updated its own prediction as time passed, comparing the target's flight path with previous guesses. A feedback network converged on guesses that minimized this error.[9]

By June 1941 Wiener and Bigelow had designed an electrical filter to perform this prediction and presented it to Bell Labs. Bode, Lovell, and their group, who were working on similar problems with their new electrical directors (the T-10 and T-15), were favorably impressed. But the Bell Labs group needed something immediately, whereas Wiener was pursuing the longer-range goal of optimal prediction. Still, Wiener was cheered by the two groups' "similarity of approach," although by this he meant no more than "the identical concepts of realization by electrical means" as opposed to mechanical methods.[10] Through the remainder of the year, Wiener worked out the theory behind his statistical approach in detail, scribbling on a blackboard as Bigelow took notes.

Warren Weaver, the mathematician turned science manager, retained an active interest in the project; he and Wiener got along well. Wiener's work "probably represents about the ultimate that could be accomplished in designing a predicting system," Weaver noted.[11] Weaver thought the "constant altitude assumption," fire control's Achilles heel, was about to yield to the NDRC's analytical approach.

Weaver let a D-2 contract, Project 29, for Wiener to write up his theoretical results. His report, *The Extrapolation, Interpolation, and Smoothing of Stationary Time Series with Engineering Applications*, was published by the NDRC as Report to the Services 370 for restricted circulation on 1 February 1942. In it Wiener explicitly brought together statistics and communications theory and sought to unify the different branches of electrical engineering. "Power engineering," he wrote, "differs from communication engineering only in the energy levels involved and in the particular apparatus used suitable for such energy levels, but is not in fact a separate branch of engineering from communications."[12] Building on his own work in harmonic analysis and operational calculus, Wiener constructed a general theory of smoothing and predicting time series. He did not limit himself to aircraft tracking but addressed any problem that could be expressed as a discrete series of data, including economic and policy issues. The report became known as the "Yellow Peril," alluding to the common propaganda phrase and the volume's yellow cover and difficult mathematics.

While gesturing at electric power, servo design, and communications theory, Wiener did not explicitly address any previous work in feedback theory. The report includes a general mathematical introduction, a treatment of linear prediction, an algorithm for minimizing the prediction error, a method for synthesizing filters to accomplish optimal prediction, and an extension of prediction to multiple time series. The final chapter details relevant examples, including the problem of deriving rates from noisy signals, so common to fire control and "of vital importance to all designers of servomechanisms."[13] Among the paper's numerous contributions was its demonstration that a feedback system could be made to optimize not only position or velocity but an arbitrarily chosen measure of "goodness," in this case the statistically defined root mean square (rms) error.

Wiener's paper certainly had an impact within the NDRC. A number of researchers, including Ralph Phillips, George Stibitz, John R. Ragazzini, and John Russell, took up and expanded on Wiener's work. Wiener's frequently cited paper formed the basis for postwar work of optimal estimation, smoothing, and signal processing, much of it intimately tied to military applications.[14]

### Innocence Abroad: Wiener's Termination

Influential as the work would prove in later years, Wiener's scheme had insurmountable problems in practice, and his sponsors harbored doubts. "It is not at all clear," Weaver wrote, "that this study will result in a design practicable for large scale production."[15] The algorithm assumed, for instance, an infinite or very long period in the past on which to base its prediction. A real target, by contrast, could be tracked for only a few seconds before the prediction was needed. Starting and stopping the system in a finite time interval introduced noise spikes at the ends of the time series, which corrupted the prediction. Furthermore, by minimizing rms error, Wiener's approach gave progressively less value to a miss based on the square of its distance from the target. But rms error did not accurately describe antiaircraft fire: if the shell did not explode within about 10 yards of the target, it was worthless, no matter how far away it was, so the error function had to fall off more steeply than the square of the distance. Wiener's optimal predictor also involved an extensive and complex network of electronics.

In July 1942 Wiener and Bigelow demonstrated their predictor to Weaver, Poitras, Fry, and Stibitz. The D-2 members were impressed with the performance. "It gave me the feeling of having my mind read," Stibitz observed when he operated the device. But questions remained, in Weaver's view, "whether this is a useful miracle or a useless miracle."[16] Norbert Wiener, after all, was trying to build a machine to predict an uncertain future, one under the control of an enemy pilot trying to escape with his life.

Wiener and Bigelow believed their predictor was limited by their statistical knowledge of pilot behavior and flight paths, so they wanted to collect data on actual human tracking operators and pilots. The two set out on a tour of sites to do research in antiaircraft fire control. By this time Weaver had become fed up with what he saw as Wiener's naïve faith in an ideal analytical solution; the project had been under way for nearly two years with no practical applications to show. The NDRC, despite its early emphasis on fundamental work and scientific approaches, liked to see concrete results. During those two years Bell Labs, the Radiation Lab, and the Servo Lab had radically transformed the practice, if not the theory, of fire control.

Weaver vented his frustration in a memo that conveys how poorly the awkward Wiener fit into the NDRC's secret and chummy world of control:

> [Wiener and Bigelow] have gaily started out on a series of visits to military establishments, without itinerary, without any authorizations, and without any knowledge as to whether the people they want to see (in case they know whom they want to see) are or are not available. WW [Weaver] is highly skeptical about this whole business. . . . Inside of twenty four hours my office begins to receive telegrams wanting to know where these two infants are. This item should be filed under "innocents abroad."[17]

While "abroad," Wiener and Bigelow visited the army's Aberdeen Proving Ground and the Frankford Arsenal and met with the Anti-Aircraft Artillery Board at Camp Davis, North Carolina. The two also visited Tufts and Princeton Universities, Fort Monroe, and the Foxboro Company in Massachusetts, all of which were conducting studies of human-operator performance under D-2 contracts. Wiener and Bigelow returned to MIT and prepared an experiment to collect statistics on human operators in the laboratory in which a subject would try to track a dot of light as it traveled along a random path on the wall while the movements were recorded on paper tapes.

Throughout 1942 Wiener retained his confidence in the program. He developed a typical wartime ambition, for "a very considerable expansion" and a staff to help with servo and radar problems.[18] But Bigelow became discouraged. After the trip, Weaver recorded that Bigelow was convinced that Wiener's statistical method "has no practical application to fire control at this time" and that the young engineer "seriously doubts that W[iener] will be able to bring himself to make this statement." Bigelow's pessimism stemmed from his observation of human operators and their subtle, nonlinear behavior.[19]

Wiener's contract concluded in late 1942, as D-2 was transforming into Division 7. The project did not survive the transition. At the new division's first meeting Weaver reported that Hendrik Bode's work on "curved flight prediction" for the new T-15 director held more promise than Wiener's predictor. In a quantitative comparison of methods for predicting actual recorded target tracks, Wiener's optimal method proved only marginally more effective than Bode's design, which was much simpler. At its second meeting Division 7 decided to terminate Wiener's work. Project 6 ended in January 1943, and Bigelow left to join a statistical fire control group at Columbia.[20] The termination of Wiener's contracts, just as D-2 became Division 7, although somewhat coincidental, reflected the NDRC's turn away from fundamental studies toward more industrial and applied projects.

Still, the shift does not fully explain Wiener's termination, for Weaver and D-2 surely recognized the profound import of Wiener's ideas. "When this war is over," Weaver reported in 1944, "the theory and mechanization of smoothing will be one of the outstanding contributions of the NDRC fire control group."[21] Indeed, just after the war, Weaver and the Rockefeller Foundation supported Wiener's early work on cybernetics. Wiener and Bigelow's two NDRC contracts cost just over $30,000, a paltry sum, and less than one-third of 1 percent of the total outlays. The $2,000 for Wiener's *Extrapolation, Interpolation, and Smoothing* was the single smallest fire control contract. Thus, it is odd that such important work costing so little money was terminated, immediate application or no. Perhaps Wiener's inability to conform alienated him from the chummy culture of the NDRC. Perhaps the committee distrusted the Jewish, left-leaning professor who disdained the constraints of secrecy.[22]

Wiener was disappointed by his failure to produce a practical device for the war effort. Still, he plunged into elaborating his work in other arenas. The previous spring, Wiener, in collaboration with the physician Arturo Rosenblueth and the physiologist Walter Cannon, had begun addressing feedback issues in physiology and neurology. Cannon's 1932 book, *The Wisdom of the Body,* explored biological homeostasis and even made analogies with social and industrial "organisms," although not with regulating mechanisms or control systems. In the spring of 1942 Wiener first mentioned the idea of the human operator as a feedback element and an integral part of the system. He discussed the "behaviorist" implications of his work in control, saying that "the problem of examining the behavior of an instrument from this [behaviorist] point of view is fundamental in communication engineering."[23] The stimulus-response model of behaviorist psychology seemed suited to the input-output orientation of communications.

This period, the last few months of Wiener's NDRC program, marked the conception of the "cybernetic vision," which would make Wiener famous after the war. Wiener's understanding of the feedback mechanisms of control and communication in both humans and machines lay at the core of cybernetics. His postwar program would seek to extend that understanding to biological, physiological, and social systems.

## The NDRC's "Cybernetic" Program

My goal here is not to dispute the origins of cybernetics nor to replace Wiener's account with other origin stories. But Wiener's own origin stories have made their way into historical accounts, leading to an overly intellectual view of the convergence of control and communications. Consequently, errors pepper the literature on cybernetics. Wiener's biographer argues, for example, that he originated the analogy between feedback mechanisms and human behavior, "in which information from the eyes or proprioceptors is processed by the nervous system to control the hand." Another recent history of Cold War computing cites "Behavior, Purpose, and Teleology" as the origin of "negative feedback," "the theory of feedback control," and "the basis for servomechanism design." Cultural critics' enthusiasm for "cyborg history" also centers on Wiener and "the World War II regime" as origin.[24] These perspectives, based as they are on Wiener's own accounts, do not consider the broader landscape within which he traveled. Taking a wider view does not invalidate cybernetics or its analysts but rather connects them to longer histories of engineering, manufacturing, and human-machine interactions.

Consider, for example, the persistent problems surrounding human-machine interaction in fire control. The army was only too aware of the difficulties of pointer matching and "human servomechanisms" in the Sperry computers. It reported to the NDRC that any new director designs must consider "the capabilities and limitations of the human operators."[25] In May 1941 Harold Hazen and the psychologist Samuel Fernberger visited Sperry Gyroscope "to explore the possibilities of motor psychophysiology in relation to instrument design." They inspected the company's tightly coupled control systems: ball turrets, hand grips and foot controls, and cathode-ray indicators for flight instrumentation. Despite the company's decades of experience designing controls, engineers at Sperry had no systematic way of determining the best configurations for their human operators, but simply "settled on that combination that feels good to the designer."[26] The NDRC came upon an industry experienced at combining people and machinery, but one seeking a systematic basis for its work.

Hazen also visited Fort Monroe, where he witnessed the army's Sperry antiaircraft directors. It was after this visit that he wrote his memo to Weaver concerning "the human operator as a fundamental link in automatic control systems." Hazen suggested that the NDRC study "the fundamental mechanical parameters of the human operator." Frequency-response methods borrowed from communication engineering, he wrote, could characterize human reactions under varying conditions. He also proposed optimizing "the nature of the device by which the operator expresses his reactions," that is, the wheels, knobs, levers, or other means of human interaction. "This whole point of view of course makes the human being . . . nothing more or less than a robot," Hazen added, "which, as a matter of fact, is exactly what he is or should be." New approaches based on servomechanism and communications theory might better match machines to their operators, Hazen argued, and so amplify the human capacity for judgment, memory, and extrapolating patterns into the future.[27]

Weaver found Hazen's ideas intriguing and circulated the memo to the rest of the committee. Preston Bassett, president of Sperry Gyroscope, thought Hazen "had done a service merely in formulating the problem," that is, the issue of "the human element in automatic control systems."[28] Samuel Fernberger suggested that emotional stability, the group behavior of machine operators, and the effects of battle stress on human control were also worthy, if difficult, objects of laboratory study. D-2 thus initiated a program of research into the human being as an element in feedback loops, particularly in the critical operations of ranging and tracking, under the direction of Fernberger (and later Bassett), sometimes in conjunction with the NDRC's Applied Psychology Panel.[29] These efforts were part of a larger trend of American psychologists' contributing to the war effort, from the famous studies of group dynamics and "men under stress" to psychoacoustics and studies of "human engineering" and the "man-machine unit." Novel demands of military aviation also elicited studies of training, selection, fatigue, and human performance.[30] By the time Wiener and Bigelow made their tour in late 1942, then, the NDRC had an extensive program of "cybernetic" research.

Under D-2 and Division 7, psychologists at a variety of universities were examining the "human element" in fire control. Princeton University set up a special laboratory for man-machine interfaces at Fort Monroe in Virginia, a stop on Wiener and Bigelow's tour. Engineers at the Foxboro Company, a well-known manufacturer of control systems and another stop on Wiener and Bigelow's tour, looked at the effects of inertia, friction, and gear ratio on hand and foot controls, as well as the effectiveness of data displays on visual recognition. The computer innovator John Atanasoff conducted experiments at Iowa State College on tracking with small knobs instead of handwheels to achieve finer control with finger muscles than would be possible with coarser hand and back movements (Project 12). A group at Columbia worked on electronic simulations of responses of human system operators. Another battery of tests tried to quantify the effects of diverse factors on operator performance, including gender, exercise, practice, stereo acuity, pupil size, and drugs. Other studies exposed machine operators to bells, loud noises, and electric shocks. Only fatigue appeared to have any consistent, quantifiable effects.

The strangest of the human-performance studies gave new meaning to the concept of stability in a control system: psychologists searched for ways to determine whether an individual would become emotionally unstable under fire. In 1941, Division 7 brought six British seaman to the Princeton Laboratory in Virginia. The men were fire control operators on the ship HMS *Dido*. Two of them had "broken up" in heavy combat off Crete, while the rest had stayed at their positions. Without being told who was who, researchers tried to distinguish the unreliable men from the "stable" ones. Rorschach tests, optical exams, electric shocks, psychoanalysis, and a number of other scientific indignities all failed to detect which of the men had "broken." Electroencephalography and electric shocks did seem to select the "abnormal" individuals, but with heavy reliance on skilled interpretation of data that seemed to disqualify the test for broad use. In the end, the researchers deemed the sample size too small to be conclusive.[31]

All of these studies articulated analogies between human operators and servomechanisms and saw the human-machine combination as a feedback system. The engineer Enoch Ferrell, of Bell Labs, for example, equated the human role in a control system to that of a negative feedback amplifier: "The difference in azimuth between the output shaft, as marked by the telescope cross-hairs, and the target azimuth is detected by a human eye and brain, amplified by human muscles, and passed through a handwheel and gear-train to the output shaft in such a polarity as to reduce the observed difference. This is a negative feedback system." Ferrel even used Nyquist's criterion to determine the stability of the human-machine combination: "If the higher frequency components are transmitted around the loop with improper phase relations then oscillations may occur and jerky tracking may result."[32]

When Wiener made the analogy between people and machines, then, he was reacting to and building on an evolving understanding, pervasive among engineers and psychologists, that the boundary between humans and machines affected the performance of dynamic systems and was a fruitful area of research. Unlike Wiener, however, NDRC researchers remained bound by military secrecy and busy with contractual obligations.

## Cybernetics as a Civilian Science

Wiener's 1943 paper, written with Rosenblueth and Bigelow, "Behavior, Purpose, and Teleology," allies servomechanisms with the "behavioristic approach" to organisms and classifies behavior by level of prediction. The paper's philosophical tone and biological metaphors reflect not only Wiener's alliance with the life sciences but also the strictures of secrecy surrounding his prior work. In *Cybernetics* Wiener acknowledged the role fire control and prediction played in his thinking, but beginning with "Behavior, Purpose, and Teleology," he also recast military control in a civilian mold. For Norbert Wiener, in the midst of the technological war, cybernetics became a civilian enterprise.

Most indicative of this alienation and reconstruction is Wiener's consistent failure to acknowledge the multiple traditions of feedback engineering that preceded him. In all his writing on cybernetics he never cited Elmer Sperry, Nicholas Minorsky, Harold Black, Harry Nyquist, Hendrik Bode, or Harold Hazen. All had published on the theory of feedback before 1940; all were recognized as important to the field; all had speculated on the human role in automatic control; some had even written on the merger of communications and control or on philosophies of feedback. But Wiener only rarely cited any theory later than Maxwell's 1867 paper "On Governors."[33] Wiener called this paper fundamental, but it lacks the basic idea of a feedback loop, which Wiener himself found so central.[34]

The omissions are striking. Wiener must have been aware of his predecessors: he advised Vannevar Bush's research program in the 1930s; he worked with MIT's Servo Lab and Radiation Lab during the war; he corresponded with Hendrik Bode during the early 1940s. Still, he wrote, *"I think that I can claim credit for transferring the whole theory of the servomechanism bodily to communication engineering."*[35] Wiener's chapter "Cybernetics in History," from *The Human Use of Human Beings,* refers only to Leibniz, Pascal, Maxwell, and Gibbs as "ancestors" of his new discipline.[36] Wiener gave cybernetics an intellectual, scientific trajectory, divorced from the traditions of technical practice from which it sprang.

Wiener's reformulation had ideological implications, especially in light of his own estrangement from military research. After Hiroshima and Nagasaki, Wiener became critical of the American military's dominance of the country's engineering efforts. Yet in the early 1940s he had been anything but a pacifist. Wiener had suggested to the army filling antiaircraft shells with flammable gasses to burn enemy planes from the sky; he had pondered what types of forested areas and grain crops were most susceptible to fire bombing. Weaver remembered him as "at least at times, about as savage a fighter as anyone who ever appeared on the front."[37] Still, the atomic bombs, and perhaps his disappointing NDRC project, changed Wiener's attitude toward military research. His primary substantive contact with what he later called "the tragic insolence of the military mind" occurred under NDRC auspices and ended in January 1943.[38] Pesi Masani, Wiener's colleague and biographer, argues that Wiener's wartime experience did not influence this estrangement because his final NDRC report suggests further research into military prediction.[39] The wartime writings certainly contain none of the criticism that appeared after 1945. Still, George Stibitz wrote in his final report on Wiener's NDRC project that "Professor Wiener has asked that no mention of his name be mentioned in connection with any War work."[40] Galison perceptively argues that Wiener, by elevating his prediction circuit to the "symbol for the new age of man," enshrined an oppositional military metaphor into the civilian science of cybernetics and its descendants.[41] In light of Wiener's wartime work, however, the survival of the oppositional model is also ironic, as Wiener's writings effectively formulated cybernetics as a specifically nonmilitary, scientific endeavor.

Wiener's own contributions were hardly insignificant. The mathematics of *Extrapolation, Interpolation, and Smoothing* were of the greatest import; he introduced statistics into the field of control; his efforts to bring an understanding of communications and control to broad communities of physiologists, physicians, and social scientists are well documented.[42] Through the informal "Teleological Society," the series of Macy conferences, and a growing identity as a public intellectual, Wiener elevated his thinking on control and communications to a moral philosophy of technology and enjoyed enthusiastic response. He recognized the industrial implications of widespread automation and contributed to a public discourse of technology and society that continues to this day.

However influential, Wiener's ideas about control, communication, and human-machine interaction arose within the context of wartime research. This context would be irrelevant were it not for the subtle and significant histories of other projects and ideas and, as we have seen, the technological trajectories that emerged from their own working through of problems in communications and control. In fact, after terminating Wiener's work the NDRC supported several other projects that can, in retrospect, be called "cybernetic." While several of those projects addressed the feedback behavior of the human operator, others examined how best to represent the world in a machine.

### Improving the Analog Infrastructure

Division 7 supported differential analyzers, numerical fire control computers, and relay computers but rejected a proposal to build an electronic, digital machine. The division's projects in computing both drew on and questioned the analog approach developed at MIT in the 1930s. At the start of the war, four Bush-style differential analyzers had become mechanized calculating facilities. The original one at MIT had 6 integrators and spent the war running ballistics calculations for the Naval Proving Ground at Dahlgren, Virginia. The Moore School of Engineering at the University of Pennsylvania built a copy of Bush's machine with 14 integrators and also made a 6-integrator machine for the Ballistics Research Laboratory (BRL) at the army's Aberdeen Proving Ground in Maryland. The Penn machine, in addition to having the largest number of integrators, and hence the ability to solve the most complex equations, was also the most refined. Engineers there had embellished Bush's original design with new servos and automatic curve followers (improvements they also applied to the BRL machine).[43] Both machines computed army ballistics tables during the war. G.E. built its own differential analyzer, based on the Penn machine, with 14 integrators, which was ready in 1943.[44] The Rockefeller Differential Analyzer, at MIT, had 18 integrators and was also completed that year. When operational, it ran firing tables for BuOrd.

Firing tables became a bottleneck in the army's ability to field new ordnance. Every combination of guns and fuzes required new range tables, accounting for a variety of parameters, such as range and temperature. As a result, pressure mounted to increase the capacity of the differential analyzers. On 1 December 1942, Division 7 let a contract, Project 62, to Penn for "Improvement of Differential Analyzers," meaning faster throughput and greater precision. The project would add new types of torque amplifiers and make a new recording device to log several values simultaneously with regard to more than one variable. Penn would also improve input and output devices, study mechanical slip in the integrators, and interface the machines to punched-card equipment. The Neiman torque amplifiers would be replaced with updated servos, reflecting the results of recent wartime work.[45]

## Claude Shannon: Channel to Bell Labs

The NDRC supported differential analyzers as computing facilities because their ability to produce firing tables, as well as data for ballistic cams, made them a legitimate part of fire control research. Firing tables, indeed, were numerical solutions to the machine-control problem of aiming a gun; producing and using them required a series of conversions between continuous curves, numerical outputs, and mechanical motions (although not in real time).

Indeed, the differential analyzers had a technical similarity to mechanical fire control computers. Whereas the analyzers used feedback loops (Hazen's back coupling) to solve equations, gun directors used feedback to solve the coupled problems of ballistics and prediction (Earl Chafee's "cumulative cycle of correction and recorrection," and the Ford Rangekeeper's "regenerative tracking"). Both made extensive use of differential gears, integrators, and servos, now the standard building blocks of mechanical computing. Early in its program, D-2 sought to apply the academic knowledge about the analyzers, including theories of feedback, to the industrially produced military machines.

Claude Shannon made the connection. After completing his Ph.D. at MIT in 1940, he went to Princeton for a postdoctoral year; he was at Princeton when the NDRC was established. On Thornton Fry's suggestion, D-2 let a contract, Project 7, to Shannon for "Mathematic Studies Relating to Fire Control." The NDRC asked him to look at Sperry's antiaircraft director and at another designed by the Frankford Arsenal and suggest how to improve the smoothness of tracking. Shannon analyzed the calculating mechanisms of the gun directors, especially their smoothing circuits, as feedback networks. Using stability and transient analysis, Shannon treated both the gun directors and the differential analyzers as dynamic systems. He wrote five papers on the topic.

Shannon compared the devices to electric circuits, found "the use of electrical analogues very useful in understanding these devices," and devised a circuit-like notation to describe them. He drew on network theory and his own relay algebra to define "analysis and synthesis" for differential analyzer setups, as though they were electrical

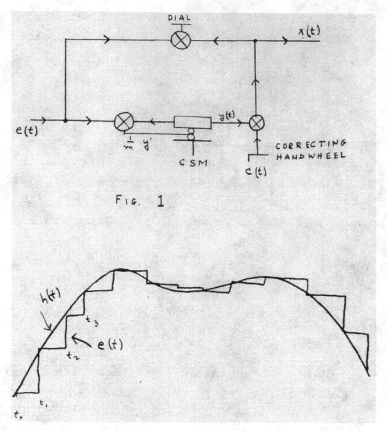

Fig. 11.1. Claude Shannon's height data smoothing mechanism, showing electrical-type schematic symbols, a feedback loop, and smoothing of jerky tracking data. Reprinted from C. E. Shannon, "A Height Data Smoothing Mechanism," 25 May 1941, Claude E. Shannon Papers, MIT Archives.

circuits. For the first time in a detailed technical study, Shannon made explicit the equivalence that Bush and Ford Instrument had intuited years before: the mechanical calculators created at universities and the fire control computers built for the military were analytically similar machines (Fig. 11.1).[46] Soon after completing this work, in 1941, Shannon joined Thornton Fry's mathematics department at Bell Labs, thus making an institutional and intellectual link between MIT's differential analyzers and Bell Labs' new work in fire control.

## MIT's "Arithmetical" Computers

Ballistics also raised the question how best to represent data in the machine—continuously, as with the Bush analyzers, or numerically, as in a firing table. Before the war, Bush himself considered numerical techniques; he had circulated memoranda outlining a "rapid arithmetical machine" based entirely on electronic switching. The device represented numbers as electronic pulses, not with the physical analogs he and his students had developed. The architecture included keyboard inputs, a control unit, an arithmetical unit, memory storage, and recording outputs. After Bush's departure, Sam Caldwell pursued his ideas, supervising the research assistants William H. Radford and Wilcox P. Overbeck under the sponsorship of the National Cash Register Company. Wartime projects distracted the staff before any full-scale hardware was built.[47] Yet the Rapid Arithmetical Machine had some influence: Norbert Wiener wrote a memo about it, and it brought Caldwell into contact with the idea of representing numbers as electronic pulses. His opinions would heavily influence D-2 and Division 7 policy on computing.

A project that stemmed from the arithmetical machine also made the link to fire control. When Claude Shannon completed his doctorate and left MIT, another student, Perry O. Crawford, replaced him at the Center for Analysis. Crawford did some work on the Rapid Arithmetical Machine project, read Shannon's thesis on relay circuits, and became interested in arithmetical computing. Because the differential analyzers were heavily involved in ballistics work, and because Caldwell became increasingly devoted to the NDRC's fire control committee, Crawford saw the connection between arithmetical computing and fire control. His cryptically titled 1942 master's thesis, "Automatic Control by Arithmetic Computation," sketched a design for a fire control computer based on electronic pulses. It was not a complete system, only the mathematical architecture required for target prediction, but Crawford's study employed components from the Rapid Arithmetical Machine and algorithms based on the Sperry Gyroscope and Ford Instrument systems.[48] Several years later, when Jay Forrester and the Servo Lab began working on a generalized flight simulator, it was Crawford who suggested they build a digital machine.

MIT's work with arithmetical calculators, while building on extensive experience with differential analyzers, had not produced any functional hardware. But working on these projects were engineers who saw the potential of applying this new way of representing the world to problems of control. These engineers' connections to the NDRC and the navy assured these techniques a place on the research agenda for fire control.

## Electronic Fire Control at RCA

MIT's electronic projects were not unique. At least one other group investigated electronic numerical computing for fire control: the television pioneer Vladimir Zworykin and his colleague Jan Rajchman at the Radio Corporation of America (RCA), working under contract with BuOrd. They too studied "computing devices in which variables are represented by discrete impulses," and they examined the implications for individual components, coordinate systems, and "the manner in which these elements are coupled together" for a new fire control system. Zworykin and Rajchman realized that a critical aspect of any pulse system would be the means by which physical quantities were translated to and from the electronic pulses. Hence they concentrated their effort on electromechanical "coders" to turn continuous fire control inputs into discrete signals. They also worked on a *computron,* a vacuum tube that incorporated elements for a 10-bit counter into a single tube (similar to one investigated for the Rapid Arithmetical Machine at MIT).[49] Individual vacuum tubes failed regularly, so the computron, they argued, by reducing the number of tubes, would correspondingly improve the reliability of any electronic pulse machine.

Despite these intriguing ideas, the project's sponsor, BuOrd, was not accustomed to such long-term research. The bureau's chief, William Blandy, pressed by the urgency of the antiaircraft situation, wanted RCA to produce an electronic gun director by the end of the contract period, April 1942. He soon realized that the bureau "was perhaps too sanguine in its hopes."[50] Blandy reluctantly acknowledged that progress toward electronic pulse computing would be slow and deliberate, so he requested that the NDRC take over the RCA project. D-2 agreed, especially since Thornton Fry had recommended to Weaver that they support long-range development of calculation techniques.[51]

In response to the navy's request, Weaver informed Zworykin of D-2's existing work in electronic computation. He mentioned Caldwell's work at MIT, the Bell Labs analog electronic director (the T-10), as well as Wiener's prediction network. Weaver also distinguished, however, between electronic analogs, which he considered immediately useful, and electronic pulses, which seemed far off. "The present state of the art," he wrote, "as regards impulse electronic computing devices, is not sufficiently advanced to warrant the attempt, at this time, to incorporate such devices into an over-all design for a predictor." D-2 believed that RCA should concentrate only on "the essential computing elements themselves."[52] Weaver suggested that Zworykin improve the computron tube rather than build a complete system.

Weaver's note also drew an institutional boundary. RCA's existing contract with BuOrd, typical for navy support, was for the delivery of a complete fire control system.[53] The NDRC had a different definition of a research contract: feel free to look into components, Weaver seemed to say to RCA, but systems are our terrain.

## Electronic Fire Control Computers

Still, Weaver had the sense that electronic computing might be a significant path for the NDRC to pursue. To assess its potential, Weaver surveyed the research groups working on related technologies—MIT, Bell Labs, RCA, and an optical group from Eastman Kodak. At this stage, he did not distinguish between analog and numerical or digital techniques but focused simply on electronics. In April 1942 Weaver called a conference, "Electronic Fire Control Computers," in New York City, to review the survey results and ideally to merge existing expertise in computing with that in control systems. Attending the meeting were representatives from the army's Frankford Arsenal and M. Emerson Murphy, head of BuOrd's antiaircraft fire control section. Researchers included Zworykin from RCA, Caldwell and Overbeck from MIT, and others from Bell Labs and Eastman Kodak. From the NDRC itself, Thornton Fry, Duncan Stewart, and George Stibitz also attended.

Weaver asked the group about the viability of electronic computers for fire control. Could they be fielded within a 24-month time frame? Or should the NDRC view them as "basic but very long range research and development" and hence assign them "low priority?" It was just five months after Pearl Harbor; Weaver was already drifting away from his early interest in fundamental work, giving highest priority to research that would produce usable systems in a short time.

The meeting began with an extended discussion on the advantages of electronic computation, and here the researchers distinguished between analog and numerical methods. Analog, whether mechanical or electronic, suffered from difficulties of scaling: How large could the electrical quantities in the machine be? Did 1 mph of target speed correspond to 1 volt? to 10 volts? The answers depended on precision. Given some minimum amount of noise or uncertainty in the signal, scaling depended on how precise the machine needed to be. If it could resolve differences of a hundredth of a volt, for example, tracking a target with a speed of 300 mph required processing signals up to 30 volts (the M-9 director, for example, used hundreds of volts to represent data). By contrast, participants noted, numerical techniques had no such scaling difficulties; arbitrarily high precision was obtainable by adding extra "digits" to the numbers. This, of course, increased the size and complexity of the machine, but how much, and with what effects, remained unknown.

The precision required, however, depended on the application. For scientific uses a high degree of precision was often desirable. But in fire control the overall system was only as good as its crudest component. Tracking inputs, heavy servos to drive guns, and even the guns themselves had limited precision. Current computers would benefit from a two- to threefold increase in computation precision, the engineers agreed, but more could not be justified. This point had critical implications for D-2's interest in numerical computers. "It is important to remember that it is impractical and indeed useless," ran the conference report, "to carry the accuracy of the computer beyond a certain point . . . either from the point of view of the input data with which the computer must operate, or from the point of view of the accuracy with which the output of the computer can be utilized."[54] Because of their concern for the material dimensions of a control system, this group did not see a pressing need for numerical computations. To them, a computer was only as accurate as its ability to translate from instruments of perception into machine representations and to articulate those representations as mechanical motion.

Similarly, the advantages of speed turned on perspective; again D-2 insisted on seeing electronics as part of a larger system of fire control. The army complained that Sperry M-4 directors had too much delay in their computation and felt that electronic computers would be faster. But NDRC engineers pointed out that a number of different speeds together determine speed in fire control solutions. The slowness of mechanical directors resulted less from mechanical techniques than from the feedback involved in approximating the calculation (this was, in part, the issue Claude Shannon had been hired to examine). The speed of raw computation was thus distinct from the amount of time the circuit took to converge on the solution. Other time delays, due to data smoothing and the time of flight of the shell, inhered physically in the problem and could not be reduced with electronics. Electronic computing improved speed for only part of the system; when considering the inputs and outputs of control systems, the speed advantage seemed less obvious. The group reached "no definite conclusion" on the speed of electronic computation and how it would affect fire control systems.[55]

Manufacturing mattered as well. Electronic computation did seem "to be of a character well suited to large scale production," but there had been so little experience with electronic computation that the topic was not discussed in detail. The issue of reliability and maintenance, however, proved more contentious. Electronic computers would use a yes-or-no signal, and "electrical circuits for such signals can be made highly reliable and insensitive to small variations." But many within the services, as well as in research, still distrusted the reliability of electronic equipment in the field. The issue reduced to one of familiarity. George Stibitz noted that "people with mechanical experience think all electronic devices full of troubles, and correspondingly reverse opinions [were held] by the others."[56]

In a follow-up memo on the meeting George Stibitz clarified his own thinking on the subject. For him, the important distinction was not between mechanical and electronic but between analog and numerical, as well as between continuous and discrete time. The key characteristic of numerical machines, Stibitz added, was that analog machines shared the same dynamics as the problems they represented, whereas digital computers did not. Indeed one advantage of numerical techniques was that they decoupled the structure of the computer from that of the calculation. Still, he acknowledged that even numerical algorithms had internal dynamics that could imitate analog feedback loops. Stibitz suggested that in the distinction between analog and "pulse," or numerical, computers, the latter be replaced with the term *digital*.[57]

For D-2, this early meeting clarified the issues surrounding electronic and numerical computing. Numerical techniques might improve fire control systems, but only to the point where other components became limiting factors. The appeal of such machines varied, based on whether one was building a large, central computation facility or a mobile, reliable, field-deployed system. D-2 was primarily interested in the latter, so RCA had not made a strong case for numerical electronic computing.

Still, D-2 kept its options open. After the meeting its representatives again visited RCA, where they saw a prototype of its electronic fire control computer. This device used a "resistive function matrix," designed by Jan Rajchman, that stored firing tables for the guns in an electronic grid, a replacement for the Sperry ballistic cam. (A derivative of Rajchman's function matrix would eventually be included in the ENIAC.) Several months had passed since RCA finished its navy contract, and the company pressed the D-2 members for a prompt decision on further moneys. D-2 acted promptly and appropriated funds for RCA to "carry the work forward" in electronic computing.[58]

But the support was cautious and qualified. Weaver emphasized to Zworykin that because of other errors in the system, fire control would not benefit from the improved precision of electronic computers. Two possibilities could change this situation, Weaver added, and generate a demand for more accurate computing: guns with much longer ranges or radically reduced errors from dispersion, ballistics, and fuzes. This judgment bore directly on D-2's definition of the type of research RCA would conduct, for it defined the work as "long-range future, rather than immediate present interest."[59] The NDRC would support the RCA work for a few months, Weaver concluded, and then decide on next steps. But as Norbert Wiener was also about to learn, within the increasingly goal-oriented NDRC, defining a project as "long-range future" amounted to killing it.

Indeed, the NDRC soon dropped RCA's electronic numerical computing project. Weaver explained that neither of D-2's major systems, Bell Labs' T-10 and T-15, could incorporate numerical techniques. He did, however, recognize the scientific potential of the work. "We arrived at this decision [to cancel the project] reluctantly," he wrote to Zworykin, "because we all have lively personal and scientific interest in seeing this computron project continued to its successful conclusion."[60] Weaver sincerely tried to find other sponsors within the NDRC to take up the project, but found no takers. "With genuine scientific regret," Weaver recommended that the project be dropped "because beautiful as the work is it does not appear at present to have any real prospect of being directly useful in the war."[61] Harold Hazen, by then taking over Division 7, also tried to find other NDRC divisions willing to support the project but to no avail.[62] The contract terminated on 31 March 1943, just weeks after Norbert Wiener's. The NDRC was narrowing its horizons.

### Rejecting the ENIAC

Defining computers within control systems and defining electronic digital computers as long-range research shaped Division 7 policy throughout the war. No episode illustrates the ambiguous effects of that policy as well as the division's response to the idea that became the ENIAC, the project that some historians hail as the first modern computer. Just as the RCA project was ending, a group from the Moore School at Penn, including John Brainerd and John W. Mauchly, were inspired by John Atanasoff's work at Iowa State. They proposed an "electronic diff. analyzer" to do ballistics calculations, purposely using the abbreviation *diff.* to stand for both "differential" and "difference."[63] The proposal built on Penn's experience with its differential analyzer and on electronics experience that Mauchly had gained at the Radiation Lab. Later the machine was renamed the Electronic Numeric Integrator and Calculator, or ENIAC.

The term *integrator* in the title reflects the importance of integration in the mathematics of the time, as well as the centrality of integrators in Penn's differential analyzer. Arthur and Alice Burks, both of whom worked on the ENIAC, described the machine as an electronic equivalent, unit for unit, of a Bush machine, a functional and structural replacement. Other members of the team contested the depth of the similarity because the ENIAC had separate control units, but ENIAC certainly drew on the machine culture of differential analyzers.[64]

Harold Hazen did not think ENIAC could become operational before the end of the war, which he assumed to be within five years. He discussed the proposal with Sam Caldwell, who emphasized the availability of the new Rockefeller Differential Analyzer, which was just coming on line. Soon Caldwell acted forcefully to scuttle the project, writing to Warren Weaver, who was now on the Applied Mathematics Panel but still an influential member of Division 7, of "a certain amount of agitation," from Penn for an "electronic differential analyzer." Wary of such a "huge undertaking," Caldwell doubted that the project could be finished until five years after the end of the war. Division 7 thus decided not to fund the ENIAC project.[65]

Some historians have argued that the conservatism of the NDRC leadership, combined with their "personal commitment to different technologies," blinded them to the value of digital techniques. Herman Goldstine, a mathematician who worked on the ENIAC while in the army, suggested that the NDRC was caught between Caldwell's mechanical analog machines and Stibitz's digital approach based on relays.[66] Indeed, Hazen and Caldwell, MIT professors and Bush disciples, had a clear stake in the success of the Rockefeller machine, as did its private sponsor, Warren Weaver. Hazen, who had written years before that numerical methods "have an artificiality irksome to the physically minded," was indeed an apostle of the analog art. Division 7 may have seen the Penn project as potential competition, but none of the confidential diaries indicate that this was the case. Mostly, they saw it as time-consuming and difficult to build. Caldwell, tired of the drawn-out, costly Rockefeller project (it was years late becoming operational), wished to avoid another such headache.

Institutional politics also played a role. The ENIAC found willing support through the BRL at Aberdeen, a branch of the Army Ordnance Department. Aberdeen indeed posed a threat to Division 7. Hazen reported that certain Aberdeen members were antagonistic and resentful toward the NDRC, which they believed was siphoning scientific talent from the army. According to Hazen, they threatened not to accept any new NDRC technology into the army.[67] In addition, the NDRC had technical concerns about the Moore School. Division 7 was already funding a project at Penn to improve the differential analyzer; the project was moving slowly and had technical troubles, including stability problems with the servos. When Weaver and Caldwell visited the university to evaluate the project in October 1943, they reported "a depressing day. Initiative and candor were entirely lacking."[68] Division 7 also found the electronics of Penn's proposal to be less sophisticated than those of NCR and RCA. Jan Rajchman visited the Moore School group in 1943 and found their technical ideas "extraordinarily naïve."[69]

The Rockefeller Differential Analyzer used a digital crossbar switch and a punched paper tape for reconfiguration and programming. By contrast, the ENIAC was rewired with cables, a process as difficult and time-consuming using a digital computer as mechanical reconstruction was using an analog one. "No attempt has been made," the ENIAC designers wrote, "to make provision for setting up a problem automatically." Their attitude toward programming reflected their experience with the mechanical differential analyzer: "It is anticipated that the ENIAC will be used primarily for problems of a type in which one setup will be used many times before another is placed on the machine."[70] The ENIAC used innovative electronic pulses to represent the world but relied on traditional methods to route those representations through its computational paths.

Division 7's interests in electronic digital computing always existed within its overall work on fire control. With the war under way, Weaver and Hazen did not fund generic technology research but rather explored all avenues that would help them achieve pressing, immediate, and short-term goals. When they received the ENIAC proposal, they had just shut down at least two projects (Wiener's and RCA's) for being too "fundamental" and long-range. Indeed, the ENIAC did not become operational until the war was over. Weaver and Hazen certainly saw the scientific and intellectual value of digital-computing research; their regret over cutting off the RCA project was sincere, and their efforts to find it another sponsor are well documented. (The record shows no similar effort regarding the Penn proposal.) Immediately after the war, in fact, Weaver provided funding from the Rockefeller Foundation for an electronic digital computer at MIT.[71]

Already by 1943, the NDRC's fire control research was moving out of a period of radical innovation and into a period of refinement, incremental improvement, and system integration. Other elements in the system simply could not benefit from more accurate computing. More than speed and accuracy, military control systems required reliability, ruggedness, and compactness, characteristics decades away in digital computing. Furthermore, Penn did not propose such a field-deployable system but rather a university calculating center akin to the differential analyzer. For the differential analyzers, the NDRC only sponsored improvements to existing machines, facilities that were already cranking out data. Division 7 might support fundamental work in fire control computers, but not a machine to produce firing tables, an application already peripheral to its charter.

Such a view does not diminish but rather underscores the radical nature of the early proponents' faith in digital techniques, despite great difficulties of reliability, size, and complexity. These problems, however, made electronic digital computing unsuitable for Division 7 support. The NDRC's failure to pursue such work, despite recognizing its scientific importance, suggests the limitations of the wartime research paradigm, which became focused on short-term, practical devices rather than long-term fundamental research.[72] Wartime research in control systems achieved success, but within, and because of, the narrow goals it defined for itself.

## Topological, Not Metric: Relay Computers

Still, D-2 and Division 7 did build digital computers, but they were computers that met the key qualifications of institutional position, rapid construction, and immediate application. George Stibitz, of Bell Labs, had been instrumental in shaping these criteria, and not by coincidence he satisfied them all. Stibitz's computers stood between control and communication. They were testers.

In late 1941 Barber Coleman's dynamic tester, a mechanical analog machine, began redefining the performance of fire control systems in the laboratory. It quickly became standard; contractors and fire control vendors, including Sperry Gyroscope and Ford Instrument, wanted their own machines to prepare for the army's rigorous acceptance tests. But like the machines it evaluated, the dynamic tester was difficult to reproduce. Different flight profiles, simulating differing paths of attacking airplanes, were "programmed" by specially machined two-dimensional cams. Changing the cams would change the flight profile, from dive bombing to evasive action, for example, but in a difficult and time-consuming process.[73] Creating these cams required a great deal of labor and skill, as well as precision machining that itself introduced errors. Mechanical cams and analog mechanisms could not provide the uniform, reproducible testers D-2 needed to maintain its authority in fire control.

### Digital Mechanisms: The Tape Dynamic Tester

George Stibitz found that he could reproduce and transfer that authority using paper tape. At Section D-2's first formal gathering, at the American Mathematical Society meeting at Dartmouth in September 1940, Stibitz had demonstrated his binary computer made out of telephone relays. The "Complex Number Computer," or Model I, worked with the imaginary numbers familiar to communications and electric power engineering, and employed Shannon's relay algebra in its design.[74] Thornton Fry, one of D-2's first members, was Stibitz's supervisor at Bell Labs and had encouraged his efforts. From this early date relay computers had a high profile within the division, a profile that Stibitz used to great success. In late 1942, D-2 let a contract, Project 60, to Western Electric and Bell Labs for a "Punched Tape Dynamic Tester." The original appropriation was $2,500. By the end of the war Division 7 had spent more than $500,000 on Stibitz's three major computers.[75]

Stibitz's envisioned his machine as "a simplified form of dynamic tester which can be duplicated readily." The comparatively inexpensive machine consisted of parts common in the telephone network. Instead of mechanical cams to program flight profiles, it used paper tapes that could be easily, cheaply, and exactly duplicated for multiple installations. "Typists replace machinists," Stibitz wrote, capturing the shift of skills that would accompany the material transition from analog to digital computing.[76] Machinists were male, highly skilled, and difficult to find in wartime. Typists were female, required less training, and were plentifully available (though the innovative women programmers working on the ENIAC soon proved that programming required much more than typing skills).

Like the earlier dynamic tester, Stibitz's machine simulated the handwheel inputs to a gun director. The new machine, however, performed a digital-to-analog conversion, using a "tape controlled servo" (Fig. 11.2). A rotating shaft connected to a series of electrical contacts and sensed the actual shaft position. These switches fed a set of relays that read the desired position off holes on the punched paper tape. Five holes coded (in binary) the desired position, or which of 32 positions the shaft ought to assume. A relay network compared the desired position with the actual position and drove the shaft one way or the other to make the error, or the difference between the two positions, equal to zero. To synthesize the relay network to perform this comparison, Stibitz used Shannon's relay algebra.[77] He termed the tape servo a "sampled" data system because it operated in discrete rather than continuous time intervals. Using a

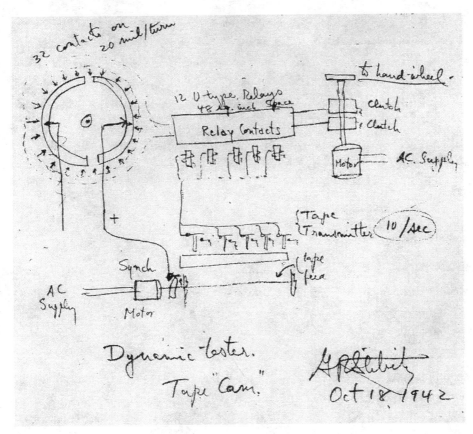

Fig. 11.2. George Stibitz's sketch for a "tape controlled servo." Reprinted from George Stibitz, "Proposed Dynamic Tester," 19 October 1942, OSRD7 GP, project file 60, box 44.

Nyquist diagram, Stibitz analyzed the servo as a feedback amplifier, using the "sampling period" to establish its equivalence with a continuous servo.[78]

As a data-driven servo, the tape dynamic tester resembled the numerically controlled machine tools that appeared after the war, and the application to machining was not lost on the NDRC. In 1941 Stibitz and Duncan Stewart made a tape-controlled device to mill the mechanical cams for the Barber Coleman Dynamic Tester, possibly the source of Stibitz's idea for the tape tester. In 1943 a Division 7 contractor wrote to Stibitz proposing "a means of constructing the metal cams by the use of a duplicating device on a milling machine, controlled by one of the Bell Laboratories tape controlled units." Indeed, under an NDRC contract the company adapted the tape dynamic tester to mill the mechanical cams for an antiaircraft gunsight tester at the University of Texas, the "Texas Tester."[79]

Stibitz's computers also joined communication and control. Input and output devices—tape readers, keyboards, teletypes—were all borrowed from telephone systems. The tape dynamic tester, like the complex number computer, could operate remotely over phone lines, and "the impulses could be transmitted over a single telegraph channel from the tape transmitter . . . and be reproduced in the form of motor rotation at the other end of the line."[80] Where Norbert Wiener theorized the fundamental notion of "the message" in computing systems, Stibitz implemented it in practice, turning teletype messages into mechanical movements and custom metallic parts.

### The Relay Interpolator as a Differential Analyzer

The tape dynamic tester could absorb and process so much data, however, that it drove further automation. To prepare the tapes for input to the machine, human computers had to calculate series of points that described the track of the simulated attacker. The tester needed about 20 of these points per second, all of which operators manually punched onto paper tape, a laborious and error-prone job. A typical run required six functions (three input and three output variables) of about 150 seconds in length, requiring about 20,000 points. The 60 or so courses required to thoroughly test a new director amounted to over a million points, or "about three years of a skilled [human] computer's time." Division 7 investigated a number of options for automating this process, including punched-card machinery from NCR and IBM, but these machines printed their output on paper, which still had to be transcribed to tape.[81]

In June 1942 Stibitz proposed another relay computer to generate the tapes from data given at 1-second intervals.[82] If only a single point in every 20 were calculated, the machine would fill in, or interpolate, the rest. Stibitz called the device the "Relay Interpolator" (RI), and it became Model II in his series. Soon Stibitz and Bell Labs began designing and building the relay interpolator (Project 70). Ten weeks later the machine's 500 relays began producing paper tapes for the dynamic tester.

The machine could do more than interpolate. As soon as it became operational, Stibitz began offering the RI to NDRC and military researchers as a general mathematical machine. The Applied Mathematics Panel distributed a pamphlet describing the device and its programming, announcing that "the NDRC now has a calculator of rather low native intelligence but of indefatigable energy." While limited to reading, writing, storing, and adding numbers, it could be programmed to repeat those operations in mind-numbing repetition.[83] The machine had 9 "registers," or locations for storing numbers, two of which could add numbers and store them in a third register. To manipulate these numbers, the RI used "a system of designation of the orders," in which each order was identified by two letters. The command "CA," for example, would copy a number from register C to register A, similar to modern computer assembly language. Other commands could add two registers, input data from the tape, or output to a typewriter (the machine had no branching instruction).

Stibitz refined his conception of the RI in a series of memos distributed to Division 7 in 1943. These included "Harmonic Analysis as a Smoothing Operation on the RI," and "The Relay Interpolator as a Differential Analyzer." Properly set up, he argued, the RI could solve not only ordinary differential equations but also partial differential equations, which the MIT machines could not do. While the RI could not explicitly solve ballistics problems, it could interpolate and improve the solutions.[84] Throughout, to distinguish the differential analyzer from his own machines Stibitz used his new term, *digital.*

In the most ambitious of these memoranda, "Unified Theory of the Relay Interpolator," Stibitz connected the RI to the general processing of signals. Interpolation, he recognized, was really a smoothing operation, and it "looked like the 'filtering' that communication engineers applied to noisy telephone signals." This memorandum showed how the RI could simulate electrical filters, electronic oscillators, differential equations, and Fourier analysis. These simulations, in fact, would even face the stability problems of continuous, linear systems: "Unstable programs would cause the output to oscillate with increasing amplitude, and conversely with stable programs."[85] The RI continued the NDRC's abstraction of flight profiles, now making them interchangeable, digital signals.

Soon after this project began, Stibitz had proposed yet another machine to automate testing, this time for the army's Anti-Aircraft Artillery Board. The original dynamic testers compared a gun director's actual outputs with actual and ideal outputs; the AAAB computer actually calculated the ideal responses, using firing tables stored on tape. A "master control tape" programmed the machine with the requisite formulas, whereupon "the operator pushes the start key and leaves the machine to do the rest." What took a human computer using a calculating machine 40 minutes, the relay machine could do in about 2.5.[86] Division 7 let a contract to Bell Labs in late 1943, Project 74, and the machine began its first calculation the following May. Bell Labs eventually built several similar machines, one for NACA, one for the Aberdeen Proving Ground, and a derivative, Model IV, for the NRL.[87] While the RI and the "ballistic computer" were intended for specific fire control applications, Stibitz and Bell Labs offered them as general-purpose computers; they remained in service for military applications until 1961.

### George Stibitz's Digital Approach

How did Stibitz obtain and sustain Division 7 support for his extensive computing program, when Penn and RCA could not do so for theirs? He certainly had advantages: he himself was a member of Division 7, the NDRC tended to favor Bell Labs and MIT, and Bell Labs was the site of the leading fire control project in the country, the T-10. Still, institutional position might explain a single, isolated project, but not this expensive and unique series of machines. Stibitz's success derived from his conception of digital computing and its relationship to fundamental research and manufacturing.

For Stibitz, digital computing was as much a structural as a mathematical strategy. With mechanical analogs, the complex interconnection of gears, differentials, and shafts constrained the machines to a particular form, but that was not true for numerical computers. "The electrical [relay] computer," he wrote, "was topological, not metric."[88] By this he meant that his machines separated function from physical form; if the wiring was correct, it did not matter how the computer was laid out mechanically. The design inhered not in the components, which, after all, were standard telephone relays, but in the wiring between them. Stibitz defined his digital machines by connections, and Shannon's relay algebra allowed him to manipulate and combine digital circuits as network diagrams, with mathematical notation.

Both Penn and RCA could design with similar abstraction (no evidence suggests that they employed Shannon's relay algebra, but neither used relays). Yet one key difference set Bell Labs apart: the translation from design to structure was simple, unproblematic, and proven, for it relied on standard telephone circuitry. "Switching equipment of the type used in the construction of relay computing machines," Stibitz wrote, "has been subjected to many years of investigation and improvement so that it is extremely reliable . . . [and] in large quantity production."[89] Nor was the benefit of telephone equipment limited to components. As one of Stibitz's colleagues recalled, "He stole a very large part of the circuitry out of standard telephone central offices."[90] Western Electric's thick Standard Operating Procedures manuals (SOPs) specified "how to do almost anything that could be done legitimately in the Bell System." Stibitz stuck to the SOPs when laying out his systems. Thus, Western Electric could build the machines quickly and reliably, relying on the technical culture of telephone engineering: "All parts required are in production and are available at short notice in any quantity likely to be required. Construction does not demand highly trained or scarce personnel. Design of the mechanism, once the fundamentals are sketched out for them, *is a familiar and routine matter for telephone machine switching engineers.* This group has not been drained as completely [by wartime demands] as have most skilled groups."[91] Any of thousands of Western Electric wiremen could build the machines as they built any telephone switching system. As Stibitz wrote, "The wiremen worked at the speed and precision with which they would have done had they been constructing dozens of relay computers in their careers."[92] If speed and precision were benefits of digital representation, they also characterized the mode of production.

Just as digital processing made data interchangeable, digital computers standardized construction. Rapid, reliable translation of ideas into things produced predictable project schedules and solid, maintainable hardware. It also fostered architectural innovation: Stibitz built three successive generations of computers in 18 months (late 1942 to mid-1944), each responding to new problems and building on prior experience. The ENIAC, by contrast, sought to innovate in both components and architecture and hence took more than two years to build, based on a single design.

Thus, Stibitz owed his success not only to digital computing but also to computers based on telephone relays. For him and for Division 7, the difference between his and the Penn project mapped onto that between system design and component development: the former could contribute to the war effort in short order; the latter represented fundamental research and might not pay off before the war ended. Stibitz summarized his philosophy in 1943: "Electronic methods may well be the computing means of the future, but their application at present would present a *research* as contrasted with a *design* problem."[93] Stibitz, Bell Labs, and relay computers reinforced the distinctions Weaver had imposed on RCA, which informed Division 7's rejection of the ENIAC: electronic analog and electromechanical digital machines were short-term developments, but electronic digital machines were fundamental, long-range, and hence would not contribute to the war effort.

## Conclusion

The story of the NDRC's involvement with early cybernetics and digital computing need not establish Harold Hazen, Warren Weaver, George Stibitz, nor their associates as originators of the postwar computing or systems sciences. Like Norbert Wiener, Ivan Getting, and numerous others, they responded to the technologies and organizations in their environment and contributed observations and new ideas. The role played by D-2 and Division 7, however, continues to refute the notion that communications and control merged first inside the head of the mathematician and that digital computers emerged from a single, eureka moment or project. Rather, the relationships between human and machine and between analog and digital, of such concern to prewar engineering cultures of control systems, carried over into wartime problems. An understanding of these relationships, as reformulated by Wiener and a host of others, embeds the appealing rhetoric of cybernetics and digital computing into the technical cultures that surrounded them.

Moreover, the NDRC's attitude toward digital machines further emphasizes the material substrate of the history of computing. The early digital projects at MIT and RCA sought to work from the component level, a technically ambitious choice but not one that could survive in the results-oriented world of NDRC projects. The Penn group's ENIAC sought to innovate simultaneously in components and architecture, but this meant that they had difficulty selling the project and that their architecture was not as innovative as it might have been. After all, the ENIAC project's major contribution to computing theory, John von Neumann's landmark treatise on the stored program architecture, described not the Penn machine but its proposed successor, the EDVAC, which was never built.[94]

George Stibitz succeeded in building wartime computers because he was able to base his architectural innovations on material practice, relying on an established and stable set of components, workers, and procedures to build the machines. All aimed to transcend the dependence on skilled machinists and to design computers that would be robust to variations in the skill level and quality of the workers who assembled them. What set these computers apart from the earlier mechanical machines was their reliance on large numbers of identical parts. Customization, in the form of connections, programming, and data tables, was increasingly removed from the component hardware. In the ENIAC and later computers, electronics did radically separate digital data from mechanical (but not physical) limitations. For D-2 and Division 7, however, information remained classical and Newtonian, tied to switch closures, to cam motions, and to servomechanisms.

# Conclusion

## Feedback and Information in 1945

### Philbrick's Supersimulator

In 1946 George Philbrick articulated a cybernetic vision of his own. Philbrick, a Harvard-educated engineer, had worked on industrial controls before World War II. In 1942, when he became a technical aide to the NDRC's Division 7, Philbrick observed the prevalence of dynamic feedback loops in control systems. "Such causal loops may be entirely automatic in nature," he wrote, "or may contain one or more human elements as an essential connecting link." When considering the human operator, Philbrick wrote, one must consider not only structure, muscles, and reflexes but also the random, noiselike perturbations introduced by the nervous system. "As to weapons, we are concerned here with the airplane, the projectile, and the man at the firing-key, the whole group operating as a unit." Furthermore, the critical task of "tracking" was not limited to fire control but appeared in all human activity. "People 'track' during every conscious moment," Philbrick wrote. "Alignment processes, in which the alignment error serves as datum for its own annihilation, are forever being carried out in the familiar operations of living."[1] As Wiener would do two years later in *Cybernetics*, Philbrick extended the behavior of the antiaircraft operator to human activity in general.

The OSRD closed down when the war ended, and Philbrick's essay appeared in its final publication. The 68-volume *Summary Technical Report of the NDRC* was published as a restricted document in 1946 (Division 7 members also wrote accounts for James Phinney Baxter's official history of the OSRD, *Scientists against Time*). Of the three volumes from Division 7, the first, *Gunfire Control*, edited by Harold Hazen, surveyed the NDRC's broad range of projects. Volume 2, *Range Finders and Tracking*, summarized the extensive work in optics and applied psychology undertaken to support that fire control research. The final volume, *Airborne Fire Control*, contained writings by Philbrick and others.

For Philbrick, new control systems influenced not only human performance but engineering practice as well. He elaborated at length on "simulation as an aid in development" that provided "a means for experimentation in which certain limitations are removed in comparison with the original system."[2] In his "philosophy of models" Philbrick distinguished between a developmental simulator, for creating a technology; a training simulator, for teaching the system's operators; and an educational simulator, for imparting the workings to an engineering student. In each case, the machine replicated in real time the dynamics of the system under study. A user could disturb the system, observe its responses, and develop an intuitive sense for its dynamics without the need for complex mathematics. Philbrick drew no conceptual lines between mechanical, electrical, and symbolic simulations, but he favored electrical machines based on feedback amplifiers because they could run simulations at many times the speed of real phenomena.

Philbrick then proposed a new machine, "an extremely general and flexible assembly, covering every conceivable type of system which could be adapted to any particular problem simply by the manipulation of conveniently provided organizational controls." Today we would call such a machine a computer, the "manipulation of conveniently provided organizational controls" a kind of programming. Philbrick called it a "supersimulator" and pointed to such systems already in development, which allowed engineers "to experience 'electronic' flight in the laboratory, the whole illusion being accomplished by simulative components." (Philbrick was alluding to a navy project at the MIT Servo Lab to build a fully programmable flight simulator; the analog machine was never completed, but the project produced the Whirlwind digital computer).[3] As war became increasingly mechanical and automated, Philbrick wrote, the supersimulator would be capable of modeling entire battles: "Warfare among [guided] missiles could thus be staged and observed entirely in the laboratory, possibly with statistical machinery in attendance for interpretation and assessment. Years of experience, and of trial and error, on the development of controls and dynamic components could thus be collapsed into hours." In 1947 he started a company, George Philbrick Researches, to build analog computers and components for industrial process control. The "Philbrick amplifier," a modular operational amplifier, became a common building block for analog computers.[4] "There is, in reality," Philbrick concluded of his supersimulator, "no limit at all."[5]

Philbrick's essay "Aiming Controls in Aerial Ordnance" was but one of a series in Division 7's *Summary Technical Report* that synthesized the war's experience into postwar technologies and engineering practice. Philbrick's approach, together with his supersimulator, exemplified the emerging visions of computing and control: the importance of feedback, connections to human behavior, infinitely flexible machines to simulate the entire world. It also displayed, however, the continuities between these ideas and the prewar threads of control systems. Hence it is worth reviewing the six theses with which this book began and exploring their implications for the history of feedback, control and computing after 1945.

## Interwar Engineering Cultures

The interactions between people and machinery had been on the agenda of naval gunnery since at least the turn of the century. Fire control systems enacted a cascade of representations by organizing human and mechanical components around instruments of perception, integration, and articulation. Telescopes and rangefinders were used to collect data about the enemy and the environment, which were fed to a Battle Tracer or rangekeeper, which in turn integrated the data, extrapolated into the future, and then drove servos to direct gunfire. Whether in gun directors high above the decks or deep in a plotting room, human operators gathered visual data, integrated and smoothed them into reliable information, and then articulated them to the next stage in the system.

Sperry Gyroscope excelled at these loops in a different realm: individual human-machine combinations amenable to large-scale manufacturing. Steersman observed and tended their Sperry gyropilots, while aviators like Wiley Post exchanged control with autopilots, extending their endurance and range. Connecting local human-machine combinations into larger systems required renewing the data to prevent their decay. Sperry's phantom picked off the position of a gyrocompass without affecting its subtle dynamics. Data transmitters and human follow-the-pointer operators served a similar function in antiaircraft systems. Human beings glued the systems together by feeding back data between successive stages. Sperry called them "human servomechanisms," but they also smoothed and interpreted the data, eliminating jitters and high-frequency noise.

In the telephone network too conversations decayed as they traveled down the line, and then required renewal. Harold Black designed his negative feedback amplifier as a telephone repeater, separating the telephone message from its carrier in electricity and opening the door to a host of new manipulations and combinations. Other Bell Labs engineers, seeking to transmit text, speech, and images down the same wire, defined a common language of signals. Telephone conversations were no longer simply intelligible speech but rather signals with specific characteristics, such as bandwidth, attenuation, and noise. At MIT, servomechanisms renewed data in differential analyzers, coupled together different stages of computation, and allowed extensive combinations of mechanical elements. This cascade led Harold Hazen to see feedback as a general principle across a variety of systems.

Renewing data enabled control systems to become interchangeable and flexible. In fire control, users could dynamically reconfigure a system using synchronous transmission and electrical switchboards. At MIT, the network analyzer could be reconfigured to represent any power network with a plugboard, and telephone switching equipment made the Rockefeller Differential Analyzer into a machine programmable with a paper tape.

This trend toward flexibility was no natural progression; organizational goals and engineering cultures shaped the approaches to feedback and control and evolved accordingly. BuOrd sponsored a specialized, secret cadre of contractors like Ford Instrument to supply its equipment. When G.E. entered the field, it electrified the system and integrated components from a variety of contractors. During the 1930s the Sperry Corporation accumulated subsidiaries devoted to the various components of control systems. AT&T created Bell Labs to capitalize on the success of the transcontinental line. Networks that could transmit all types of information as generic signals promised to free the company from regulatory restrictions and enter new markets. At MIT, technical flexibility enabled analog machines to be cast as scientific instruments rather than engineering problem solvers, giving engineers academic legitimacy, broadening their pool of users, and providing access to new resources like those of the Rockefeller Foundation. Between the world wars, control systems evolved in different ways according to their particular technical and institutional environments.

## Continuities of Wartime Research

Control systems also elucidate the continuities of wartime research with the prewar and postwar worlds. At the Rockefeller Foundation, Warren Weaver used private patronage to draw together disparate sciences. When World War II began, Weaver brought this approach to the NDRC. His fire control committee, Section D-2 of the NDRC, later Division 7, included members from a variety of prewar engineering cultures—Edward Poitras from the Palomar project, Thornton Fry from Bell Labs, Harold Hazen from MIT, and Preston Bassett from Sperry Gyroscope, to name but a few. Weaver also brought administrative practices designed to foster intellectual exchange and to unite these previously disparate engineering cultures.

Weaver's committee began with an interest in fundamental approaches, one related to its institutional position outside the military services. To avoid overlapping with commercial interests and military projects, the committee saw control systems as broader than any particular project or industry, considered the general aspects of the problem, and concentrated, at least initially, on fundamental work. As the war went on, the NDRC drifted toward development, but it had to carve out its role carefully, as other government programs sponsored their own programs. The Mark 14 gunsight, for example, solved a pressing problem in antiaircraft defense with a clever, manufacturable solution, embodying relationships between Draper's and Brown's laboratories at MIT, BuOrd, and Sperry Gyroscope.

An understanding of wartime research programs in fire control helps us to understand the NDRC's influence on postwar science policy in America. Section D-2 and Division 7 borrowed the techniques of prewar private patronage and employed expert program managers with the authority to shape research agendas. After the war, this model was continued by the Office of Naval Research (ONR), and nearly two decades later it was replicated by the Advanced Research Projects Agency (later the Defense Advanced Research Projects Agency, or DARPA). Founded in 1958 in the wake of *Sputnik,* DARPA, like the NDRC, relied on expert program managers rather than peer review to shape research agendas and distribute funds. Because it was attached to no particular military service but reported directly to the secretary of defense, DARPA concentrated on problems, such as computing, that did not obviously fit under any particular service. DARPA fostered numerous innovations in computer science, including computer graphics, human-computer interaction, and the now famous ARPAnet, progenitor of the Internet.[6] A detailed picture of control systems research during World War II helps connect today's world of computers, human interfaces, and networks to a larger history of military-industrial-academic interaction, not only in the Cold War but in earlier periods as well.

## The Antiaircraft Problem

Before World War II, communications and control were gradually converging, but the difficulty of shooting down high-speed aircraft accelerated and solidified their union. Army and navy projects defined "the antiaircraft problem" in the 1920s and '30s and developed adequate but unremarkable solutions by the start of World War II. Engineers who built these systems assumed that aircraft would fly straight and level until they released their bombs. In 1940 Vannevar Bush recognized the inability of these technologies to adapt either to new aircraft and tactics or to the new technology of radar. Similarly, BuOrd realized that the navy's capital ships were exceedingly vulnerable to air attack. During the war, the antiaircraft problem influenced the creation of the NDRC, dominated BuOrd's agenda, and drove control systems research in the United States.

Solving the antiaircraft problem required researchers to combine real-time calculation, mathematical prediction, and human behavior (of both the operator and the enemy) within a noisy, dynamic environment. Engineers rethought the problem around notions of signals and noise. Servos and computing elements no longer simply amplified power or calculated quantities; now they transmitted messages. Gordon Brown's Servo Lab combined their indigenous transient analysis with frequency domain techniques imported from telephony. Similarly, at Bell Labs Donald Parkinson and Clarence Lovell employed telephone amplifiers to build a gun director with electromechanical servos. They too applied Black's, Bode's, and Nyquist's theories and techniques of feedback and stability. When Ivan Getting and his group at the Radiation Lab were designing the tracking servos for their SCR-584 radar, they used communications theory to describe the noisy echoes from aircraft and to sort valuable data from damaging noise. Norbert Wiener introduced statistics and autocorrelation to the prediction problem and showed how to design optimal feedback loops for particular parameters. Despite these efforts, the antiaircraft problem proved unsolvable for systems with ballistic shells; the controls had to move inside the projectile and continue to direct it during flight. This process began with the proximity fuze during the war and continued with the development of guided missiles.

After the war, new variants of the antiaircraft problem continued to raise pressing technological questions. In the 1950s the Servo Lab built the Whirlwind digital computer, originally intended as a flight simulator (Fig. 12.1). Inspired in part by Perry Crawford's work on "arithmetical" fire control, Jay Forrester and his group soon re-

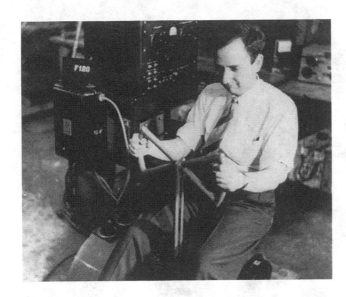

Fig. 12.1. Robert Everett, of the MIT Servo Lab, in the control seat for the Aircraft Stability and Control Analyzer, a flexible, analog simulator that transformed into the Whirlwind computer. Courtesy of MIT Museum.

cast the project as an antiaircraft computer. Whirlwind spawned the SAGE air defense system, which one Air Force colonel called "a servomechanism spread over an area comparable to the whole American Continent."[7] SAGE brought fire control into the world of digital electronics, information processing, and national systems. It still had to solve the old problems of tracking targets, smoothing signals, predicting positions, and directing weapons—problems made more difficult by ballistic missiles and supersonic bombers.[8] SAGE operators used "light guns" to identify targets, recalling the pointer matching and pip matching of earlier generations (Fig. 12.2). Indeed, the problems of tracking, predicting, and shooting down aerial attackers are with us still, though in different forms. From Ronald Reagan's Star Wars dream, to the controversies over the Patriot missile in the Gulf War, to current debates about ballistic missile defense, to the new threat of terrorists turning commercial aircraft into guided bombs, what Vannevar Bush called "the antiaircraft problem" continues to influence our technological world.

## The Diversity of Post-war Practice

Despite their breakthroughs and the synthesis of disparate threads, World War II's control systems projects did not culminate in a single, unified vision. Engineers took numerous paths. Ivan Getting spent the 1950s as a vice president at Raytheon and then

Fig. 12.2. Operators at consoles for the SAGE air defense system in the late 1950s. Note the light guns, used for target identification, similar to the pip matching of World War II. Courtesy of MITRE Corporation Archives.

became the first president of the Aerospace Corporation, an organization that grew out of the defense contractor Thompson, Ramo, & Woolridge (TRW) in 1960 to do system engineering for the air force. There he envisioned a system of satellites for radio navigation and created the Global Positioning System (GPS), which grids the earth today.[9] Bell Labs' fire control group, which included Hendrik Bode and Walter Mac-Nair, built the Nike series of antiaircraft missiles, which was based in part on their wartime work. Bode played a prominent role on Cold War scientific advisory committees and finished his career as a professor of systems engineering at Harvard.[10] The Servo Lab's work with the Whirlwind computer and MIT's Digital Computer Laboratory and its numerous descendents, such as the SAGE Air Defense System Lincoln Laboratories, and the MITRE Corporation, are well known. Albert Hall also started his own descendent of the Servo Lab: the Dynamic Analysis and Control Laboratory, which designed and simulated analog control systems for guided missiles. Gordon Brown, as dean of engineering at MIT, became convinced that the Radiation Lab's research style, with its heavy emphasis on physicists, was superior to that of his own Servo Lab, which was run by engineers. Hence he ushered in the "university polarized around science,"[11] which defined MIT's science-based approach to engineering education for the rest of the century. Perry Crawford left the navy in 1952 for IBM, where he spearheaded SABRE, an adaptation of real-time command and control systems to automate American Airlines' national ticketing operations. In the navy, the gun club became the nuke club: BuOrd chief William Blandy, with aid from Horacio Rivero, in 1946 directed Operation Crossroads, in which the navy tested its first atomic bombs.[12]

Several textbooks published soon after the war also defined control from diverse perspectives. In 1944 Warren Weaver asked LeRoy MacColl, of Bell Labs, to approach the subject as a mathematician, and in his 1945 book (which included an ugly, misogynistic foreword by Weaver) he rigorously applied Bode's and Nyquist's techniques to electromechanical servo design. Other texts emerged from the Radiation Lab (Nichols, James, and Phillips), the process industries (Ahrendt and Taplin), and the Servo Lab (Brown and Campbell). Professional activities followed a similar trend, as numerous different engineering societies created subgroups on automatic control. Because of these diverse homes, control engineers became increasingly concerned with standardizing language and terminology.[13] The theoretical insights of the 1930s and '40s coalesced into "classical," linear control theory, as feedback, control, and computing informed a wide variety of engineering practices. It also continued to evolve, particularly into nonlinear realms, optimal control, and "state-space" methods—directions heavily influenced by Cold War priorities for controlling missiles and spacecraft.[14]

Control engineering was but one of several descendents of wartime work in feedback, control, and computing. Fields like cybernetics, systems engineering, automation, systems dynamics, inertial guidance, and command and control typified the multiple, overlapping legacies of wartime control systems research. They joined operations research, systems analysis, game theory, automata theory, and a host of others as the postwar systems sciences. Each emphasized, in its own way, feedback, systems, dynamics, computing, and modeling of complex systems. Following Wiener's exhortations, in the decades after the war not only engineers but also economists, anthropologists, and social scientists took up the lessons of feedback, stability, and interconnected systems, applying them to everything from global ecology to urban transportation. Common to these arenas was not so much a particular theory or methodology as the idea that various aspects of the world could be understood as systems and modeled as flows, feedbacks, and human-machine interactions. The language of feedback, control, communications, and information proved enormously flexible and adaptable; the historian Vyacheslav Gerovitch calls it "cyberspeak," comparing it to the "newspeak" language of Soviet Stalinism.[15] As Ivan Getting realized when he proposed the Radiation Lab as a "systems integrator," seeing the world at this high level of abstraction bought these engineers and systems scientists purview over broad arenas.

Popular images appeared as well, in part as the result of proselytizing by former NDRC researchers like Norbert Wiener, Gordon Brown, Warren Weaver, and Louis Ridenour. A famous cover of *Time* magazine portrayed cybernetics as an anthropomorphic automatic computer in a military uniform, reading its own output. Debates over the allocation of tasks between people and machines crystallized into the term *automation,* which arose to capture the potentials of self-acting machinery and its social and industrial implications. As David Noble has pointed out, numerically controlled (N/C) machine tools, developed by the MIT Servo Lab, among others, brought computers and control into age-old struggles between managers and skilled machinists. Kurt Vonnegut's first novel, *Player Piano,* explored the social dislocations that might result from large-scale use of digital computing and industrial automation.[16] In Stanley Kubrick's *Dr. Strangelove,* control rooms set the stage for a play about the end of the world in which erratic humans battled unerring automatic systems for control of nuclear arsenals. In music, the solid-body electric guitar emerged in the 1950s to reduce feedback through an amplifier, but musicians began incorporating the unusual sounds into their work. Jimi Hendrix brought feedback into the counterculture with his performance of "The Star Spangled Banner" at Woodstock in 1969.[17]

## Control Systems and the History of Computing

From at least the 1920s, control mechanisms called computers connected people and machines, coalesced into larger projects, and integrated broad networks. This history of control systems, rather than framing computers solely as logic machines, reattaches them to the problems, engineers, and materials from which they sprang. Historians tend to write of George Stibitz's relay computers, for example, as the Model I, followed by the Model II, which led to the Model III, and so on. By seeing computers as control systems, we find him building the complex number computer, followed by the tape dynamic tester, the Relay Interpolator, and the Antiaircraft Artillery Board computer. The more evocative names of his machines direct our attention to the fact that they solved particular historical problems rather than just running numbers in a laboratory. Similarly, MIT's Whirlwind computer did not emerge, Athena-like, from a laboratory previously committed to other technologies. Rather, Whirlwind's engineers had been trained in the Servo Lab designing gun mounts and servos as part of fire control systems—the source of their emphasis on feedback, real-time operation, and human-machine interaction.

Embedding computers in their technical cultures also draws our attention to their material basis, for the designs of machines determine who will build them, in what factories, with what skills: from the manufacturing issues that beset World War I startups like Sperry Gyroscope and Ford Instrument, to the Ford Motor Company's attempts to build mechanical antiaircraft computers, to George Stibitz's use of telephone wiring techniques in his relay computers. The story continues after World War II, with the introduction of transistors, printed circuits, modular circuit elements, and of course integrated circuits. Indeed, what is Moore's Law—the idea that the power of computers doubles every 18 months—but a statement that progress in computing depends on progress in manufacturing? A history that includes materials, techniques, and industries reveals modern computing as part of a larger story of technology and culture, rather than the product of a discontinuous break between old and new.

## Analog and Digital

With his idea of "technics" Lewis Mumford captured the idea that machines create and manipulate representations of the world. Indeed, the very term *analog* implies that a physical quantity stands for something else. Similarly, *digital* alludes to counting the world on our fingers. These two methods of representation reappear throughout the history of control systems, countering the view of the transition from analog to digital computing as a simple succession of superior technology. Analog and digital, in fact, emerged together, in a subtle and lengthy evolution of engineering techniques for modeling the world and its dynamics.

Between the world wars, engineers developed methods for interchanging data, as both analog quantities and digital symbols. Vannevar Bush, for example, employed the common principle of the circuit: any system that flowed could be represented by electrical diagrams. For modeling such circuits, Bush and Hazen saw continuous electrical computation as an improvement over the numerical methods employed in business machines. Yet to make their machines flexible, Bush and his students employed switches—manual plugboards and telephone relays—to reconfigure the computations.

Theorists like Harry Nyquist and Ralph Hartley studied the translations between the telephone's waves and the telegraph's discrete states, establishing the equivalence between the continuous world and the 1s and 0s of electrical pulses. Similarly, Claude Shannon, when studying how to set up an analog problem with a punched paper tape and a series of relay closures, synthesized network theory and Boolean algebra into a theory of switching. The switch designer George Stibitz immediately took up the ideas and, leaving the continuous components aside, built calculators and interpolators out of telephone switches. Although designed for fire control applications, these machines adapted to a variety of mathematical tasks. Again, flexible machines enabled flexible organizations.

Gradually, researchers articulated the dichotomy between "continuous" representations and "numerical" or "arithmetic" ones. The terms *analog* and *digital* appeared nearly simultaneously, as NDRC members debated their relative merits, along with those of mechanical, electromechanical, and electronic representations. Which combination worked best for any given application depended not only on technical logic but on a host of practical and material questions: How large, and how reliable would the machine be? What kinds of components would the machine require? What kinds of skills were required to make it? How fast, and how accurate, were the input and output mechanisms? For George Stibitz, "speed and accuracy" described not only the workings of digital machines but the workers who built them as well.[18] Questions of analog and digital in computers concerned the practical processes by which the world was represented in machines and by which those representations were made effective in the world.

A complete history of analog computing has yet to be written, but it would show that analog computers matured in parallel with digital computers, not before them. The heyday of analog computing was the 1950s and '60s and focused on real-time simulation.[19] Indeed, the philosophy of analogs survives today—whenever we run a simulation on a digital computer, or compose thousands of bytes into an image, or move a joystick to control a vehicle (or drive a mouse, for that matter). Modern software packages for engineering simulation, for example, have menus of building blocks that resemble those in George Philbrick's analog computers, and even evoke the mechanical circuit elements of Bush's differential analyzer.[20]

## Data Smoothing and Prediction

One final example illustrates these six theses and exemplifies the convergence of communications and control. Complementing Philbrick's notes on simulation, another paper in Division 7's *Summary Technical Report* discussed the translation of the world into transmissible signals. "Data Smoothing and Prediction in Fire-Control Systems," by Richard B. Blackman, Hendrik Bode, and Claude Shannon of Bell Labs, examined the problems of tracking and predicting moving targets and made explicit the emerging connections between feedback, control, and computing.

For Blackman, Bode, and Shannon, fire control raised a problem of electrical communications and hence required an analysis "couched entirely in electrical language." The authors acknowledged the importance of Norbert Wiener's work and devoted significant effort to summarizing his statistical approach. Ultimately they rejected it, however, due to problems in applying the rms-error criterion to fire control, as well as its assumptions about statistical behavior of human pilots. Instead, the paper used terms from communications engineering to analyze fire control, drawing heavily on Bode's work in feedback theory. "There is an obvious analogy," they wrote, "between the problem of smoothing the data to eliminate or reduce the effect of tracking errors and the problem of separating a signal from interfering noise in communications systems." Hence tracking was ultimately a problem of filtering: separating the signal "or true [flight] path" from the noise of "tracking errors." While noting that "this analogy . . . must of course not be carried too far," the paper considered inputs and disturbances in fire control systems as signals in the frequency domain.

As Wiener had done in *Extrapolation, Interpolation, and Smoothing of Stationary Time Series*, Blackman, Bode, and Shannon broadened the relevance of their study beyond fire control, treating it as "a special case of the transmission, manipulation, and utilization of intelligence." Like Wiener, they pointed out that predicting time-series data had broad application to "weather records, stock market prices, production statistics, and the like."[21]

Just a few years later Claude Shannon expanded these ideas in his 1948 paper, "A Mathematical Theory of Communication," which famously created information theory.[22] Following Nyquist's and Hartley's analyses years earlier, Shannon defined the act of communication as transferring a given message, or a series of symbols, from one place to another (from one person to another or from one machine to another) through a noisy channel of finite bandwidth. He provided a measure of channel capacity, which led to a theory of efficient coding, and techniques for maximizing the available capacity of a channel. After initial publication in the technical literature, Shannon's theory was published as a small book that included a popular introduction and explication of information theory by Warren Weaver. Weaver not only explained the theory but also suggested wide implications beyond electrical communications.[23] Shannon had thought about this topic since at least 1939, he drew on his own experience in cryptography during the war, and he cited Wiener's paper. But as Weaver's introduction and the article by Blackman, Bode, and Shannon show, Shannon's theory also carried the traces of fire control.[24] Information now became the common currency that AT&T had sought in the 1930s, carrying generalized signals and messages through the continuous world as discrete pulses.

PHILBRICK'S supersimulator and Shannon's information theory highlight the state of feedback, control, and computing at the end of World War II. It is not a story of firsts or of exclusivity; no single engineer, organization, or invention made all the difference. Others did similar, important work in other countries and contexts. Rather, the significance of this history depends on its coherence and continuity, and on broad, diverse communities' remaining connected over long periods. When Norbert Wiener wrote in *Cybernetics* that in 1942 "it had already become clear to Mr. Bigelow and myself that the problems of control engineering and of communication engineering were inseparable, and that they centered not around the techniques of electrical engineering but around the much more fundamental notion of the message," he was one of many engineers and scientists coming to such conclusions.[25] They had their own interests before the war, they attacked problems with both analog and digital elements, and they sought to develop techniques for representing the world as signals, manipulating those signals, and then translating them back into the world. Whether conceiving a general-purpose simulator, designing a digital control computer, or compressing signals into transmission channels, by the end of World War II engineers had begun to describe the world with the language of signals, noise, and information.

This history of feedback, control, and computing before cybernetics not only chronicles these engineering developments but also reconfigures our historical categories. Until now, historians of computing have concentrated on hardware, biographies, institutions, and cultural representations. Now we might address modeling and simulation, machine representation, or the role of manufacturing. Other topics suggest themselves as well: the importance of training, the evolution of user interfaces, the creation of the human operator. This book begins to examine these topics and to suggest the relationships between them, but each extends beyond the present scope. More remains to be done.

TODAY we sit in front of screens, manipulating words, images, even sound and moving pictures, and transmitting them as equivalent data over common channels. As Mumford noted, our technics serve to "dissociate" these representations from their referents in the world. That dissociation is a historical process, carried on at particular times for particular reasons. Our computers retain traces of earlier technologies, from telephones and mechanical analogs to directorscopes and tracking radars.

As users, too, we inherit the legacies of machine operators from earlier ages. When we articulate a mouse to direct a machine, do we not resemble Sperry's pointer-matching human servomechanisms? When we interpret glowing images and filter out signals from noise, do we not resemble a pip-matching radar operator? The "user," "computer operator," or even "net surfer" of today is no recent invention but a historical, technological descendent, an aggregate that includes pilot, machinist, human computer, telephone operator, radar tracker, fire control officer, and antiaircraft gunner.

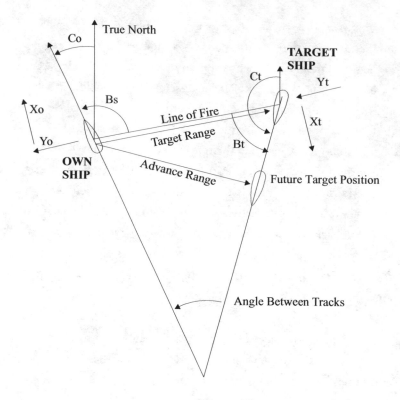

Co - Own Ship's Course  Xo - Component of Own Ship's Speed Normal to Line of Fire
So - Own Ship's Speed  Yo - Component of Own Ship's Speed Along Line of Fire

Ct - Target's Course  Xt - Component of Target Ship's Speed Normal to Line of Fire
St - Target's Speed  Yt - Component of Target Ship's Speed Along Line of Fire

Bt - Target Angle  Bs -  Relative Target Bearing
 (Target's course w/
  respect to line of fire)

Fig. A.1. Measurement of angles in the Ford Rangekeeper.

# Appendix A

# Algorithm of the Ford Rangekeeper Mark 1

The own ship has a course and speed *(Co* and *So)*, as does the target *(Ct* and *St)* (Fig. A.1). The essential problem is to determine the course and speed of the target ship. Consider an imaginary line, the *line of fire*, that connects the two ships, which could also be called the *line of sight.* The length of this line, of course, is the *target range.* The course of the target with respect to the line of fire is known as the *target angle,* or *Bt.* Consider, then, that each ship has two velocity components relative to the line of fire, *Xo* and *Yo* for the own ship, *Xt* and *Yt* for the target ship. *Y,* for both ships, indicates the component of the ship's motion along the line of fire (in the direction of range). The sum of *Yt* and *Yo* is equal to the time it takes the two ships to close or open in range; hence it is the *rate of change of range. Xo* and *Xt* are the components of the ships' motion normal to (at right angles to) the line of fire, and their sum is known as *deflection.* To hit a moving target, one would not aim the gun directly along the line of sight, but at some angle ahead of it, and that is *deflection.*

Figure A.2 shows the basic layout of the Ford Rangekeeper's data flow and algorithm.* Starting at the top, the course and speed of the own ship come into the machine from a gyrocompass repeater and a revolution counter, respectively. Then initial guesses for the target's course and speed are cranked in by hand, usually taken from a Battle Tracer or hand plot (1); the own ship's course is subtracted from the target bearing to give target course (2). The two *component solvers* resolve these data into their components relative to the line of fire, that is, *Xt, Yt,* and *Xo, Yo.* Differential gears subtract these components from each other to give *dR,* the change in range, and *RdB,* the change in deflection (3). Then *dR* goes into an integrator, which produces *R,* or

---

*This description of the Ford Rangekeeper is compiled from data in Ford's original proposal for the device, Hannibal Ford, "Ford Range and Deflection Predictor," 15 May 1915, RG 74, E-30, box 696, subject file 30199, as well as from H. M. Terril, "Notes on the Theory of the Ford Range Keeper," print issued by the USS *New Mexico* Gunnery Department, ca. 1919, RG 38, entry 178, box 3, Conf. 59 (65) folder, courtesy John Tetsuro Sumida and Christopher Wright, and from "Rangekeeper, Mark I, Mod. 3," in U.S. Naval Academy, *Fire Control Installations,* 1939.

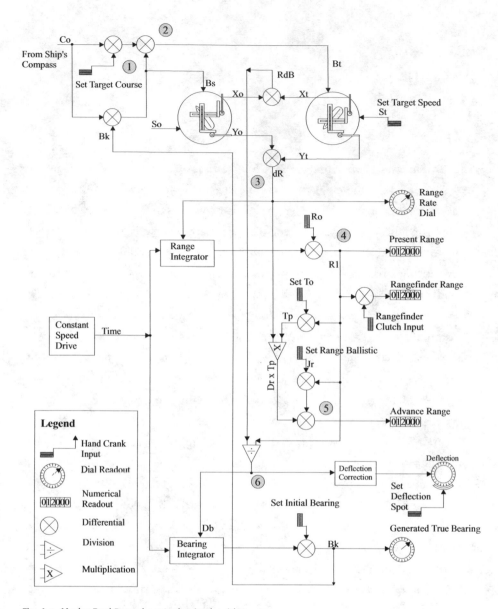

Fig. A.2. Mark 1 Ford Rangekeeper, basic algorithm.

range, which varies linearly at a rate determined by $dR$. But this $R$ is only an incremental range, so it needs to be added to an initial observed range, $Ro$, to produce an accurate *present range*, which is read off a numerical dial (4). To calculate *advance range*, the range of the target at some time in the future, the range rate, $dR$, needs to be multiplied by that time interval. The time interval, $Tp$ (time of prediction), consists of two factors. First is the time of flight of the shell, the amount of time it takes the shell to reach the target after firing. The Ford Rangekeeper treats the time of flight as linearly proportional to range, which is only an approximation. The second component of $Tp$ is the amount of time between when the advance range is read off the dial and when the gun is actually fired, or the *transmission interval*. This quantity includes delays in data transmission to the turret, loading the shell, elevating the gun, and so on. The transmission interval, $To$, is cranked in by hand. The sum of the time of flight and the $To$ is then multiplied by the range rate, which is then added to the previously computed present range to derive the advance range (5). The advance range can be adjusted up or down manually depending on spotting corrections. Thus, if the spotter sees that a shot falls 100 yards short, $Jr$ can be manually entered to subtract from the calculated advance range. $RdB$, or the rate of change of deflection, is then divided by $R$, the calculated range, to derive the change of bearing, $dB$ (6), which is then integrated to produce bearing, which, when added to initial bearing, calculates a *generated true bearing*.

A final mechanism derives the deflection. Deflection itself is expressed in knots, as the speed at which the target ship is sailing perpendicular to the line of fire. For a given range, this can be converted to an angle, the amount off the line of sight the gun needs to aim, and depends on four factors: (1) the target's change of bearing during the time of flight; (2) *drift*, the tendency of the trajectory to curve to the right due to the fact that the projectile is spinning; (3) wind; and (4) spotting and ballistics. The mechanism, a set of cams and multipliers, takes as input the present range, from which it calculates a time of flight, and the deflection rate, or $RdB$. $Jd$, the spotting correction ("left" or "right" a certain number of yards), is entered by hand. The system then calculates drift via another cam, and outputs on a dial the deflection setting for the guns.

Although elegantly executed, with the exception of the improved integrators these calculations are not qualitatively different from those in the British fire control machines. A key innovation of the Ford Rangekeeper, however, comes in the final stage of output. This setup allows direct comparison of the "guessed" quantities entered in the beginning of the calculation and the calculated quantities produced by the machine. A cartoon of the target ship appears on a rotating dial. Its angle indicates the target ship's bearing, and a small "button" within the cartoon itself indicates the speed of the ship, as entered in the initial estimation. Ford's innovation was to derive range rate back out of the calculated advance range as it changes. This he accomplished with an ingenious and subtle use of feedback, connecting the input and the output of an integrator together through a differential. If the speed of the shafts coming into the differential are the same, the output shaft does not turn at all, and the mechanism is in a kind of equilibrium. Since one of these shafts represents advance range, and the other the output of the integrator, they will only turn at the same rate when the integrator is adjusted so that the output speed exactly matches the rate at which the advance range is changing. Otherwise, the output of the differential changes the position of the balls on the integrator, moving them toward the equilibrium position. At the equilibrium point, the position of the balls is proportional to the rate of change of the advance range. Thus, by taking a shaft rotation as an input and producing an output proportional to its speed, this arrangement acts as a differentiator (like a tachometer). Through a feedback loop, Ford inverted the function of the integrator—an accomplishment not repeated in other fields until a decade later.

The ship's own speed along the line of fire, $Yo$, is then subtracted from the output of the mechanism, producing the target's speed. This *calculated target component along the line of fire*, or the rate at which the range to the target is changing, then drives a needle, called the *horizontal wire*, that overlays the indicator that reads the observed speed button. Thus, another feedback loop is set up, this one involving the human operator. The human operator looks at the needle and the button, which implicitly compares the estimated values of target speed and course, with the calculated quantities based on other observations and integration. The operator's job is to reduce the "error" indicated by the distance between the button and the needle, which he does by adjusting the estimates of target speed and course accordingly. This cycle of correction continues until the dials and needles match up. At that point the rangekeeper has converged on a solution for the target's course and speed that matches both the estimates and observations, and the predicted advance range will be accurate for setting the guns.

A similar cycle works for bearing. An integrator converts the rate of change of bearing into an incremental bearing, which, when added to the initial observed bearing, produces a generated true bearing. This reading drives a needle called the *gold pointer* on a bearing scale. On the same scale, a *white pointer* reflects the observed present bearing, as indicated by the target bearing instrument. The operator then observes the difference between these two. Because these bearings may differ only slightly, a *vertical wire* exaggerates the difference to make it easier for the operator to read. Where the horizontal wire indicates errors in $Yt$, the white pointer and the vertical wire indicate a need to correct $Xt$. It is worth noting, however, that the range comparison is based on advance range, the result of a prediction into the future, whereas the bearing comparison works off present bearing. The first model of the Ford Rangekeeper made no bearing predictions.

# Appendix B

# NDRC Section D-2 and
# Division 7 Contracts for Fire Control

Division 7 contractors developed several new gun directors, building on the Sperry systems or taking entirely new approaches (Projects 2, 30). A combined project of Division 7, the Radiation Lab, and G.E. built an integrated control system for naval fire control (71, 79, 85, 86). One project modified the Sperry directors, adding features and integrating them with radars for which they were not designed (51). The Bristol Company designed plotting boards to lay out the geometry of an engagement on paper similar to the plotting boards used in naval gunnery (64). Several projects addressed fundamental or theoretical aspects of the fire control problem, including new types of prediction (4, 11, 12, 78), simplified mechanisms (68), or controllers for new types of guns. Many studied or improved optical rangefinders for various types of directors and gunsights. The Barber Coleman Company modified the British M-5, or "Kerrison," director for easier production and put together conversion kits to update the units in the field so that they could achieve higher performance (31). Western Electric similarly modified the Sperry M-7 (51).

Section D-2 and Division 7 put a great deal of effort into instruments of perception, improving the classical methods of optical ranging and tracking. Polaroid developed a "short base" rangefinder for use with small guns or aboard an airplane (32). Barber Coleman combined tracking telescopes and rangefinders into a single unit (52). Eastman Kodak and Bausch & Lomb studied improved optics and better geometry for ranging devices. They found a major source of error to be optical distortions caused by the temperature differential of air within the sight itself. Filling the devices with helium markedly improved their accuracy; the American Gas Association provided its testing lab for this work and designed seals and pressure relief valves for containing the gas (41). Other sources of error included haze, camouflage on targets, low light levels, misalignment, and bad calibration. Some studies considered optical design, reticule patterns, illuminated reticules, and eyepieces (44, 58).

Lead-computing sights moved the gunner's reticule to automatically lead the target for gunners defending against close-in attack (61, 73). The McMath-Hulbert

TABLE B.1  NDRC Section D-2 and Division 7 Contracts

| Number and Title | Contractor | Investigator | Supervisor | Dates | Cost ($) |
|---|---|---|---|---|---|
| 1. Servomechanisms | MIT | Hazen | EJP | 11/1/40 to 9/1/41 | $6,721 |
| 2. Electrical Director | Western Electric/Bell Labs | Fletcher | DJS | 11/6/40 to 9/30/43 | 224,468 |
| 3. Methods of Improving Optical Rangefinders | Cal Tech | Bowen | TCF | 12/1/40 to 1/1/43 | 127,500 |
| 4. Geometrical Predictor | Cal Tech | Bowen | DJS | 12/1/40 to 2/1/42 | |
| 5. | | | | | |
| 6. General Mathematical Theory of Prediction and Applications | MIT | Wiener | WW | 12/1/40 to 1/31/43 | 28,209 |
| 7. Mathematical Studies Relating to Fire Control | Princeton | Shannon | WW | 12/1/40 to 10/1/41 | 3,044 |
| 8. Studies of Fire Control Equipment and Personnel | Princeton | Flood | TCF | 12/1/40 to 1/31/43 | 271,509 |
| 9. Mathematical Studies | U. of Wisconsin | Skolnikoff | WW | 2/1/41 to 8/15/41 | 11,730 |
| 10. Psychological and Physiological Factors of Importance in Fire Control | Tufts | Carmichael | TCF | 3/1/41 to 6/30/42 | 89,586 |
| 11. Fundamental Director Studies | Western Electric/Bell Labs | Fletcher | DJS | 2/10/41 to 11/30/45 | 166,062 |
| 12. Prediction Devices | Iowa State College | Atanasoff | TCF | 3/1/41 to 11/2/42 | 28,168 |

| Project | Organization | Person | Initials | Dates | Amount |
|---|---|---|---|---|---|
| 13. Height Finder (Mihalyi) | Eastman Kodak | Bishop | TCF | 3/1/41 to 5/1/43 | 39,909 |
| 14. Optically Tracked Radio Range Finder (Mickey) | Western Electric/Bell Labs | Bown | SHC | 4/4/41 to 4/4/42 | 38,324 |
| 15. Hydraulic Controls for Small Caliber Guns | United Shoe Machinery | Roberts | EJP | 2/1/41 to 3/31/42 | 50,000 |
| 16. Hydraulic Servos | United Shoe machinery | Roberts | EJP | 3/1/41 to 3/1/42 | 24,388 |
| 17. Fire Control Research | Eastman Kodak | Bishop | TCF/DJS | 6/1/41 to 11/30/45 | 24,364 |
| 18. | | | | | 1,195,604 |
| 20. Data Transmission System (Seacoast) | Western Electric/Bell Labs | Clark | EJP | 5/1/41 to 2/1/42 | 56,137 |
| 23. Statistics of Train Bombing | Princeton | Williams | WW | 1/1/43 to 8/31/44 | 123,503 |
| Bombardiers Calculator | Columbia | Williams | | | |
| | U. of California | Neyman | | | |
| 25. Dynamic Tester | Barber Coleman | Lilja | HLH | 8/10/41 to 8/31/45 | 64,779 |
| 26. Simplified Electrical Predictor | GM Labs | McMaster | DJS | 9/26/41 to 2/3/42 | 4,113 |
| 27. Servomechanisms | Barber Coleman | Lilja | EJP | 4/1/43 to 12/31/44 | 9,546 |
| 28. Intermediate Director | Barber Coleman | | | | |
| 29. Extrapolation, Interpolation and Smoothing of Stationary Time Series | MIT | Wiener | WW | 2/1/42 to 10/31/42 | 2,000 |
| 30. Electrical Director (BTL II) | Western Electric/Bell Labs | Fletcher | DJS | 11/10/41 to 11/30/45 | 600,183 |
| 31. Simplified Director (Type BC) | Barber Coleman | Lilja | HLH | 10/15/41 to 10/31/41 | 87,000 |
| 32. Short Base Range Finder | Polaroid | Land | TCF/DJS | 12/1/41 to 8/31/44 | 140,000 |

(continued)

TABLE B.1    NDRC Section D-2 and Division 7 Contracts

| Number and Title | Contractor | Investigator | Supervisor | Dates | Cost ($) |
|---|---|---|---|---|---|
| 33. Air-Borne Fire Control Equipment | Franklin Institute | McClarren | SHC | 2/1/42 to 10/31/45 | 1,070,000 |
| 34. "Pilot Model, Data-Transmission System" | Leeds & Northrup | Quereau | EJP | 2/1/42 to 6/30/43 | 29,521 |
| 35. Improvement of Servo for 37 and 40mm Guns | MIT | Brown | EJP | 2/1/42 to 4/30/43 | 41,973 |
| 36. Effects of Fatigue on Space Perception | Dartmouth | Pearson | TCF | 2/1/42 to 3/31/43 | 13,500 |
| 37. Effectiveness of Controls and Data Presentation | Foxboro | Bristol | TCF | 3/10/42 to 8/31/45 | 247,112 |
| 38. Rocket Director Development | Bristol | Bristol | DJS | 4/15/42 to 6/30/44 | 50,387 |
| 39. Computations | Franklin Institute | Allen | WW | 3/27/42 to 8/31/44 | 24,701 |
| 40. Gyroscopic Director | U. of Michigan | McMath | EJP | 5/15/42 to 11/30/45 | 75,000 |
| 41. Helium Retentivity | AGA Testing Lab | Conner | TCF/PRB | 5/1/42 to 8/31/45 | 64,391 |
| 42. Relation Between Fatigue and Tracking | Tufts | Carmichael | TCF | 7/1/42 to 7/31/43 | 204,000 |
| 43. Acuities in Telescopic Vision | Harvard | Holway | TCF/PRB | 7/1/42 to 12/31/45 | 142,994 |
| 44. Emotion in Military Performance | Brown | Graham | TCF/PRB | 7/1/43 to 8/31/45 | 38,999 |
| 45. Stereoscopic Acuity Reticle Design | Ohio State | Bridgman | TCF | 7/1/42 to 11/30/43 | 14,767 |
| 46. Servos for Medium-Caliber Guns | Westinghouse | Wolfert | EJP | 5/25/42 to 2/29/44 | 81,438 |
| 47. Air Warfare Analysis | Columbia | Hotelling | WW | 7/1/42 to 8/31/44 | 513,000 |

| Project | Company | Person | Initials | Dates | Amount |
| --- | --- | --- | --- | --- | --- |
| 48. Electronic Computing Devices for Predictors | RCA | Zworykin | DJS | 7/1/42 to 12/31/42 | 20,000 |
| 49. Fire Control Analysis Device | Stanolind Oil & Gas | Silverman | SHC | 7/25/42 to 4/30/43 | 10,127 |
| 50. Testing Plane to Plane Fire Control Equipment | U. of Texas | LaCoste | SHC | 9/1/42 to 11/30/45 | 935,000 |
| 51. Modification of M-7 Director for Field Conversion | Western Electric/Bell Labs | Fletcher | DJS | 9/1/42 to 2/1/43 | 29,945 |
| 52. Combined Tracking and Rangefinding Devices | Barber Coleman | Lilja | HLH | 9/1/42 to 8/31/44 | 12,978 |
| 53. Torpedo Director | IBM | Daly | SHC | 9/1/42 to 3/1/43 | 34,954 |
| 54. Anti-Aircraft Fire Control Testing | U of North Carolina | Ruark/Shearin | DJS | 3/10/42 to 10/31/45 | 124,880 |
| 55. Gyroscopic Computer (Pneumatic) | Wilcolator | Taplin | EJP | 10/15/42 to 6/30/43 | 4,954 |
| 56. Invar Bar for M-2 Height Finder | Bausch & Lomb | Bausch | TCF | 10/1/42 to 5/1/43 | 2,392 |
| 57. Air-Borne Gunnery Computers | G.E. | Bowman | SHC | 11/1/42 to 9/30/45 | 165,000 |
| 58. Range Finder Redesign | Eastman Kodak / Bausch & Lomb / Keuffel & Esser | Bishop / Bausch / Keuffel | TCF/TD | 2/1/43 to 3/31/45 | 222,143 |
| 59. Anti-Tank Director | Barber Coleman | Peterson | EJP | 11/1/42 to 2/28/45 | 21,174 |
| 60. Punched Tape Dynamic Tester | WE/BTL | Seibel | DJS | 11/10/42 to 10/31/45 | 376,094 |
| 61. Anti-Aircraft Computing Sight | Pitney-Bowes | Bernart | DJS | 12/1/42 to 5/31/44 | 25,000 |
| 62. Improvement of Differential Analyzers | U. of Pennsyvania | Brainerd | WW | 12/1/42 to 8/31/44 | 18,500 |

*(continued)*

TABLE B.1    NDRC Section D-2 and Division 7 Contracts

| Number and Title | Contractor | Investigator | Supervisor | Dates | Cost ($) |
|---|---|---|---|---|---|
| 63. Data Recorder | Western Electric/Bell Labs | Dow | DJS | 11/1/42 to 12/31/44 | 179,800 |
| 64. Chart Type Data Smoother and Retransmitter | Bristol | Waidelich | DJS | 1/10/43 to 8/31/44 | 57,305 |
| 65. Muzzle Velocity Instrument | Westinghouse | Hanna | IAG | 1/1/43 to 10/31/45 | 31,634 |
| 66. Tank Fire Control | Bausch & Lomb | Bausch | TCF | 2/1/43 to 8/31/44 | 11,622 |
| 67. Vector Gun Sight & Assessing Camera | Jam Handy Org. | Campbell | SHC | 4/1/43 to 9/30/45 | 115,820 |
| 68. Mechanical Director | Byrant Chucking | Rose | DJS | 7/1/43 to 10/31/45 | 63,874 |
| 69. Steering Mechanism for Torpedoes | Foxboro | Howe | EJP | 8/1/43 to 9/30/45 | 50,000 |
| 70. Relay Interpolator | Western Electric/Bell Labs | Dow | WW | 7/1/43  to 8/31/44 | 22,075 |
| 71. Gyro Unit for M56 Director | G.E. | Coutant | IAG | 8/1/43 to 8/31/44 | 48,639 |
| 72. Torpedo Director | G.E. | Coutant | IAG | 7/1/43 to 9/30/45 | 39,227 |
| 73. Course Invariant Sights | Baker Mfg. | Baker | DJS | 8/1/43 to 11/30/45 | 80,000 |
| 74. AA Board Computer | Western Electric/Bell Labs | Dow | WW | 9/1/43 to 9/30/44 | 108,220 |
| 75. Mechanism to Measure the Smoothness of Control of Aircraft Turrets | Waugh Labs | Roy | EJP | 11/20/43 to 8/31/45 | 16,117 |

| | | | | | |
|---|---|---|---|---|---|
| 76. Fire Control Electronics | Columbia | Ragazzini | SHC | 11/15/43 to 9/30/45 | 85,000 |
| 77. Redesign of Gun Directors Mk 49 | MIT | Edwards | ALR | 11/1/43 to 8/31/44 | 8,773 |
| 78. Second Derivative Curvature Attachment for M9 Director (T17) | Western Electric/Bell Labs | Fletcher | DJS | 12/1/43 to 12/31/45 | 21,179 |
| 79. Gun Director Mk 56 | G.E. | Leveen | IAG | 1/1/44 to 10/31/45 | 1,273,532 |
| 80. Aircraft Fire Control Analysis—Patuxent | Northwestern | Calvert | SHC | 2/15/44 to 10/31/45 | 485,000 |
| 81. Speed Regulator for Motors and Motor Generators | Leeds & Northrup | Lane | EJP | 2/1/44 to 8/31/45 | 12,760 |
| 82. Control Elements for Fire Control Applications (Pneumatic) | Lawrance Aeronautical | Young | EJP | 6/1/44 to 9/30/45 | |
| 83. Chronograph T4 | Westinghouse | Osbon | IAG | 7/1/44 to 10/31/45 | |
| 84. Components for Pilot-Operated Sights | Bristol | Mabey | SHC | 7/1/44 to 9/30/45 | |
| 85. Computer for Mk 56 | Librascope | | IAG | 9/30/44 to 10/31/45 | 348,247 |
| Total Cost | | | | | 11,090,595 |

*Source:* NDRC Index Card File, RG 227, National Archives, Index to Contracts.
*Note:* Some contract numbers are not used.

Observatory at the University of Michigan studied pneumatic controls for these sights (40), as did Eastman Kodak. The Bristol Company designed an antiaircraft rocket director (38), Bausch & Lomb and Barber Coleman made antitank sights (59, 66) and a stabilizer for an aerial camera, and G.E. designed a torpedo director (72).

Even at the start of the war, microwave ranging techniques (later called radar) demonstrated the potential to automate perception and replace optical tracking. Still, the technology remained in its infancy, and many feared that the enemy would develop suitable countermeasures and render microwave detection useless. D-2 let a contract to Bell Labs for a "radio ranging device" that would replace the most unreliable input to a director. The result of that project, the SCR-547 radar, was nicknamed "Mickey" because its separate parabolic antennas for sending and receiving gave it the look of mouse ears. This device determined range only, still needing telescopes for tracking (14), and was quickly superseded. Westinghouse built a radar that could measure the velocity of a shell as it left the muzzle of a gun (65, 83), but most radar work was taken over by the MIT Radiation Lab, under NDRC Division 14.

Though Division 7 did not have a laboratory of its own, two Division 7 contractors established large laboratories for specific control problems. Eastman Kodak completed a broad range of work under a single contract, including rangefinder improvements, lighting studies, and pneumatic controls (17). The Franklin Institute became a central laboratory for airborne fire control and conducted studies in torpedoing, bombing, gunnery, rocketry, and integrated systems (33).

Testing posed a major problem for all types of antiaircraft devices. At the start of the war, no quantitative comparisons could be made of the relative performance of new technologies. To redress this problem, the Barber Coleman Company built a dynamic tester, which generated perfect inputs for gun directors and compared their outputs with ideal solutions (25). George Stibitz, at Bell Labs, made three digital relay computers for testing, using easily changed paper tapes as the sources for target aircraft trajectories (60, 63, 54). The University of Texas built a simulation facility for airborne devices (50), and another lab developed a means for measuring the smoothness of a turret's motion (75) for manufacturing testing.

Division 7's most lasting research concerned the integration components of the control system, particularly in the areas of mathematics and computation. Norbert Wiener, of MIT, studied a statistical method for predicting the future trajectory of an airplane based on its past performance (6, 29). As a part of the testing program, George Stibitz built computers that interpolated intermediate points into trajectories and calculated the ideal output for a fire control system (70, 74). RCA studied the feasibility of electronic computing methods (48). Engineers at the University of Pennsylvania's Moore School continued to improve their Bush-style differential analyzer (62). Differential analyzers at the BRL, MIT, and Pennsylvania did computations for a variety of studies, under a contract with the Franklin Institute (39). Division 7 also referred a few projects to the NDRC's Applied Mathematics Panel, such as Columbia University's work on bombing statistics (23) and one for the general analysis of aerial combat (47).

Division 7 also funded the articulation component of control, letting 16 contracts for investigations in servomechanisms. The MIT Servo Lab studied fundamental theory and designed a number of servos (1, 35), several of which were put into large-scale production (46). Barber Coleman did research in clutch-type servos (27). The United Shoe Machinery Corporation developed boosters to aid gunners in moving machine guns aboard bombers (15) and did fundamental research into hydraulic servos for the gun mounts (16). Other projects, at Leeds & Northrop, for example, developed motor regulators for use aboard aircraft (81). In two projects that extended the methods of antiaircraft gunnery to coastal defense, the units of the control system were separated by long distances and required devices to transmit data back to a base station (20, 34).

Combining perception, integration, and articulation led to an overall view of the system. Harold Hazen suggested studying the human operator "as an integral component of an automatic control system" during the development and design process. Fire control spawned seminal studies on what today would be referred to as human factors in automation. Seven contracts studied a broad array of psychological and physiological factors in rangefinding and tracking performance (10, 43, 45, 37). All except fatigue produced negative results, showing no effects on ranging or tracking (36, 42). Other work sought standards for selection of rangefinder operators, including height, vision, intelligence, mechanical ability, interpupillary distance, and coordination. This work sought to put what had previously been an ad hoc informal process, namely, matching the capabilities of the human to the characteristics of the machine, onto a scientific, psychological, and physiological foundation.

# Appendix C
# Algorithm of Bell Labs' T-10 Director

As in the Sperry machines, the T-10 director takes three inputs: azimuth ($\alpha$), elevation ($\epsilon$), and range *(r)*. It produces three outputs for the guns, azimuth ($\alpha_p$), elevation ($\epsilon_p$), and the fuze setting or time of flight ($\Delta T$). Box I converts the slant-range input to a voltage, and box II combines slant range with elevation to derive its height component. Box III combines the target height with azimuth to derive the target position in rectangular coordinates ($x$, $y$, and $v$ for vertical height). Box IV performs the actual prediction, deriving the target velocities (i.e., differentiating the position components with respect to time), multiplying the velocities by the time of flight ($\Delta T$) and adding them to the original positions. The time-of-flight parameter closes a feedback loop around the prediction calculation. The output of Box IV, then, is the predicted position of the target, $x_p$, $y_p$, and $v_p$. Blocks V, VI, and VII then convert this set of three voltages, representing rectangular coordinates, back to polar coordinates, represented now by angular shaft positions. The servomotors do both angular conversion (multiplying by a sine or cosine) and electrical-to-mechanical translation (Figs. C.1 and C.2).

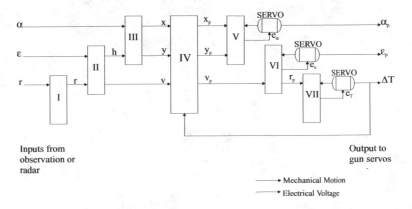

Fig. C.1. Simplified block diagram of the T-10 director. Reprinted from "Final Report: D-2 Project #2c Study of Errors in T-10 Gun Director," OSRD7 GP.

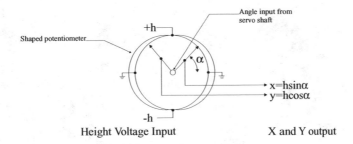

Fig. C.2. Coordinate conversion with a sinusoidial potentiometer driven by a servo shaft. Reprinted from "Final Report: D-2 Project #2c, Study of Errors in T-10 Gun Director," OSRD7 GP.

# Notes

*Abbreviations*

| | |
|---|---|
| ATT | AT&T Archives, Warren, New Jersey |
| EAS Papers | Elmer Ambrose Sperry Papers, Hagley Museum and Library, Wilmington, Delaware |
| NARA | National Archives and Records Administration, College Park, Maryland |
| NARL | National Archives, Radiation Laboratory Records, Waltham, Massachusetts |
| NWCL | Naval War College Library, Newport, Rhode Island |
| NW Papers | Norbert Wiener Papers, MIT Archives, Cambridge, Massachusetts |
| OSRD | Record Group 227, Office of Scientific Research and Development, National Archives and Records Administration, College Park, Maryland |
| OSRD7 | Record Group 227, Division 7 Records, Office of Scientific Research and Development, National Archives and Records Administration, College Park, Maryland |
| OSRD7 GP | Record Group 227, Division 7 General Project Files (E-86), Office of Scientific Research and Development, National Archives and Records Administration, College Park, Maryland |
| RAC | Rockefeller Archive Center, Sleepy Hollow, New York |
| RG 74 | Record Group 74, U.S. Navy, Bureau of Ordnance Records, National Archives and Records Administration, College Park, Maryland |
| SGC Papers | Sperry Gyroscope Company Papers, Hagley Museum and Library, Wilmington, Delaware |

*In the notes to chapters 7 through 11 the names of D-2 and Division 7 members are abbreviated as in the original memos.*

| | |
|---|---|
| CSD | Charles Stark Draper |
| DJS | Duncan J. Stewart |
| EJP | Edward J. Poitras |
| GAP | George A. Philbrick |
| GRS | George R. Stibitz |
| GSB | Gordon S. Brown |
| HLH | Harold L. Hazen |

| IAG | Ivan A. Getting |
| KTC | Karl Taylor Compton |
| NW | Norbert Wiener |
| PRB | Preston R. Bassett |
| SHC | Samuel H. Caldwell |
| TCF | Thornton C. Fry |
| WW | Warren Weaver |

## Chapter 1 Introduction

1. Mumford, *Technics and Civilization*, 247.

2. Ibid., 228.

3. Lindbergh, *Spirit of St. Louis*, 486.

4. Smith, "Selling the Moon," 189; Ward, "Meaning of Lindbergh's Flight."

5. Mumford, *Technics and Civilization*, 3, 14.

6. Ibid., 33.

7. Ibid., 25.

8. Ibid., 36.

9. Mumford, *Myth of the Machine*.

10. Gibson, *Neuromancer*.

11. Wiener, *I Am a Mathematician*, 265.

12. Heims, *John von Neumann and Norbert Wiener*; idem, *Constructing a Social Science for Postwar America*.

13. Licklider, "Man-Computer Symbiosis"; Norberg and O'Neill, *Transforming Computer Technology*; Haraway, "Cyborg Manifesto," in *Simians, Cyborgs, and Women*.

14. Wiener, *I Am a Mathematician*, 265. For a similar account and a similar claim, see idem, *Cybernetics*, 8.

15. Gerovitch, *From Newspeak to Cyberspeak*; Heims, *John von Neumann and Norbert Wiener*; Galison, "Ontology of the Enemy"; Edwards, *Closed World*; Kay, *Who Wrote the Book of Life?* Keller, *Refiguring Life*, chap. 3; Aspray, "Scientific Conceptualization of Information."

16. Gitelman, *Scripts, Grooves, and Writing Machines*; Rabinbach, *Human Motor*; Dagognet, *Etienne-Jules Marey*; Daston and Galison, "Image of Objectivity"; Kern, *Culture of Time and Space*; Nelson, *Frederick W. Taylor*; Taylor, *Principles of Scientific Management*; Hounshell, *From the American System to Mass Production*, chap. 6; Eksteins, *Rites of Spring*. For earlier manifestations of this history, see Channell, *Vital Machine*; and Schaffer, "Babbage's Intelligence."

17. *Control theory* refers to a body of mathematical concepts that quantitatively describe the behavior of dynamic systems. *Control engineering* is practice and technique that employs control theory to design such systems, including the professional development of a discipline of control engineering, with its own journals, professional societies, and career tracks.

18. Mackenzie, *Inventing Accuracy*, 31; Hughes, *Elmer Sperry*; idem, *Networks of Power*.

19. Pickering, "Cyborg History and the World War II Regime."

20. Reingold, "Vannevar Bush's New Deal for Research"; "Science Agencies in World War II"; Owens, "Counterproductive Management of Science in the Second World War."

21. Edwards, *Closed World.*

22. Hughes, *Rescuing Prometheus;* Edwards, *Closed World;* Hughes and Hughes, *Systems, Experts, and Computers;* Kay, *Who Wrote the Book of Life?*

23. Ceruzzi, *History of Modern Computing,* ix. Ceruzzi was referring to the material on analog computing in Aspray, *Computing before Computers.*

24. Owens, "Where Are We Going, Phil Morse?" begins to address the historical relationship between analog and digital.

25. Levin, "Contexts of Control."

26. Beniger, *Control Revolution.*

27. Mayr, *Liberty, Authority, and Automatic Machinery in Early Modern Europe;* Bennett, *History of Control Engineering, 1800–1930* and *History of Control Engineering, 1930–1955.* Bennett leaves it "to others to delve into the complex relationships between the technology and its social and economic consequences," by which he means "unemployment, economic growth, removal of degrading and onerous work, and de-skilling" (*History of Control Engineering, 1930–1955,* viii).

28. See Mackenzie's concept of "gyro culture" in *Inventing Accuracy,* 31.

29. Constant, *Origins of the Turbojet Revolution.*

30. Bennett, "Development of Process Control Instruments"; idem, "Industrial Instrument."

31. Oppelt, "On the Early Growth of Conceptual Thinking in Control System Theory"; Gerovitch, "Speaking Cybernetically"; Bennett, *History of Control Engineering, 1930–1955;* Porter, "Servo Panel."

32. Galison, *Image and Logic,* 61.

33. Mackenzie, *Inventing Accuracy;* Bijker, Hughes, and Pinch, *Social Construction of Technological Systems.* For recent refinements, see Alder, *Engineering the Revolution;* and Hecht, *Radiance of France.*

34. Turing, "On Computable Numbers"; Hodges, *Alan Turing;* Hutchins, *Cognition in the Wild,* 356–70. See a similar account in Edwards, *Closed World,* 250–56.

35. Benjamin, "The Work of Art in the Age of Mechanical Reproduction," in Benjamin and Arendt, *Illuminations.*

36. See, e.g., Shapin and Schaffer, *Leviathan and the Air-Pump;* and Galison, *Image and Logic.*

37. Latour, *Science in Action;* idem, "Drawing Things Together."

38. Latour, *Science in Action,* 241; Serres and Latour, *Conversations on Science, Culture, and Time;* Serres, *Parasite.*

39. Hutchins, *Cognition in the Wild,* 65, 95, 170.


*Chapter 2 Naval Control Systems*

1. For the "gun club," see Rosenberg, "Officer Development in the Interwar Navy."

2. Arthur Hungerford Pollen, in *Land and Water,* 17 August 1916; Sumida, "Quest for Reach," 77–78.

3. BuOrd to Commander in Chief, Atlantic Fleet, 11 July 1916, RG 74, E-30, box 747, subject file 30309, Folder Without Line Numbers, 1915–1917.

4. Commander in Chief, Atlantic Fleet, to Chief of Naval Operations, 11 August 1916, and Ford Instrument Company to Chief of BuOrd, 20 September 1916, RG 74, E-30, box 743, subject file 30309, folder 1–50.

5. Captain H. J. S. Brownrigg, "'Ford,' Fire Control System: Interviews with Representatives of Ford Instrument Co. of New York," January–June 1919, Naval Library, Ministry of Defence, London. Courtesy John Tetsuro Sumida. See also Sumida, *In Defence of Naval Supremacy*, 314–15.

6. McBride, *Technological Change in the United States Navy*. For a broader view of the interwar period, see Murray and Millett, *Military Innovation in the Interwar Period*, esp. chaps. 9 and 10.

7. For the historical origins of military discipline, see McNeill, *Pursuit of Power*, esp. chap. 4; Smith, introduction to Smith, *Military Enterprise and Technological Change*; and Alder, *Engineering the Revolution*.

8. Crary, *Techniques of the Observer*, 147. See also idem, *Suspensions of Perception*; and Virilio, *War and Cinema*, 14–15, 83.

9. Hutchins, *Cognition in the Wild*, 65.

10. My emphasis on perception and articulation derives from Gille Deleuze's essay on Foucault, "A New Cartographer." His distillation of Foucault, that "all knowledge runs from a visible element to an articulable one, and vice versa," echoes the translation performed by the human operators in control systems (Deleuze, "A New Cartographer," in Deleuze and Hand, *Foucault*, 23–44).

11. Bush, *Pieces of the Action*, 183; Jahn, "Employees Honor Hannibal C. Ford"; Ward, "Hannibal Ford, Sperry Pioneer." See also "Hannibal Ford"; and Ford Instrument Company, Division of the Sperry Corporation, "News Release: For Release in the Event of Mr. Ford's Death." For an anecdotal account of Ford Instrument, see Reynolds and Rowe, *Operation Success*, 153–60.

12. For problems of the magnetic compass, see British Admiralty, Technical History Section, *The Development of the Gyrocompass prior to and during the War*, pamphlet TH 20 (London, October 1919), 3. Courtesy John Sumida. For early gyrocompass history in Germany, see Mackenzie, *Inventing Accuracy*, 34–35.

13. Hughes, *Elmer Sperry*, 284–85. To avoid confusion, I use *Sperry* or *Sperry Gyroscope* to refer to the company and *Elmer Sperry* to refer to the man, except where syntax makes *Sperry* clearly the man.

14. "Sperry: The Corporation."

15. Richardson, *Gyroscope Applied*, chap. 2; Hughes, *Elmer Sperry*, 146; Elmer Sperry, "Gyroscopic Navigation Apparatus," U.S. Patent 1,255,480 (17 September 1918) and "Repeater System for Gyro-Compasses," 1,296,440 (4 March 1919).

16. Hughes, *Elmer Sperry*, 241.

17. Padfield, *Aim Straight*, chap. 5; Morison, "Gunfire at Sea," in *Men, Machines, and Modern Times*; Morison, *Admiral Sims*, 83, 178, 145; Padfield, *Guns at Sea*, chap. 29.

18. In this context, *director* can mean the actual mechanism whereby the main gunnery officer

aims his telescopes and transmits orders to the turrets, or it can refer to that officer himself. *Director fire* tends to refer to the whole system as set up with a main officer in the foretop controlling fire. Later, in U.S. Navy terminology, the mark series assigned to directors had a variety of meanings; the Mark 1, for example, was really a computer, whereas the Mark 6 was a gyroscopically controlled stable element. A given fire control system could be composed of many different elements, referred to as directors, each with a distinct function.

19. Padfield, *Guns at Sea,* 245.

20. For one, possibly biased chronology of gunnery in the Royal Navy during this era by a participant, see Dreyer, *Sea Heritage,* 43–46. For an excellent firsthand summary of fire control development in the U.S. Navy in the years 1915–20, see William R. Furlong, "Development of Fire Control," memorandum, [1920?], William R. Furlong Papers, Library of Congress, Washington, D.C., box 6, Ordnance—American folder. For Percy Scott and director firing, see Padfield, *Guns at Sea,* 245–46; and idem, *Aim Straight.* Scott remained a firm believer in spotting, and while he favored director fire, he was not as sanguine about the new fire control technologies (Sumida, "Quest for Reach," 55).

21. Padfield, *Aim Straight,* 190–91.

22. Padfield, *Guns at Sea,* 225. See also Sumida, *In Defence of Naval Supremacy,* 74–75. Sumida lists three major weaknesses in this method: poor visibility could obstruct the necessary range readings, the range rate itself was inaccurate, and data had to be transferred manually out of the Vickers clock because its output was too weak to drive a data transmitter ("The Quest for Reach," 72).

23. Sumida, *In Defence of Naval Supremacy.* It is worth noting that Sumida follows Pollen's terminology and uses *change of range rate* to mean the rate of change of the range (velocity). For clarity, I avoid *change of range rate* because to the modern reader it might also suggest the rate at which the change of range is changing (acceleration), thereby causing confusion.

24. Ibid., 133; idem, "Quest for Reach," 62–66.

25. Sumida, *In Defence of Naval Supremacy,* 333, 217–19; Padfield, *Guns at Sea,* 226.

26. Fire Control Board report, 1905, quoted in Friedman, *U.S. Naval Weapons,* 28, emphasis added.

27. Van Auken, *Half Century of United States Naval Ordnance,* 20; Padfield, *Guns at Sea,* 247–50; Furlong, "Development of Fire Control"; Friedman, *U.S. Naval Weapons,* 27.

28. Furer, *Administration of the Navy Department in World War II,* 205–6.

29. U.S. Navy, Bureau of Ordnance, *Navy Ordnance Activities,* 151.

30. Elmer A. Sperry to BuOrd, 14 March 1914, RG 74, E-30, box 587, subject file 29758, folder 1–50. See also Sperry, "Repeater System for Gyro-Compasses," U.S. Patent 1,296,440, which alludes to but does not mention fire control.

31. Reginald E. Gillmor to Sperry, 16 June 1916, and Sperry to J. Strauss, Chief of BuOrd, 17 July 1916, including excerpts from Gillmor's letter, RG 74, E-30, box 586, subject file 29758, folder 101–150.

32. "Mr. Gillmor's Report—Result of his Investigations," RG 74, E-25, box 586, subject file 29758, folder 110.

33. Sperry to Lieutenant Commander F. C. Martin, 31 August 1916, EAS Papers, box 32, Lt. Logan Cresap Fire Control Correspondence folder.

34. Sperry was also a friend of Admiral Bradley Fiske, who is credited with an early director-firing invention. The two men had patented a device for "Automatic Gun Pointing" in 1917 (U.S. Patent 1,238,503, 28 August 1917).

35. Hannibal C. Ford, "Battle Tracer," U.S. Patent 1,293,747. See also Elmer Sperry's "System of Gunfire Control," U.S. Patent 1,356,505 (19 October 1920).

36. W. R. Van Auken to Sperry, 4 March 1917, EAS Papers, box 32, W. R. Van Auken folder.

37. "The Sperry Fire Control System," Bulletin 304, 1916, SGC Papers, box 33. See also Ford, "Battle Tracer," U.S. Patent 1,293,747.

38. See also the following patents by Elmer Sperry: "System of Gunfire Control," U.S. Patent 1,356,505; "Plotting-Indicator," U.S. Patent 1,215,425 (13 February 1917); and "Multiple-Turret Target-Indicator," U.S. Patent 1,296,439 (4 March 1919).

39. "Turret Control Equipment," Bulletin 303, 1916, SGC Papers, box 33.

40. Captain Albert Gleaves (USS *Utah*) to Secretary of the Navy and BuOrd, 1 June 1915, RG 74, box 587, subject file 29758, folder 1–50.

41. Lieutenant Palmer, Commanding Officer, USS *New York,* to BuOrd, 14 July 1915, and Commander, *New York,* to Commander, Battleship Division 6, 2 August 1916, ibid., folder 101–150.

42. U.S. Navy, Bureau of Ordnance, *Navy Ordnance Activities,* 152. In the late 1920s the Arma and Ford Instrument Companies began producing Dead Reckoning Tracers, which performed roles very similar to those of the Battle Tracer, and some of the Ford machines are still in use on warships today ("Report of Test of Arma Dead Reckoning Tracer Equipment for U.S.S. Texas," 13 April 1927, RG 74, S-71, box 1740).

43. Palmer to BuOrd, 14 July 1915; Commander, *New York,* to Commander, Battleship Division 6, 2 August 1916; "Mr. Gillmor's Report—Result of his Investigations."

44. Ford Instrument Company, "Report on Organization and War Activities of the Ford Instrument Company, Inc.," June 1919, RG 74, E-25, box 2740, subject file 36276, folder 110, mentions that naval interest "necessitated the abandonment of commercial projects then underway, including the manufacture of the Carrie Gyro Compass, and other work in which the company was engaged." This document was the source for the information on Ford Instrument in U.S. Navy, Bureau of Ordnance, *Navy Ordnance Activities,* 159. The book does not mention the Carrie compass, but it has been the source for the majority of the scant historical material on Ford Instrument. For a comparison of the Carrie Gyro Compass and the Sperry model, see British Admiralty, Technical History Section, *Development of the Gyrocompass prior to and during the War,* 7.

45. The navy purchased more than 400 Vickers Clock Mark 2s during World War I (U.S. Navy, Bureau of Ordnance, *Navy Ordnance Activities,* 152).

46. BuOrd to Secretary of the Navy, 3 June 1915, RG 74, E-30, box 587, subject file 29758, folder 1–50.

47. License agreements between Ford and Sperry for the Battle Tracer, 4 May 1915, and for the range clock, 7 July 1915, SGC Papers, AC 1915, box 67, Ford Instrument Company Patents folder;

H. H. Thompson to Sperry, 27 February and 1 December 1919, SGC Papers, box 32, Fire Control Patents and Hannibal Ford Interference folder. Contrast this situation with that of Carl Norden, who quit Sperry Gyroscope in 1913 to manufacture bombsights for the navy. "Sperry took Norden's resignation as a personal affront, beginning a half century of conflict between the Sperry Company and Norden. Sperry felt he had taught Norden everything he knew about gyroscopes and therefore should share in any of Norden's future patents" (McFarland, *America's Pursuit of Precision Bombing,* 50).

48. BuOrd to Secretary of the Navy, 3 June 1915.

49. Ford Instrument Company, "Announcement," n.d., received by BuOrd 4 December 1915. See also Jahn, "Employees Honor Hannibal Ford."

50. "Range Keeper Mark I," Sperry Gyroscope album 659, photo 86.273, SGC Papers, box 34. Sperry to Martin, BuOrd, 8 May 1916; Naval Inspector of Ordnance to BuOrd, 11 May 1916 (about Sperry Gyroscope rangekeeper); and Naval Inspector of Ordnance to BuOrd, 11 May 1916 (about Ford Instrument rangekeeper), RG 74, E-30, box 587, subject file 29758, folder 51–100. The naval inspector for these trials was assigned primarily to the E. W. Bliss Company, a manufacturer of torpedoes in New York. At this point neither Sperry Gyroscope nor Ford Instrument had enough navy work to justify its own inspector, but soon thereafter a permanent inspector was assigned to each company.

51. BuOrd to Commander in Chief, Atlantic Fleet, 11 July 1916.

52. BuOrd to Battleship *New York,* 2 August 1916, and Sperry to Martin, 2 December 1916, RG 74, E-30, box 586, subject file 29758, folder 101–150. Sperry Gyroscope missed the yard period of the *New York* in August, which was why it had to wait until December to demonstrate the device. Battleship captains were willing to help BuOrd evaluate new technologies, but not if it meant holding their ships in port. A photograph in the SGC Papers shows that the Sperry Gyroscope instrument lacked both the design flair of Ford Instrument's and the feedback loop for continuous correction of its solution ("Range Keeper Mark I," Sperry Gyroscope album 659, photo 86.273).

53. Commanding Officer, *Louisiana,* to Commander, Battleship Force One, "Pollen Fire Control Instruments," 31 March 1919, RG 74, box 345, subject file 28499. The British comparison is quoted in Friedman, *U.S. Naval Weapons,* 33.

54. Lieutenant H. M. Terril, "Notes on the Theory of the Ford Range Keeper," issued by the USS *New Mexico* Gunnery Department, ca. 1919, RG 38, entry 178, box 3, Conf. 59 (65) folder, courtesy John Tetsuro Sumida and Christopher Wright.

55. Ibid., 2.

56. Kelvin probably suggested the device to Pollen personally in 1904, as he served on the board of Pollen's Linotype Company and as a scientific mentor to Pollen (Sumida, *In Defence of Naval Supremacy,* 78). On the relative advantages of different types of integrators, see Clymer, "Mechanical Integrators," 20–22; idem, "Analog Computers of Hannibal Ford and William Newell"; and Bromley, "Analog Computing Devices." Clymer was an engineer at Ford Instrument.

57. Hannibal C. Ford, "Mechanical Movement," U.S. Patent 1,317,915 (7 October 1919).

58. The production device became the Ford Rangekeeper Mark 1, mod 1, the mod number in-

dicating minor modifications made after the *Texas* tests. These modifications mostly entailed removing the automated input from the target bearing transmitter and replacing it with a follow-the-pointer operation, thus making the machine "entirely independent of all other apparatus. . . . In other words, the machine is self-contained" (Martin, BuOrd, to D. C. Bingham, fleet gunnery officer on the staff of Admiral Mayo, 14 May 1917, RG 74, E-30, box 743, subject file 30309, Folder Without Line Numbers, 1918). For the navy's order, see ibid. For the Mark 2 "Baby Ford" Rangekeeper, see Hannibal C. Ford, "Range-Keeper," U.S. Patent 1,370,204 (1 March 1921), which covers the Mark 2 machine; no individual patent covers the first rangekeeper. See also U.S. Naval Academy, *Notes on Fire Control*, chap. 6, "Secondary Battery Rangekeepers," for the Baby Ford; and Martin's comments on the Baby Ford in Martin to Bingham, 14 May 1917. For BuOrd's comments on reliability, see BuOrd to Ford Instrument Company, August 1917, and BuOrd to Commander in Chief, Atlantic Fleet, 3 August 1917, RG 74, E-30, box 747, Subject 30309, Folder Without Line Numbers, 1915–1917.

59. Ford Instrument Company, "Status of Contracts," 19 April 1918, RG 74, E-30, General Correspondence 1915–1926, box 747, subject file 30309 (Ford Rangekeeper), folder 201–250.

60. For the emergence of mass production and its origins in military arsenals, see Smith, "Army Ordnance and the 'American System' of Manufacturing"; Hounshell, *From the American System to Mass Production*. Philip Scranton details the close-knit relationships between specialty producers in *Endless Novelty*, although the book does not address military production. See also Brown, *Baldwin Locomotive Works*; and Piore and Sabel, *Second Industrial Divide*.

61. Brownrigg, "'Ford,' Fire Control System." See also Sumida, *In Defence of Naval Supremacy*, 314–15.

62. Ford Instrument Company, "Report on Organization and War Activities of the Ford Instrument Company, Inc.," 2–6.

63. R. C. Wilber, Ford Instrument Company, to Naval Inspector's Office, 28 February 1918, RG 74, E-30, General Correspondence 1915–1926, box 747, subject file 30309 (Ford Rangekeeper), folder 100–150; Ford Instrument Company, "Report on Organization and War Activities of the Ford Instrument Company, Inc."

64. Roosevelt to Ford Instrument Company, 30 April 1918, RG 74, E-30, box 747, subject file 30309, folder 201–250.

65. R. K. Davis, Office of the Judge Advocate General, to Admiral Furlong, "Bureau of Ordnance, New Patent Clause," 6 April 1938, RG 74, box 1740, Ford Instrument Company folder.

66. Jules Breuchaud to Ralph Earle, 28 May 1917, RG 74, E-30, box 743, subject file 30309, folder 1–50; Earle to Breuchaud, 1 July 1917, RG 74, E-30, box 747, subject file 30309, Folder Without Line Numbers, 1915–1917. As a member of the Naval Consulting Board, Elmer Sperry had access to the Ford machine, and after the board visited BuOrd late in 1916 Ford wrote that "[I] have no doubt that Mr. S. was all eyes when examining the instrument" (Ford to Martin, 4 October 1916, RG 74, E-30, box 743, subject file 30309, folder 1–50).

67. Elemer Meitner, "Range Predicting Apparatus," U.S. Patent 1,387,551 (applied 14 May 1915, granted 16 August 1921); Hannibal Choate Ford, "Range Bearing and Keeper," U.S. Patent 1,450,585

(applied 19 June 1918, granted 3 April 1923); Elmer A. Sperry and Elemer Meitner, "Director Firing System," U.S. Patent 1,755,340 (applied 9 April 1917, granted 22 April 1930).

68. McFarland, *America's Pursuit of Precision Bombing,* 50–60; Barth is quoted on 58–59.

69. I borrow the idea of private arsenals from Sapolsky and Gholz, "Restructuring the Defense Industry."

70. Sperry Company memorandum, 5 June 1920, EAS Papers, box 32, Capt. William McEntree, Naval Constructor folder; U.S. Navy, Bureau of Ordnance, *Navy Ordnance Activities,* 152.

71. Sperry to Martin, 31 August 1916; Sperry to Gillmor, 20 October 1916, EAS Papers, box 32, Capt. William McEntree, Naval Constructor folder; Sperry to Gillmor, 13 February and 25 July 1917, ibid., Comdr. F. C. Martin folder.

72. Chafee to BuOrd, 26 January 1915, and Commander, Battleship Force, to Commander in Chief, 25 January 1918, RG 74, E-30, box 585, subject file 29758, folder 1–50.

73. Commander, Battleship Force, to Commander in Chief, 5 October 1916, RG 74, box 586, subject file 29758, folder 101–150. See also Commander in Chief to Commander, Battleship Force, 22 September 1916; Commanding Officer, *Utah,* to Commander, Division 7 Battleship Force, 25 August 1916; Commander, *New York,* to Commander, Battleship Force, 6 August 1916; and Commander, *New York,* to BuOrd, 14 August 1915, ibid.

74. "Turret Control Equipment."

75. Sperry to Martin, 6 June 1917, and Sperry Gyroscope Company to BuOrd, 1 December 1917, RG 74, E-30, box 586, subject file 29758, folder 151–200. E. A. Sperry, "Synchronous Transmission System," U.S. Patent 1,468,330 (18 September 1923); E. A. Sperry, "Self-Synchronous Transmission System," U.S. Patent 1,850,640 (22 March 1932); E. A. Sperry, "Synchronizing Mechanism," U.S. Patent 1,656,962 (24 January 1928).

76. Earle to Naval Inspector of Ordnance, Sperry Gyroscope Company, 16 February 1918, SGC Papers, summarizing a letter to Earle from the commanding officer of the *Texas,* 22 January 1918, RG 74, E-30, box 585, subject file 29758, folder 301–350; Naval Inspector of Ordnance, Sperry Gyroscope Company, to Sperry Gyroscope Company, 25 February 1918, emphasis in the original, ibid. Folders 151–200, 251–300, and 450–500 contain numerous complaints about ruggedness and synchronization in the Sperry systems. For the review of the system by the National Bureau of Standards, see Earle to Sperry Gyroscope Company, 1 May 1918, ibid., folder 351–400; and technical note to Earle, handwritten, 12 March 1918, ibid., folder 301–350.

77. Sperry to Van Auken, 18 December 1917, SGC Papers, box 32, Capt. William McEntree, Naval Constructor folder. For a more sanguine view of Sperry's 1918 reorganization, see Hughes, *Elmer Sperry,* 210–11. Sperry Gyroscope's new organizational chart, as published in Sperry's public newsletter, does not list the secret fire control division.

78. Gillmor to Assistant Secretary of the Navy, 13 June 1918, RG 74, E-30, box 584, subject file 29758, folder 450–500.

79. Hughes, *Elmer Sperry,* 295–303.

80. Sperry to Strauss, 17 July 1916. BuOrd to Sperry Gyroscope, 14 August 1916, RG 74, E-30, box 586, subject file 29758, folder 101–150. Naval Inspector of Ordnance, Sperry Gyroscope, to BuOrd,

21 and 23 March 1916; Strauss to Sperry Gyroscope, 23 March 1916; and D. M. Mahood to BuOrd, 29 March 1916, RG 74, E-30 box 587, subject file 29758, folder 51–100.

81. Friedman, *U.S. Battleships*, 182–83.

82. Furlong Papers, box 4; for Furlong's hiring of electrical engineers, see box 1, General Correspondence folder.

83. Hewlett, "Selsyn System of Position Indication."

84. Wilbur R. Van Auken, "Adoption of General Electric Fire Control System," 23 July 1929, RG 74, box 1740, General Electric Fire Control—General folder; see also Hammond and Pound, *Men and Volts*, 356–58, 370–71. For the "all electric ship," see McBride, "Strategic Determinism in Technology Selection."

85. Furlong to Admiral Larimer, 8 January 1932, Furlong Papers, box 4, General Correspondence, Military File. This letter supports a commendation by the U.S. Navy of Hewlett for his work on the Selsyn fire control system and narrates the development process. See also General Electric Company, "Report of Expenditures up to October 1920 for Development of Fire Control Apparatus as Shown by the Following Special Manufacturing Orders Since February 14, 1919," RG 74, box 3251, subject file 39117, folder 1–50.

86. "Elements of the Synchro System," in U.S. Naval Academy, *Fire Control Installations*, 139.

87. G.E. and Sperry eventually settled out of court. For BuOrd patent policy, see O. G. Murfin, memorandum for file, "Patent Rights for Fire Control Material," 12 August 1933; and Davis to Furlong, "Bureau of Ordnance, New Patent Clause," 6 April 1938. For Sperry's position regarding G.E., see Sperry Gyroscope Company, "An Analysis of the Fire Control Patent Situation in the U.S. Navy," [ca. 1920], and associated correspondence, in RG 74, E-30, box 2924, subject file 37186. This folder contains a handwritten note, dated 1922, saying that "G.E. reports that it has arranged with Sperry, Hammond, Ford to use apparatus. This is covered by clause in contract allowing not over $9000 for patent rights."

88. In these "master gun" setups a gunnery officer aimed at the target with a single, instrumented gun mount (equipped with a Mark 2 "Baby Ford" rangekeeper), and gunners on the other mounts "followed the pointer" to track the master's movements. For a good summary of fire control in the early 1920s, see Commander R. C. Hyatt, U.S. Navy, "Discussion of 'Modern Fire Control System,'" Lecture to Naval War College, 2 February 1925, NWCL, XOGF-44. For a list of G.E.'s fire control work in 1920, see Naval Inspector of Ordnance, General Electric, to BuOrd, "*Maryland* and *Colorado* Contract #2992, Fire Control, General Information of Progress of Work," 15 November 1920, RG 74, E-30, box 3251, subject file 39117, folder 1–50. Friedman, *U.S. Naval Weapons*, 35, appraises fire control in the twenties. For a technical description of one of these systems, see "Main Battery Fire Control System, U.S.S. *West Virginia*," chap. 4 in U.S. Naval Academy, *Fire Control Installations;* and U.S. Naval Academy, *Notes on Fire Control*, 250–70.

89. Brittain, *Alexanderson*, 204, 219–22, 237–42.

90. Ibid. See also U.S. Naval Academy, Department of Ordnance and Gunnery, *Naval Ordnance and Gunnery*, vol. 1, *Naval Ordnance*, chap. 10, sec. D, "Amplidyne Follow Up System," 221–26; and Bennett, *History of Control Engineering, 1930–1955*, 10–12.

91. Commander G. L. Schuyler, "The Present Status of Ordnance Developments in the U.S. Navy," Lecture to Naval War College, 9 March 1928, RG 15, NWCL. Annual lectures on ordnance by BuOrd personnel to the Naval War College trace the development of fire control during the twenties. For a clear, diagrammatic example of the switchboard system, see U.S. Navy, Bureau of Ordnance, *Main Battery Fire Control System*, OP 1387 (Washington, D.C., 14 June 1948), 5.

92. U.S. Navy, Bureau of Ordnance, *Surface Fire Control*, OP 1701, [mid-1940s?]; idem, *Main Battery Fire Control System*.

93. Davy, "Case Study: The American Bosch Arma Corporation" (courtesy John Tetsuro Sumida). S. J. Davy, a fire control engineer at Arma, compiled this history from annual reports and interviews with Arma employees.

94. Ibid.; Hyatt, "Discussion of 'Modern Fire Control System,'" 8.

95. Mahood to Chief of BuOrd, 22 December 1931, RG 74, E-25, box 1740, Arma folder.

96. Sperry Gyroscope to BuOrd, 19 August 1915, RG 74, E-30, box 587, subject file 29758, folder 51–100; Gillmor to Sperry, 17 January 1917, SGC Papers, box 32, Comdr. F. C. Martin folder; Cresap to Sperry, 11 February 1918, ibid., Lt. Logan Cresap Fire Control Correspondence folder; Earle to Naval Inspector of Ordnance, Sperry Gyroscope, 17 March 1918, RG 74, E-30, box 585, subject file 29758, folder 301–350; Sperry to Earle, 15 April 1918, and Earle to Sperry, 27 April 1918, ibid., folder 400–450; Bruno A. Wittkuhns to BuOrd, 2 July 1931, RG 74, S-71, E-25, box 1741, Sperry Gyroscope Company folder.

97. The Mark 2, as we have seen, was the "Baby Ford," attached to the gun directors themselves. Mark 3 was another main battery director. Mark 4 attached to an antiaircraft director. Mark 7 was used in cruisers.

98. "Rangekeeper, Mark 8," in U.S. Naval Academy, *Fire Control Installations*, 7.

99. U.S. Navy, Bureau of Ordnance, *Rangekeeper Mark 8 and Mods*, OP 1068 (Washington, D.C., 1949). This was an instruction manual for the Mark 8.

100. Rowland and Boyd, *U.S. Navy Bureau of Ordnance in World War II*, 373.

101. U.S. Navy, *Administrative History*, 79:17.

102. C. C. Badger, "Memorandum for Chief of Bureau," RG 74, E-25, box 1740, Fire Control—General folder. William Newell, interview by author, 12 May 1995, Mount Vernon, N.Y.

103. C. C. Bloch to Chief of Naval Operations (Director of Chief of Fleet Training), 13 May 1924. Schuyler to Chief of Bureau of Ordnance, 22 September 1926, reports a visit by a Vickers representative who described the British efforts. Both in RG 74, E-25, box 1801.

104. See the chronology for the Mark 19 supplied in the Dugan patent suit, *Joseph Dugan v. United States*, 1933, RG 74, S-71, E-25A, S-71, General Correspondence 1926–44, box 14.

105. U.S. Navy, *Administrative History*, 79:137–45. For antiaircraft procedure on which the Mark 19 was based, see United States Fleet, Battle Fleet Anti-Aircraft Board, "Procedure for Antiaircraft Fire Control," 29 September 1926, RG 74, S-71, box 1801; Newell interview, 12 May 1995.

106. For summaries of the state of naval fire control in 1940, see U.S. Navy, *Administrative History*, 79:137–45 (by a participant); and Jurens, "Evolution of Battleship Gunnery," 259–60.

107. U.S. Navy, *Administrative History*, 79:146–47.

108. "Production of Major Ordnance Items, 1 July 1940 through 30 June, 1945," RG 74, Office of History, Records Relating to Official History, box 47, 87.

109. U.S. Navy, Bureau of Naval Personnel, *Naval Ordnance and Gunnery,* 429.

110. Allison, *New Eye for the Navy,* 115–17.

111. U.S. Navy, Bureau of Naval Personnel, *Naval Ordnance and Gunnery,* 470.

112. U.S. Navy, *Administrative History,* 79:148.

113. "Type Tests—6″/47 Caliber Triple Mount—Determination of the Accuracy of the Transmission and Operation of the Automatic Control," 1 August 1937, RG 74, S-71, box 1769, 3.

114. Ibid., 19.

115. These data are taken from the results of battle practices, compiled into tables in Jurens, "Evolution of Battleship Gunnery," 261–63. Despite these numbers, the variety and variability of these assumptions, coupled with minimal combat experience between the world wars, meant that little agreement existed regarding the accuracy of these systems. Gunnery officers did not even agree that long-range gunnery *should* be accurate. Strange as this may seem, the debate came down to whether one should think of long-range guns as rifles, where every shot aims to hit (they did indeed have rifled barrels), or as shotguns, where the very pattern of shots improves one's chances of hitting. The guns themselves had a natural dispersion, meaning that two identically fired shots would not hit at exactly the same spot but rather would be distributed within a circle of some determinable size. This problem got worse as the guns were fired and wore down the linings, until some point at which they needed relining. The debate was far from academic, for it had implications for where the navy put its resources in refining guns, powder, mounts, training, and fire control instruments. Obviously, there was no point in improving the accuracy of fire control beyond the point of natural gun dispersion, and systems for surface fire did indeed stabilize at this level of maturity (see, e.g., Schuyler, "Note on Salvo Dispersion," and idem, "Distribution of Shots in Long Range Salvos," for the "shotgun" case; and Chase, "Accuracy of Fire at Long Ranges," for the "rifle" case). As Donald Mackenzie points out in his superb analysis of inertial guidance, the technology itself did not necessarily follow a trajectory of ever-increasing accuracy. In fact, accuracy itself, that seemingly obvious technical quantity, does not necessarily have a stable meaning, nor is it always subject to agreed-upon standards of measurement (Mackenzie, *Inventing Accuracy,* chap. 7).

## Chapter 3 Taming the Beasts of the Machine Age

1. Crimi, *Crimi,* 147–57; "Mechanical Brains."

2. Introduction to Sperry Company history, [1942?], SGC Papers, box 40. Similarly, Thomas Morgan, Sperry Corporation president, wrote that "the primary value of Sperry's military products is that they extend the physical and mental powers of the men in the Armed Forces enabling them to hit the enemy before and more often than the enemy can hit them" (Sperry Corporation, annual report, 1943, Annual Reports Collection, Baker Library, Harvard Business School, Harvard University, Cambridge, Mass.).

3. Elmer Sperry, quoted in Hughes, *Elmer Sperry,* 173. Sperry's words have a biblical ring: "Of

all vehicles on earth, under the earth and above the earth, the airplane is that particular beast of burden which is obsessed with motions, side pressure, skidding, acceleration pressures, and strong centrifugal moments . . . all in endless variety and endless combination." Elsewhere Sperry used more technical language: "The steering of ships, torpedoes, airplanes, dirigibles, etc., where the rudder is aft, presents several problems, some of which, at least, are due to the fact that the controlling of the direction of such a craft is in constant unstable equilibrium" (Sperry, "Automatic Steering").

4. Hughes, *Elmer Sperry.*

5. Ibid., 275–79.

6. *A True Story of the Devil,* n.d., SGC Papers, box 1.

7. Sperry, "Automatic Steering," 54.

8. Sperry Gyroscope Company, *The Gyro-Compass and Gyro-Pilot: Their Operating Principles, Construction, and Uses,* Publication no. 17–1610 (Brooklyn, N.Y., n.d.), SGC Papers. See also Sperry, "Automatic Steering"; and Mayr, *Feedback Mechanisms,* 102–4.

9. Advertisement, n.d., SGC Papers, box 3.

10. "Sperry: The Corporation."

11. Bennett, "Industrial Instrument."

12. *Automatic Steering for Naval Vessels,* Publication no. 19–3 (1932), SGC Papers, box 2, emphasis in the original.

13. Sperry, "Automatic Steering," 61.

14. Ibid.

15. Hughes, *Elmer Sperry,* 269.

16. Sperry, "Description of the Sperry Automatic Pilot"; McFarland, *America's Pursuit of Precision Bombing,* 36–39; Sperry Gyroscope Company, *A-3 Gyropilot,* Catalog Collection, National Museum of American History, Smithsonian Institution, Washington, D.C.; Nealey, "Integration of the Automatic Pilot System and the Norden Bombsight."

17. J. H. Conger, quoted in Mohler and Johnson, *Wiley Post,* 5; Will Harris, in *Oklahoma City Times,* 16 August 1935, 1, quoted in ibid., 116.

18. Post and Gatty, *Around the World in Eight Days,* 26.

19. Ibid.

20. Bassett, "Servomechanisms," 22.

21. *New York Times,* 23 July 1933, 1; Post, "Destination—New York." Post's Sperry autopilot and the *Winnie Mae* are on display at the Smithsonian Air and Space Museum in Washington, D.C. Events dramatically demonstrated the importance of the Sperry autopilot's reducing Post's fatigue. The day after Post landed, two Italian aviators were critically injured landing in New York after a transatlantic flight. The cause of the crash: pilot fatigue (*New York Times,* 24 July 1933).

22. *New York Times,* 24 July 1933, 2.

23. "Sperry: The Corporation"; Lambright, *Powering Apollo,* 22–29.

24. Sperry Corporation, annual report, 1937, Annual Reports Collection, Baker Library, Harvard Business School.

25. Data taken from Sperry Corporation annual reports, 1933–40, ibid. Research and develop-

ment spending at the company actually went down as a percentage of sales during the war, when sales increases radically. See also Sanderson, "Sperry Corporation."

26. Advertisement, *Sperryscope* 10, no. 5 (1944): back cover.

27. "The Sperry Corporation."

28. "The Products of The Sperry Corporation," [early 1930s], SGC Papers.

29. Crowell, *America's Munitions*, 135–47.

30. J. B. Rose to Sperry, 11 August 1917, EAS Papers, box 32; Sperry to T. Wilson, Frankford Arsenal, 10 July 1925, ibid., box 33. See also Scott, *Naval Consulting Board*, 195–96.

31. "Ordnance History of the Sperry Gyroscope Company, Inc.," SGC Papers, box 33, Fire Control folder.

32. Sperry Company memorandum, [Preston R. Bassett?], "Development of Fire Control for Major Caliber Anti-Aircraft Gun Battery," 2, SGC Papers, box 33.

33. Sperry to Wilson, 10 July 1925. For the Sperry bombsights, see also McFarland, *America's Pursuit of Precision Bombing*.

34. Sperry Company memorandum, [Bassett?], "Development of Fire Control for Major Caliber Anti-Aircraft Gun Battery."

35. Ibid. See also Sperry Company form 1607, "Sperry Universal Director: Information to be Furnished by Customer," SGC Papers, box 3. This document was intended for foreign governments who wanted Sperry to customize directors to different types of guns.

36. See, e.g., Wells, "New Fire Control for Divisional Weapons," which explicitly suggests extending Sperry's antiaircraft fire control system to standard artillery. Sperry also did some work in seacoast artillery fire control in 1929 and 1938, although one report admitted that "our primary contribution to the seacoast fire control problem has been our sustained interest in a very slow moving field" (Roswell Ward, "Gun Data Computer, M1," 31 January 1944, SGC Papers, box 40; see also Green, Thompson, and Roots, *Ordnance Department*, 344).

37. Sperry Gyroscope Company, *Anti-Aircraft Gun Control*, Publication no. 20–1640 (Brooklyn, N.Y., 1930), 7, in SGC Papers, box 4.

38. Earl W. Chafee, "Study of the Requirements for a Satisfactory Antiaircraft Fire Control System," 15 February 1943, Fire Control Design Division, Frankford Arsenal, Philadelphia, SGC Papers, box 33. This was the final report from the "Chafee Inquiry," which Chafee conducted for the NDRC Fire Control Division in order to assess the importance of radar in antiaircraft fire control systems. It includes, however, a detailed history of Sperry's fire control development written by Chafee.

39. Sperry Gyroscope memorandum, "History of Anti-Aircraft Director Development (Army Ordnance)," [fall 1935?], SGC Papers, box 4, 9–16.

40. Sperry Gyroscope Company, *Anti-Aircraft Gun Control*. This document does not list an author, but its language and explanations are quite similar to those in an article published by Chafee, "A Miss Is as Good as a Mile," in *Sperryscope,* the official Sperry Company organ, in April 1932. See also Earl W. Chafee, Hugh Murtagh, and Shierfeld G. Myers, "Apparatus for the Control of Gunfire," U.S. Patent 2,065,303 (22 December 1936).

41. Paul Ceruzzi writes in "When Computers Were Human" that before the 1940s the term *computer* meant a person who performed mechanical calculations. Paul McConnel notes that "calling their calculating devices 'computers' appears to have been an accepted practice with aviators as early as 1926" ("Some Early Computers for Aviators"). Chafee's use of the term for antiaircraft directors probably derives from its use in aviation.

42. Sperry Gyroscope Company, *Anti-Aircraft Gun Control*, 21.

43. Robert Lea, "The Ballistic Cam in Dean Hollister's Lamp," SGC Papers, box 33.

44. Sperry Gyroscope Company, *Anti-Aircraft Gun Control*, 32.

45. Ibid., 15.

46. Ibid., 25.

47. Ibid., 24, emphasis added.

48. Ibid., 24–25.

49. Ibid., 36.

50. *Anti Aircraft Defense* (Harrisburg, Pa., 1940), SGC Papers, box 33. This book reprints the manuals for the Sperry M-2 director and discusses the mechanical problems that can be caused in this generation of Sperry directors by contradictory input data.

51. Sperry Company memorandum, "Power Controls," 7 February 1944, SGC Papers, box 40.

52. Sperry Gyroscope Company, *Universal Director and Data Transmission System,* Publication no. 14–8051 (1 August 1932), 6, SGC Papers, box 2 (essentially a specification for the T-8); idem, *Anti-Aircraft Gun Control*, 18.

53. Sperry Gyroscope memorandum, "History of Anti-Aircraft Director Development." For an explanation of the Sperry velocity servo, see Bromley, "Analog Computing Devices," 190.

54. Noble, *Forces of Production*.

55. G. B. Welch, "History of the Fire Control Sub-Office," trimonthly report covering period ending 31 March 1944, Frankford Arsenal, Philadelphia, April 1944, NARA, RG 156, Records of the Office of the Chief of Ordnance, Executive Division, Historical Branch, Publications and Reports, box 3. For Sperry's relationships with subcontractors in general, see Morgan, "Sperry and Subcontracting."

56. Welch, "History of the Fire Control Sub-Office."

57. Sperry Company memorandum, "M-5 and M-6 Directors," [1942–43?], SGC Papers, box 33, Fire Control folder. For photographs of internal mechanisms and a block diagram, see Bromley, "Analog Computing Devices," 186–91; and for an explanation of the device's use and British perspective on it, see Routledge, *Anti-Aircraft Artillery,* 52–55.

58. Wittkuhns to Chafee, Bassett, and Thompson, 15 June 1936, SGC Papers, box 33.

59. *Sperryscope* listed the company's products inside the front cover. This list comes from vol. 9, no. 2 (1940). Because of the developmental nature of many of Sperry Gyroscope's products, dates of introduction are open to interpretation. The dates of introduction of Sperry's 1940 products are compiled from a chart in *Sperryscope* 7, no. 7 (1935): 16, and a "family tree" of Sperry products from "The Story of the Sperry Corporation," [1943?], SGC Papers, box 40. The "family tree" simply omits the year 1916, when the Battle Tracer and the fire control system were introduced.

60. Varian, *The Inventor and the Pilot.*

61. "Bomber Defense from a Little Black Box."

62. Sperry Company report, "Aircraft Fire Control," [1946?], SGC Papers, box 22; Vinson, "Aircraft Computing Sights," ed. Roswell Ward, 8 February 1944, SGC Papers, box 40; Roswell Ward, "Aircraft Turrets: Description of Product Development and History," 16 February 1944, ibid.

63. For a technical comparison of the G.E. and Sperry systems, see J. B. Russell, "Aerial Gunnery," in Hazen, *Airborne Fire Control,* 177–245. The G.E. system had great problems in practice.

64. Data taken from Sperry annual reports, 1939–45. Also see the "family tree" in "The Story of the Sperry Corporation," and page 18 of that booklet, for photos of these devices. Project histories can be found in SGC Papers, box 40, Renegotiation Documents folder. The author was born across the street from the Lake Success factory, in the Long Island Jewish Hospital.

65. Introduction to Sperry Company history, emphasis added.

66. Ibid.


*Chapter 4 Opening Black's Box*

1. Black, "Inventing the Negative Feedback Amplifier". On the early morning ferry rides see George Stibitz's memoir, "Zeroth Generation," 54.

2. For a recent textbook, see Dorf, *Modern Control Systems.* For other accounts of Black's invention, see Bode, "Feedback"; "Career of the 1957 Lamme Medalist Harold S. Black"; and Mabon, *Mission Communications,* 39–40. Among historian's accounts the most thorough is Bennett, *History of Control Engineering, 1930–1955,* chap. 3, "The Electronic Negative Feedback Amplifier." See also O'Neill, *Transmission Technology,* chap. 4, "Negative Feedback"; Kline, "Harold Black and the Negative-Feedback Amplifier"; and the film, *Communications Milestone: Negative Feedback.*

3. Bode, *Synergy,* 138–40.

4. Brooks, *Telephone,* 142.

5. Brittain, "Introduction of the Loading Coil." See also Espenschied, Jackson, and Brainerd, "Discussion of James G. Brittain, 'The Introduction of the Loading Coil'"; and Wasserman, *From Invention to Innovation.*

6. Carson, *Electric Circuit Theory and the Operational Calculus;* Bush, *Operational Circuit Analysis.* For background on Heaviside, see Bennett, *History of Control Engineering, 1800–1930,* 195–200; Nahin, *Oliver Heaviside;* and Yavetz, *From Obscurity to Enigma.*

7. Late in his career, Steinmetz did work on transient phenomena and made important contributions to the understanding of transients in electric power systems (Kline, *Steinmetz,* 138–49).

8. Gherardi, "Dean of Telephone Engineers"; Fagen, *Early Years,* 32–35, 44. Ironically, in a consolidation of research Carty closed Western Electric's Boston engineering department, which had been investigating Lee de Forest's audion for use as an amplifier. Hugh Aitken argues that the closing of the lab may have cost the company several years toward making a practicable telephone amplifier (Aitken, *Continuous Wave,* 75–78). By contrast, other historians, including Lillian Hod-

deson and Leonard Reich, see the move to a single department in New York as progressive toward industrial research. Hoddeson, "Emergence of Basic Research," 530, notes that the phrase *fundamental research* began to appear in the company's rhetoric about 1907; Coon, *American Tel & Tel,* 197, also makes the point about fundamental research. See also Reich, *Making of American Industrial Research,* for the defensive aspect of early industrial research.

9. Shaw, "Conquest of Distance by Wire Telephony," reprints Carty's original proposal for the transcontinental line. Reich quotes Carty's understanding of the strategic importance of the repeater: "A successful telephone repeater . . . would not only react most favorably on our service where wires are used, but might put us in a position of control with respect to the art of wireless telephony, should it turn out to be a factor of importance" (quoted in Reich, "Industrial Research and the Pursuit of Corporate Security," 512). The organization chart of the AT&T/Western Electric engineering departments in 1905, 1907, 1909, 1911, 1915, and 1925 is reprinted in Shaw, "Conquest of Distance by Wire Telephony," 400–406; and Fagen, *Early Years,* 43–55.

10. Frank Baldwin Jewett to Robert Millikan, n.d., quoted in Fagen, *Early Years,* 258. Jewett and Millikan had boarded together at Chicago, and Jewett was the best man at Millikan's wedding (Millikan, *Autobiography,* 52–53, 116–17). Millikan remained a consultant in long-distance telephony, and his testimony helped settle the protracted suit between G.E. and AT&T over the vacuum tube (ibid., 120–22).

11. Hoddeson, "Emergence of Basic Research," 533. In *Science in Action,* 125–26, Bruno Latour uses Jewett's appropriation of the electron as an example of machines as abstract apparatus for tying together interested groups. See Shaw, "Conquest of Distance by Wire Telephony," 376–79, for a detailed discussion of mechanical and mercury arc repeaters. See also Fagen, *Early Years,* 241–56.

12. Shaw, "Conquest of Distance by Wire Telephony," 375, 379–82. In Aitken's view, the distance between telephony and wireless delayed the Bell System's adoption of the audion for a number of years (Aitken, *Continuous Wave,* 244–45, 546).

13. Colpitts, "Dr. H. D. Arnold." Actually, mechanical repeaters initially carried the transcontinental line, but they were quickly replaced with electronic ones. Gradually, more repeaters were added and the number of loading coils reduced; the coils reduced the bandwidth of transmission and the speed of signal propagation, which led to problems with echoes. Shaw, "Conquest of Distance by Wire Telephony," 389–92, provides a detailed technical description of the transcontinental line. The line was not permanent but was "built up by switches" when needed, as was the New York–Denver line (Fagen, *Early Years,* 263–64).

14. Nance and Jacobs, "Transmission Features of Transcontinental Telephony," 1062.

15. Martin, "Transmitted Frequency Range for Telephone Message Circuits"; Martin, "Transmission Unit and Telephone Transmission Reference Systems"; Borst, "Decibel"; Hartley, "TU Becomes 'Decibel'"; Johnson, "Thermal Agitation of Electricity in Conductors"; Nyquist, "Thermal Agitation of Electric Charge in Conductors."

16. The transcontinental line so solidified the alliance technically that loading coils were gradually removed from the network. The transcontinental line was fully unloaded in 1920, more than

tripling the velocity of transmission, which reduced echo effects and improved the sense of nearness of the speakers (Shaw, "Conquest of Distance by Wire Telephony," 396).

17. In 1926 Bell Labs opened a laboratory in Whippany, New Jersey, for radio research. In 1941 it moved most of its operations to a new campus in Murray Hill, New Jersey. Fagen, *Early Years,* 54–55, compares Bell Labs with the old AT&T and Western Electric engineering organizations. For its similarity to the initial Bell Labs organization, outlined below, see the organizational charts in Shaw, "Conquest of Distance by Wire Telephony," 406.

18. Hallenbeck, "Inspection Engineering Department"; Lyng, "Development of Apparatus"; Findley, "Apparatus Development Department"; idem, "Mathematical Research"; Reich, *Making of American Industrial Research,* 213; Fry, "Industrial Mathematics."

19. Findley, "Research Department," quotation on 164.

20. Findley, "Systems Development Department."

21. Arnold, "Systematized Research," 313.

22. Charlesworth, "General Engineering Problems of the Bell System."

23. Stone Stone, "Practical Aspects of the Propagation of High Frequency Electric Waves along Wires," described high-frequency multiplex telephony as "identical with that of the new continuous wave train" radio, and he included the Alexanderson alternator as an element of telephone design. See also Espenschied, "Application of Radio to Wire Transmission Engineering"; and for "wired wireless," see Fagen, *Early Years,* 282.

24. Colpitts and Blackwell, "Carrier Current Telephony and Telegraphy," contains a detailed history of carrier methods in telephony, as well as an elegant explanation of carrier modulation and transmission.

25. In 1924 the "C" system went into service, incorporating lessons from the more experimental A and B systems. C carrier systems were so successful that the last one was not removed from service until 1980 (O'Neill, *Transmission Technology,* chap. 1, "The State of the Technology [1925–1930]," 3–14).

26. Charlesworth, "General Engineering Problems of the Bell System."

27. See the table in O'Neill, *Transmission Technology,* 3.

28. This account is based on Black, "Inventing the Negative Feedback Amplifier"; and Harold S. Black to A. C. Dickieson, 16 June 1974, ATT. For a typical effort to design linear vacuum tube amplifiers, see Kellogg, "Design of Non-Distorting Power Amplifiers."

29. Harold S. Black, "Translating System," U.S. Patent 1,686,792 (9 October 1928).

30. Even if one needs to string several feedback amplifiers together to get the desired amplification, the gains accumulate by multiplying, whereas the distortions accumulate by adding. Two amplifiers used as in this example would combine to produce the original "open-loop" gain of 10,000, but with a distortion of 1/100 of the original distortion, which is still a 98 percent reduction of distortion for the same gain.

31. Black, "Inventing the Negative Feedback Amplifier."

32. Ibid., 59–60.

33. Black to Dickieson, 16 June 1974; Harold S. Black, patent application 298,155 (8 August 1928); "File History of Black Application Serial No. 298,155," ATT.

34. Friis and Jensen, "High Frequency Amplifiers."

35. Dickieson's recollections of Black's conflict with H. D. Arnold, and intimations of constant conflict with his superiors correlate with Black, "Inventing the Negative Feedback Amplifier," 59–60.

36. Dickieson to M. J. Kelley, 6 July 1972, ATT, emphasis added.

37. Friis and Jensen, "High Frequency Amplifiers," 204.

38. Dickieson to Kelley, 6 July 1972.

39. "When many amplifiers are worked in tandem . . . it becomes difficult to keep the overall circuit efficiency constant, variations in battery potentials and currents, small when considered individually, adding up to produce serious transmission changes in the overall circuit," wrote Black. "Stabilized Feedback Amplifiers" was presented at the winter convention of the AIEE, New York City, January 1934, and it was published in *Electrical Engineering* in the same month (see the discussions in *Electrical Engineering* by F. A. Cowan [April 1934], G. Ireland and H. W. Dudley [March 1934], and Harry Nyquist [September 1934]).

40. Affel, Demarest, and Green, "Carrier Systems on Long Distance Telephone Line," 384. The third author of this paper, C. W. Green, was Harold Black's boss.

41. Black, patent application 298,155, 8 August 1928; "File History of Black Application Serial No. 298,155" ATT.

42. Harold S. Black, "Wave Translation System," U.S. Patent 2,102,671 (21 December 1937), 2, emphasis added.

43. In 1921, for example, Colpitts and Blackwell, in "Carrier Current Telephony and Telegraphy," wrote that singing in a carrier system could arise when the gain was greater than 1 and when there existed "sufficient unbalance" between the circuits (313).

44. Harvey Fletcher analyzed the howling telephone as a dynamic electrical system to understand the relationship between impedance, frequency, and the tendency to break into the oscillation in "The Theory of the Operation of the Howling Telephone with Experimental Confirmation." Fletcher's paper does not employ the term *stability* or *feedback*, although it does analyze electro-acoustic circuits, which greatly resemble canonical feedback systems. (Shaw, "Conquest of Distance by Wire Telephony," 382–83). For the problems of handset howling, see Fagen, *Early Years*, 146–50.

45. Bennett, *History of Control Engineering, 1930–1955*, 77; Foster, "Reactance Theorem."

46. Johnson, "Transmission Regulating System for Toll Cables." George Ireland made this observation in his discussion of Black's paper, "Stabilized Feedback Amplifiers."

47. Stoller, "Synchronization and Speed Control of Synchronized Sound Pictures"; Stoller and Morton, "Synchronization of Television"; Stoller "Speed Control for the Sound-Picture System."

48. Stoller, "Speed Control for the Sound-Picture System." Stoller also published on voltage regulators (see Stoller and Power, "Precision Regulator for Alternating Voltage").

49. Bode, "Harry Nyquist, Obituary," April 1977, ATT.

50. For a detailed account of the Morristown trial, see Clark and Kenall, "Carrier in Cable"; and O'Neill, *Transmission Technology*, chap. 5, "Carrier on Cable." Making the system work as planned proved no simple matter, but such was the purpose of an engineering trial. Repeater amplifiers

did not pose the only problems: cable design (the number, size, and shielding of each of the many wire pairs) proved especially critical as well. Shielding, grounding, and interference between signals plagued the system. Because of the Depression, AT&T changed its emphasis from creating new systems to improving capacity with the existing plant. Bell Labs engineers thus had several years to refine the results of the Morristown trial and to work on ways of compressing more transmission onto existing wires. The Morristown trial formed the basis for the type "K" carrier system, introduced in the late 1930s, which carried 12 voice channels on cables at frequencies from 12 to 50 kHz for distances of up to 4,000 miles. K carrier furnished 70 percent of the increased capacity in the country (which doubled from 1940 to 1947) and remained in service until at least 1980. K carrier also included a pilot-wire transmission regulation scheme, with an automatic self-balancing regulator and a self-synchronizing motor (Green and Green, "Carrier Telephone System for Toll Cables").

51. Harry Nyquist, "Constant Current Regulation," U.S. Patent 1,887,599 (15 November 1932), and "Phase Regulating System," U.S. Patent 1,683,725 (11 September 1928), applications filed in 1928 and 1926, respectively. Hamilton, Nyquist, Long, and Phelps's paper "Voice-Frequency Carrier Telegraph System for Cables," which erroneously gives Nyquist's first initial as N., also includes a discussion of the precision governor required for generating carrier frequencies for this telegraph system, which suggests that Nyquist had been exposed to regulation before his 1932 paper on feedback, "Regeneration Theory."

52. Nyquist, "Discussion of H. S. Black, 'Stabilized Feed-Back Amplifiers.'"

53. Nyquist, "Regeneration Theory"; Bennett, *History of Control Engineering, 1930–1955*, 82–84.

54. Nyquist, "Regeneration Theory." In 1934 Bell Labs engineers compared Nyquist's criterion with E. J. Routh's test from his 1877 Adams Prize paper on stability in dynamic mechanical systems. Finding the two stability analyses to be compatible, they linked the new feedback theory to the older work on dynamic stability. Despite this link, however, their work makes no mention of applying feedback amplifier theory to other dynamic systems (Peterson, Kreer, and Ware, "Regeneration Theory and Experiment").

55. Bennett, *History of Control Engineering, 1930–1955*, 83; H. Nyquist, "Regenerative Amplifier," U.S. Patent 1,915,440 (27 June 1933).

56. Millman, *Communications Sciences*, 16–17; O'Neill, *Transmission Technology*, 204–8. For a good summary of the work on network theory in the 1920s and 1930s, see Wildes and Lindgren, *Century of Electrical Engineering and Computer Science*, chap. 9.

57. Bode, "Feedback."

58. Bode, "General Theory of Electric Wave Filters."

59. Espenschied and Strieby, "Systems for Wide-Band Transmission over Coaxial Lines"; Strieby, "Million-Cycle Telephone System"; O'Neill, *Transmission Technology*, chap. 6, "Coaxial Cable," esp. 131–39. The system Bode worked on became known as the L1; it was tested on a line from New York to Philadelphia in 1936–38 and put into service just before World War II.

60. Bode, "Variable Equalizers." Black acknowledged this use of feedback paths in "Feedback Amplifiers."

61. H. W. Bode, "Design Method for Feedback Amplifiers—Case 19878," 1 May 1936, ATT.

62. Bode, "Relations between Attenuation and Phase in Feedback Amplifier Design." For other discussions of this paper, see Bennett, *History of Control Engineering, 1930–1955,* 84–86; Millman, *Communications Sciences,* 29–30; and O'Neill, *Transmission Technology,* 68–70. In later years, Bode displayed some aversion to Black's version of events. He wrote to Dickieson in 1974, after reviewing Black's account: "This is not exactly how one ordinarily writes formal technical history [interestingly, Bode had some notion of "formal" technical history]. . . . In a less personalized account, it might be possible to present basic technological issues in a more satisfactory way" (Bode to Dickieson, 17 September 1974, ATT).

63. Bode, "Relations between Attenuation and Phase in Feedback Amplifier Design," 426–35.

64. Bode, "Feedback," 117.

65. Bode, *Network Analysis and Feedback Amplifier Design,* iii–iv.

66. "Spinal Cord of a Nation."

67. Mills, "Communication with Electrical Brains," 52; "A Mechanical Brain"; AIEE Committee on Communication, "Annual Report: Recent Advances in the Communication Art"; idem, "Annual Report: Electrical Communication." The annual reports of all AIEE committees track a wide range of electrical technologies during this period. For a social history of automatic switching, see Lipartito, "When Women Were Switches"; Joel, *Switching Technology;* Chapuis, *One Hundred Years of Telephone Switching;* Brooks, *Telephone,* 193; and Dahl, "Improved Equipment for Information Service."

68. Hochheiser, "What Makes the Picture Talk."

69. Findley, "Research Department"; Hartley and Fry, "Binaural Location of Complex Sounds."

70. "Western Electric Wartime Developments, 1917–1918," ATT.

71. Fletcher, "Nature of Speech and its Interpretation"; idem, "Physical Measurements of Audition and their Bearing on the Theory of Hearing"; idem, "Useful Numerical Constants of Speech and Hearing"; McGinn, "Stokowski and the Bell Telephone Laboratories"; Millman, *Communications Sciences,* 93–102. The tone and loudness controls on modern audio equipment emerged from this research. For the cultural landscape of sound reproduction in the early twentieth century, see Thompson, *Soundscape of Modernity.*

72. Black to Dickieson, 14 August 1974, ATT. Stokowski, Bode, and Fletcher maintained a long collaboration (Leopold Stokowski to Bode, 8 and 30 March 1940, Hendrik Bode Papers, box 1, folder 1, Harvard University Archives, Cambridge, Massachusetts; *New York Times,* 13 April 1933, quoted in McGinn, "Stokowski and the Bell Telephone Laboratories," 59). Stokowski manipulated the controls with such enthusiasm that he sometimes irritated those who preferred the orchestra's own volume variations.

73. Dudley, "Carrier Nature of Speech."

74. Rydell, *World of Fairs,* 126–27; Steinberg, Montgomery, and Gardner, "Results of the World's Fair Hearing Tests."

75. Coon, *American Tel & Tel,* 203; "Bell System Exhibit at the Century of Progress Exposition." Multiplexing of telegraph signals had gone on for many decades; Bell was working on a telegraph multiplexer when he invented the telephone.

76. Nyquist, "Certain Factors Affecting Telegraph Speed"; idem, "Certain Topics in Telegraph

Transmission Theory." See also the discussion of the latter paper by Nyquist's son-in-law, John C. Lozier, "The Oldenberger Award Response: An Appreciation of Harry Nyquist."

77. Nyquist's measure, that a wave must be sampled at twice its bandwidth to be transmitted without distortion, is frequently referred to as the "Nyquist rate." A modern compact disc player, for example, samples the music at 44 kHz in order to reproduce music in the audible band of about 20 kHz.

78. Hartley, "Transmission of Information." See brief discussions of Nyquist and Hartley in Cherry, "History of the Theory of Information"; and Pierce, "Early Days of Information Theory." Comparing Nyquist's and Hartley's work on transmission, Pierce wrote, without elaboration, that "it is [Claude] Shannon's feeling, and mine, that Nyquist's work was more fruitful." Cherry found Hartley's work more general than Nyquist's, and noted, "Hartley's has a very modern ring about it . . . [it] may be regarded as the genesis of the modern theory of the communication of information." See also Segal, "Théorie de l'information."

79. Shannon, "Mathematical Theory of Communication," reprinted in Shannon, *Claude Elwood Shannon,* 5–83.

80. Jewett, "Electrical Communication, Past, Present, and Future."

*Chapter 5 Artificial Representation of Power Systems*

1. The concept of "feedback culture" expands on Donald Mackenzie's idea of "gyro culture" in *Inventing Accuracy,* 31.

2. Mayr, "Maxwell and the Origins of Cybernetics"; Fuller, "Early Development of Control Theory."

3. Bennett, *History of Control Engineering, 1800–1930,* 74–88.

4. Minorsky, "Directional Stability of Automatically Steered Bodies"; Bennett, "Nicholas Minorsky and the Automatic Steering of Ships."

5. Minorsky, "Directional Stability of Automatically Steered Bodies," 282–83. See also Bennett, *History of Control Engineering, 1800–1930,* 142–47.

6. Minorsky, "Principles and Practice of Automatic Control"; Bennett, "Development of the PID Controller."

7. For another discussion of the nature and significance of Hazen's work, see Bennett, "Harold Hazen and the Theory and Design of Servomechanisms." The chapter "Theory and Design of Servomechanisms," in Bennett's *History of Control Engineering, 1930–1955,* presents a similar discussion and makes the observation of the familiarity of Hazen's language for a modern control engineer (110).

8. Hazen, "Extension of Electrical Engineering Analysis."

9. Layton, *Revolt of the Engineers,* 68; Noble, *America by Design.*

10. Jordan, *Machine-Age Ideology,* 121–22.

11. Kline, *Steinmetz.*

12. For the history of the interconnection of electrical power networks, see Hughes, *Networks of Power;* and Hunter and Bryant, *History of Industrial Power in the United States,* 364–67. For a

highly technical history of the control of electric power systems from the point of view of the Leeds and Northrup Company, see Cohen, "Recollections of the Evolution of Realtime Control Applications to Electric Power Systems."

13. See AIEE Committee on Power Transmission and Distribution, "Annual Report." This committee included Ralph Booth and Vannevar Bush. For a general review of the subject, see Fortescue, "Transmission Stability." Stuart Bennett noted the connection between power system stability and control engineering but concluded that "the problems of power-system stability although recognized early did not lead to any theoretical developments in control systems" (*History of Control Engineering, 1800–1930*, 170–71).

14. Fortescue, "Transmission Stability," 987.

15. Kline, *Steinmetz*, 148–49.

16. Cohen, "Recollections of the Evolution of Realtime Control to Electric Power Systems."

17. Carlson, "Academic Entrepreneurship and Engineering Education"; Pang, "Edward Bowles and Radio Engineering at MIT"; Lecuyer, "Making of a Science Based Technological University"; Owens, "MIT and the Federal 'Angel'"; Servos, "Industrial Relations of Science."

18. For Gordon Brown's recollection, see interview by Richard R. Merz, 27 January 1970, Computer Oral History Collection, National Museum of American History.

19. Bush, "Gimbal Stabilization"; Wildes and Lindgren, *Century of Electrical Engineering and Computer Science,* chap. 4. For biographical background on Bush, see Owens, "Straight-Thinking"; and Zachary, *Endless Frontier.*

20. Lecuyer, "Making of a Science Based Technological University"; Servos, "Industrial Relations of Science."

21. Carson, "Theory of Transient Oscillations in Electric Networks"; Nahin, *Oliver Heaviside.*

22. Bush, *Operational Circuit Analysis,* 1–2.

23. Bush and Booth, "Power System Transients."

24. Ibid., 229.

25. Owens, "Straight-Thinking," 127.

26. Bush and Booth, "Power System Transients," 232.

27. Fortescue, "Discussion of Bush and Booth, 'Power System Transients.'" This discussion, by six commentators, provides a good overview of the state of the stability problem in 1925.

28. Harold Hazen, interview by Richard R. Merz, 15 December 1970, Computer Oral History Collection, National Museum of American History.

29. Terman's *Radio Engineer's Handbook* became a standard text for electronics. For Terman's later work at Stanford, see Leslie, *Cold War and American Science;* Leslie and Hevly, "Steeple Building at Stanford."

30. Terman, "Characteristics and Stability of Transmission Systems," 1.

31. Ibid., 274.

32. Edgerton, "Abrupt Change in Load on a Synchronous Machine"; idem, "Benefits of Angular-Controlled Field Switching"; Wildes and Lindgren, *Century of Electrical Engineering and Computer Science,* 145–47.

33. Brown and Germeshausen, "Effect of Controlled Field Switching"; Green, "Static Study of

the No Load Flux Distribution"; Constantine Barry, " Parallel Operation of Alternators"; Wang, "Study of Synchronous Machines Not Running at Synchronous Speed."

34. Hughes, *Networks of Power,* 23. For a typical example at the time, see Woodward, "Calculating Short-Circuit Currents in Networks." Edwin Harder would employ power system simulations and analog computers to study servo and regulator problems for many years after World War II (see Aspray, "Edwin L. Harder and the Anacom").

35. Hazen and Spencer, "Artificial Representation of Power Systems"; Wildes and Lindgren, *Century of Electrical Engineering and Computer Science,* 99; Harold Hazen, interview by Marc Miller, 2 March 1977, Belmont, Mass., Computers at MIT Oral History Collection, MIT Archives, Cambridge, Mass., 9. See also Hazen interview, 15 December 1970, which has many similar comments by Hazen.

36. Hazen and Spencer, "Artificial Representation of Power Systems," 24–31. For a discussion of models and simulations in engineering, see Holst, "George A. Philbrick and Polyphemus," 144; and Ferguson, *Engineering and the Mind's Eye.*

37. Hazen interview, 2 March 1977, 12; Hazen and Spencer, "Artificial Representation of Power Systems," 25.

38. Hazen interview, 2 March 1977, 7.

39. Ibid., 20, emphasis in the original.

40. Ibid.

41. Hazen, "Memoirs," 65.

42. Hazen interview, 15 December 1970.

43. Schurig, Hazen, and Gardner, "M.I.T. Network Analyzer"; Wildes and Lindgren, *Century of Electrical Engineering and Computer Science,* 96–105.

44. Wildes and Lindgren, *Century of Electrical Engineering and Computer Science,* 103.

45. Hazen, "Electrical Water-Level Control."

46. Carson, "Theory of Transient Oscillations in Electric Networks."

47. Bush, Gage, and Stewart, "Continuous Integraph." For this and his other 1927 paper, with Harold Hazen, the Franklin Institute awarded Bush its Levy Medal in 1928.

48. For histories of mechanical integrating machines, see Bromley, "Analog Computing Devices"; Horsburgh, *Handbook of the Napier Tercentenary Celebration;* Clymer, "Mechanical Analog Computers of Hannibal Ford and William Newell"; and idem, "Mechanical Integrators."

49. Bush, Gage, and Stewart, "Continuous Integraph," 65.

50. Bush, "New Type of Differential Analyzer."

51. Thompson, "Mechanical Integration of the Linear Differential Equations"; Paynter, "Differential Analyzer as an Active Mathematical Instrument"; Bromley, "Difference and Analytical Engines."

52. Ford, "Mechanical Movement," U.S. Patent 1,317,915; Clymer, "Mechanical Integrators." Curiously, Wildes and Lindgren, *Century of Electrical Engineering and Computer Science,* 89, shows a picture of the Ford integrator, even though it was not used in any of the MIT machines.

53. Bush and Hazen, "Integraph Solution of Differential Equations," 586–88.

54. Hazen, "Working Mathematics by Machinery," 345; see also Hazen interview, 2 March 1977, 12. Lord Kelvin had also been involved in the genesis of the fire control systems for *Dreadnought*-era battleships, and he was on the board of directors of the company of Arthur Hungerford Pollen, who designed those systems (see Sumida, *In Defence of Naval Supremacy*, 78).

55. Paynter, "Differential Analyzer as an Active Mathematical Instrument," 3–7.

56. Documents on the daily operation of the machine at MIT are scarce, but see Arthur Porter's notebook on the Manchester machine for a good sense of the engineering practice of a differential analyzer (MS 474, Science Museum Library, Imperial College, London).

57. Vannevar Bush, "Proposal for the Design of an Improved Differential Analyzer," 22 April 1935, RAC 1.1, series 224, project files, Differential Analyzer, folder 22.

58. Brown interview, 27 January 1970.

59. Hartree et al., "Time Lag in a Control System"; Hartree, *Calculating Instruments and Machines*, chap. 3. For the G.E. analyzer, see Kuehni and Peterson, "New Differential Analyzer"; and Wildes and Lindgren, *Century of Electrical Engineering and Computer Science*, 92.

60. Bush, "Proposal for the Design of an Improved Differential Analyzer."

61. Bush, "Differential Analyzer."

62. Ibid., 465, emphasis added.

63. Ibid., 459.

64. Ibid., 477.

65. See, e.g., Aspray, *Computing before Computers*, 183; and Campbell-Kelly and Aspray, *Computer*, 62–63.

66. Owens, "Vannevar Bush and the Differential Analyzer," 87.

67. Bush, *Pieces of the Action*, 262.

68. Hazen, "Extension of Electrical Engineering Analysis," 4.

69. Bush, Gage, and Stewart, "Continuous Integraph," 69. For the importance of numerical devices in business practice, see Cortada, *Before the Computer*.

70. Hazen, "Working Mathematics by Machinery," 326.

71. Hazen, "Extension of Electrical Engineering Analysis," 147.

72. Hazen, Jaeger, and Brown, "Automatic Curve Follower"; Gordon Brown, telephone interview by author, 26 August 1994. In later years, other groups built automatic curve followers that proved more practicable.

73. Gordon Brown, "An Amplifier Wattmeter Combination for the Accurate Measurement of Watts and Vars," Gordon Brown Papers, MIT Archives. This paper cites Black, "Stabilized Feedback Amplifiers"; and Tellegen, "Inverse Feedback."

74. Brown, "Photocell Receiver and a Direct Current Vacuum-tube Amplifier for the Cinema Integraph"; idem, "Cinema Integraph"; Hazen and Brown, "Cinema Integraph."

75. Brown interview, 26 August 1994. See, e.g., Howard, "Measurement and Analysis of Errors in the Cinema Integraph"; Hedeman, "Numerical Solutions of Integral Equations on the Cinema Integraph"; Brown interview, 27 January 1970.

76. Brown, "Cinema Integraph," 33.

77. Hazen, "Theory of Servo-Mechanisms," 279.

78. Ibid., 284.

79. Trinks described a "relay" governor as one that used an additional source of power, such as a hydraulic valve, besides the rotating balls to regulate an engine, a definition different from Hazen's. For discrete time servos, see Jury, "On the History and Progress of Sampled-Data Systems," 17; and Bissel, "Modeling Sampled-Data Systems."

80. Hazen, "Theory of Servo-Mechanisms," 286.

81. Ibid., 315, 329.

82. Ibid., 283, 281.

83. Ibid., 283.

84. Ibid., 281.

85. Ibid., 328.

86. Hazen, "Memoirs," 3–9.

87. See, e.g., James, Nichols, and Phillips, *Theory of Servomechanisms*, 16; Gordon Brown, "Behavior and Design of Servomechanisms," OSRD 39, Report to the Services 2, November 1940, MIT, reprinted as Brown, "Transient Behavior and Design of Servomechanisms."; and Harris, "Frequency Response of Automatic Control Systems."

88. Harold Hazen to Stuart Bennett, 22 October 1975, courtesy Stuart Bennett.

89. Hazen, "Theory of Servo-Mechanisms," 282.

90. Hazen, "Design and Test of a High-Performance Servo-Mechanism."

91. Hendrik Bode Papers, box 1, folder 2.

92. Hazen to Bennett, 22 October 1975.

93. Hazen, "Discussion of H. A. Thompson, 'A Stabilized Amplifier for Measurement Purposes.'"

94. John Taplin, interview by author, 10 August 1995, Wellesley, Mass., notes in author's possession. See also Bennett, *History of Control Engineering, 1930–1955*, 90.

95. Weaver diary, 21 November 1932, RAC 1.1, series 224, project files, Differential Analyzer, folder 22. On Weaver and his machine shop, see Warren Weaver oral history, RAC, 91. See also the account in Owens, "Vannevar Bush and the Differential Analyzer," 75–79.

96. Bush to Weaver, 7 July 1933; Weaver to Bush, 10 July 1933; and Bush to Weaver, 16 November 1934, RAC 1.1, series 224, project files, Differential Analyzer, folder 22.

97. Bush to Weaver, 17 March 1936, ibid., folder 23.

98. Bush, "Proposal for the Design of an Improved Differential Analyzer."

99. Weaver diary, 17 May 1935, 14 April 1936, RAC 1.1, series 224, project files, Differential Analyzer, folder 22.

100. Weaver diary, 22 March, 9 April, and 12 April 1936, ibid., folder 23; Bush to Max Mason, 13 April 1936, ibid., folder 24.

101. S. H. Caldwell, "Present Status of the Project Design of an Improved Differential Analyzer," 1 October 1935 and 16 December 1935, ibid., folders 22 and 25, respectively.

102. Ibid., 16 December 1935; Bush to Weaver, 17 March 1936.

103. See also S. H. Caldwell, "Prospectus of a Computing and Analyzing Center at the MIT," 30 September 1937, and Weaver diary, 10 January 1939, ibid., folders 24 and 25, respectively; and Owens, "Vannevar Bush and the Differential Analyzer," 80.

104. Caldwell to Weaver, and Weaver diary, 30 March 1940, RAC 1.1, series 224, project files, Differential Analyzer, folder 25.

105. Weaver diary, 12 March 1942, ibid., folder 26. Two other computing projects at MIT dried up because of the war: Bush's rapid selector, a large-scale information-retrieval device based on microfilm and the "rapid arithmetical machine," also conceived by Bush, which sought to employ vacuum tubes to accomplish binary switching.

106. Bush, "New Type of Differential Analyzer"; S. H. Caldwell, "Progress in the Design and Construction of the New Differential Analyzer," 27 December 1937, RAC 1.1 series 224, project files, Differential Analyzer, folder 27, describes the totality of the donation, which consisted of a truckload of equipment weighing more than 17 tons.

107. Shannon, "Symbolic Analysis of Relay Switching Circuits," reprinted in Shannon, *Claude Elwood Shannon*, 713.

108. Bush, "Differential Analyzer," 448.

109. Stibitz, "Zeroth Generation," 74.


*Chapter 6 Dress Rehearsal for War*

1. Hazen, "Memoirs"; Bush, *Pieces of the Action*, 183.

2. Brown interview, 26 August 1994.

3. Edwin B. Hooper, interview by Richard T. Glasow and Nelson Wood, 22 August 1978, and interview by A. B. Christman, February 1971, Washington, D.C., Edwin Hooper Papers, Oral Histories folder and box 10, Library of Congress; Horacio Rivero, interview by John T. Mason Jr., 20 May 1975, Newport, R.I., Naval Operational Archives, Washington, D.C. See also biographies for Hooper, Rivero, Mustin, and Ward; and Hooper, interview by John T. Mason Jr., 23–26 June 1970, Washington, D.C., all in Admirals Biographies Collection, ibid.

4. This account comes primarily from Rivero's oral history, with additions from Hooper oral history, both in Admirals Biographies Collection.

5. Brown interview, 26 August 1994. See also the manuscript version of Wildes and Lindgren, *Century of Electrical Engineering and Computer Science*, 5-10 to 5-15, in Karl Wildes Papers, MIT Archives.

6. Brown interview, 27 January 1970.

7. Hooper interview, February 1971.

8. Hooper and Ward, "Control of an Electro-Hydraulic Servo Unit."

9. Rivero and Mustin, "Servo Mechanism for a Rate Follow-up System," 2.

10. Ibid., 10.

11. Rivero oral history; Brown interview, 26 August 1994.

12. Lloyd Mustin, memorandum introducing S. M. thesis upon declassification, 1971, MIT

Archives. Mustin recalled that "though Dr. Draper did not suggest any gun control applications at that time, he later acknowledged the contribution of this thesis to the development of his own concepts." For a detailed account of Draper's relationship to Sperry in the 1930s, see Dennis, "Change of State," chap. 2, and for the lead-computing gunsight. see ibid., chap. 4. For an example of MIT and Sperry Gyroscope's collaborative research, see Draper, Bentley, and Willis, "M.I.T.-Sperry Apparatus for Measuring Vibration."

13. Brown and Campbell, *Principles of Servomechanisms,* 9.

14. For a detailed technical summary, see Bennett, *History of Control Engineering, 1930–1955,* 138–39.

15. Millikan to Secretary of the Navy, 8 January 1935, RAC, General Education Board, series I.4, CIT, box 611, folder 6472.

16. McBride, "Greatest Patron of Science?"; McDowell, "Naval Research."

17. C. S. McDowell, "Final Report on the 200-Inch Telescope Project," [1938?], RAC, General Education Board, series I.4, CIT, box 612, folder 6473, 1.

18. Ibid. See also Florence, *Perfect Machine,* 296.

19. Sinclair Smith, "The Control and Drive for the 200-Inch Telescope," 26 August 1936, RAC, General Education Board, series I.4, CIT, box 612, folder 6473, 2; McDowell, "Final Report," 5.

20. Max Mason to George Ellery Hale, 2 and 8 January 1935, RAC, General Education Board, series I.4, CIT, box 611, folder 6472.

21. C. S. McDowell to Mason, 10 January 1935, ibid.

22. Mason to Hale, 5 December 1934, ibid., folder 6471; Weaver diary, 29 March 1935; Mason to Bush, 24 May 1935, RAC 1.1, series 224, project files, Differential Analyzer.

23. Weaver diary, 29 March 1935; Mason to Bush, 24 May 1935.

24. Mason to Bush, 24 May 1935; Weaver diary, 22 March 1936.

25. Mason to McDowell, 13 June 1935, RAC, General Education Board, series I.4, CIT, box 611, folder 6472. On Bush and Ford's participation, see also McDowell, "Final Report," 7.

26. Bush to Ford, 26 May 1935, RAC 1.1, series 224, project files, Differential Analyzer. Bush, *Pieces of the Action,* 183–85, seems to suggest that he met Ford.

27. McDowell to Furlong, 11 January 1938, RG 74, S71 Restricted, box 53.

28. Poitras and St. Louis, "Photoelectric Efficiency for Visible and Ultra-Violet Radiation."

29. Max Mason diary, 27 September 1938, RAC, General Education Board, series I.4, CIT, box 611, folder 6473.

30. F. H. Dean to Chief of Bureau, 27 July 1935, and McDowell to Furlong, 10 November 1937, RG 74, S71 Restricted, box 53.

31. McDowell, "The 200-inch Telescope."

32. Rule, "Electrical Features of the 200-Inch Telescope," 72.

33. Florence, *Perfect Machine,* 321.

34. Ibid.

35. Ibid., 331.

1. See, e.g., Kevles, *Physicists;* and Rhodes, *Making of the Atomic Bomb.* For discussions of wartime research in the context of modern science policy, see Forman, "Behind Quantum Electronics"; Baxter, *Scientists against Time;* and Stewart, *Organizing Scientific Research for War,* 322. Baxter's book was the "short version," of the official history. The long version was published as separate volumes, of which the most relevant to the present study are Boyce, *New Weapons For Air Warfare;* and Guerlac, *Radar in World War II.*

2. Dupree, "The *Great Instauration* of 1940"; idem, *Science in the Federal Government,* chap. 19.

3. Reingold, "Vannevar Bush's New Deal for Research"; Pursell, "Science Agencies in World War II"; Hart, *Forged Consensus,* chap. 5; Owens, "Counterproductive Management of Science in the Second World War."

4. Bush to Herbert Hoover, 27 April 1945, Vannevar Bush Papers, Library of Congress, box 51, folder 1261.

5. Owens, "Counterproductive Management of Science in the Second World War," 537n; Dennis, "Change of State," 357.

6. Bush to Hoover, 10 and 29 April 1939, Vannevar Bush Papers, box 51, folder 1261.

7. Bush to Jewett, 23 March 1939, Jewett folder, Bush file, Carnegie Institution of Washington, quoted in Pursell, "Science Agencies in World War II," 360.

8. Draft memorandum, n.d., OSRD, Central Classified File.

9. Pursell, "Science Agencies in World War II."

10. Bush, *Pieces of the Action,* 31–32, quoted in Owens, "Counterproductive Management of Science in the Second World War," 522.

11. Office Files of Karl Taylor Compton, NDRC Misc. folder, OSRD7, has several early organization charts for the NDRC that do not include any section devoted to fire control, although Owens reprints a chart Bush shared informally with colleagues in June 1940, before Roosevelt's executive order, which includes a division for fire control, in "Counterproductive Management of Science in the Second World War," 523.

12. For the official history of D-2 and Division 7, see Boyce, *New Weapons for Air Warfare,* chaps. 3–9. The sources for these chapters are the personal histories written for Boyce by the section members in OSRD7 GP, History File. Boyce published these accounts virtually intact but edited out much of the most interesting material on institutional friction, conflict, and competition. I therefore cite the original members' histories wherever possible, although only for general observations. For the actual chronologies of committee activities, I work from the original committee documentation on which the members' histories are based.

13. Warren Weaver oral history, RAC, 561–62.

14. Weaver, *Scene of Change,* 45.

15. Kohler, *Partners in Science,* chap. 10. Kohler examines Weaver's time at the Rockefeller Foundation in detail but does not include his wartime work. Reingold, "Vannevar Bush's New Deal for

Research," 325, comments that Bush's proposal for NSF contracts derived from a combination of Weaver's Rockefeller strategy and the OSRD contract but does not note Weaver's role in the two.

16. Warren Weaver, "N.S. Notes on Officers' Techniques," 11 January 1946, RAC, RG 3, series 915, box 2, folder 13, 2.

17. Kohler, *Partners in Science,* 302; Warren Weaver oral history, RAC, 435.

18. Warren Weaver, "How Do You Do, Dr. X?" August 1952, RAC, RG 3, series 915, box 2, folder 13, 2.

19. Weaver, "N.S. Notes on Officers' Techniques," 13.

20. Warren Weaver oral history, RAC, 665.

21. Hazen, "Memoirs," 3–34.

22. Reingold, "Vannevar Bush's New Deal for Research," 299.

23. Warren Weaver, interview, 2 July 1977, recording in author's possession.

24. WW diary, 9 July 1940, OSRD7 GP.

25. Ibid., 18 July 1940.

26. Ibid., 29 July 1940.

27. WW, agenda for D-2 meeting, 1 August 1940, OSRD7, box 70, Collected Diaries, vol. 1.

28. Kohler, *Partners in Science,* 298.

29. Randell, *Origins of Digital Computers,* 241–45. Stibitz's paper presented at this conference does not survive, but Randell's volume reprints a similar paper from 1940 (247–52). George Stibitz's memoir reprints the program of this meeting ("Zeroth Generation," appendix pp. I-5 to I-6).

30. WW diary, 18 September 1940, OSRD7 GP.

31. Ibid., 3 October 1940.

32. Ibid., 6 November 1940.

33. Ibid., 9, 23 October, 7, 8 November 1940; D-2 diary, 10 November 1940, OSRD7 GP, box 70, Collected Diaries, vol. 1; SHC diary, 1 November 1940, ibid. The *Quincy* sank in the Battle of Savo Island, in Guadalcanal, in 1942; in 1992 I was part of an expedition that located and photographed the wreck.

34. WW diary, 24, 25 October 1940, OSRD7 GP; S. H. Caldwell, "A History of Section D-2, NDRC," 21 February 1946, 7, OSRD7, Office Files of Harold Hazen, box 6.

35. Warren Weaver oral history, RAC, 581.

36. Diary of Section D-2 meeting, 16 October 1940, OSRD7 GP; WW to D-2, 4 November 1940, ibid.

37. Guerlac, *Radar in World War II,* 249, 252 n. 20.

38. Hazen, *Gunfire Control,* 5.

39. Stewart, *Organizing Scientific Research for War,* 322. Owens, "Counterproductive Management of Science in the Second World War," has comprehensive statistics on OSRD funding.

40. Early in 1942 Weaver became afraid that the growing size of the NDRC would mean that "it will spend a great deal of its energy solving problems which it itself creates." He expressed to

Bush his longing for the days of July 1940, when "the NDRC was small, indefinitely flexible, mobile, and unafraid" (WW diary, 19 March 1942, OSRD7 GP, box 70, Collected Diaries, vol. 3).

41. WW diary, 5 and 12 November 1942, ibid., box 72, Collected Diaries, vol. 5; WW to Heads of Divisions 4, 5, 6, 7, 14, 10 December 1942, OSRD, E-151, Applied Mathematics Panel General Records, box 16.

42. Fry, "Industrial Mathematics," 258.

43. Owens, "Mathematicians at War"; Weaver, *Scene of Change*, 87.

44. Division 7 meeting minutes, 18 December 1942, OSRD7 GP, box 72, Division 7 Meetings folder.

45. Hazen to Irvin Stewart, 31 January 1946, OSRD7, Office Files of Harold Hazen, box 6. This memo contains Hazen's personal observations on the operation and management of Division 7 and is the basis of the chart in Stewart, *Organizing Scientific Research for War*, 12.

46. For a full listing of Division 7 members, consultants, and technical aides, see Hazen, *Gunfire Control*, 168–69, Hannibal Ford had been mentioned as a possible member of D-2 and Division 7, but he played only a peripheral role. At least three Ford Instrument employees, however— Edward Poitras, J. D. Tear, and R. E. Crooke—were officially associated with the fire control division.

47. General information on Division 7 contracts comes from OSRD7 GP. Financial information is from the NDRC Index Card File, OSRD7, Index to Contracts.

48. OSRD7 GP.

49. Dupree, "*Great Instauration* of 1940," 459; Stewart, *Organizing Scientific Research for War*, chap. 13; Owens, "Counterproductive Management of Science in the Second World War," 521, 525–26.

50. Dennis, "Change of State"; Guerlac, *Radar in World War II*.

51. Hazen to Stewart, 31 January 1946. This letter was Hazen's personal history of Division 7 for Stewart's *Organizing Scientific Research for War*.

52. Merritt Roe Smith used the phrase "like bees pollinating flowers" to describe how skilled workers and supervisors transferred knowledge about precision arms manufacturing among government arsenals, private arms makers, and other industrial firms in the mid-nineteenth century in "Military Roots of Mass Production." Michael Dennis wrote that the OSRD resembled "a large scale consulting agency," serving as an advisory liaison between industry and military. But that view captures only the atypical Section T (which developed the proximity fuze) (Dennis, "Change of State").

53. EJP diary, 1 July 1941, OSRD7 GP, Collected Diaries, vol. 2.

54. Edward J. Poitras, "Tentative Calendar of Instrumental Developments Section D-2," 27 April 1942, OSRD7, Office Files of Warren Weaver, Index folder.

55. Harold Hazen, "Fire Control Activities of Division 7, NDRC," in Hazen, *Gunfire Control*, 4. Stuart Bennett has noted the "systems approach" in his comparison of British and American fire control work during the war in *History of Control Engineering, 1930–1955*, 125.

*Chapter 8 The Servomechanisms Laboratory and Fire Control for the Masses*

1. Diary of Section D-2 meeting, 16 October 1940, OSRD7 GP.

2. Caldwell to Brown, 17 October 1940, OSRD7 GP, project file 1, box 1; Brown, "Behavior and Design of Servomechanisms."

3. E. S. Smith to T. C. Fry, 10 February 1941, and attached "Membership List: Process Industries Division, Committee on Industrial Instruments and Regulators," and D-2 diary, "Conference with Mr. E. S. Smith, Chairman, A.S.M.E. Committee on Industrial Instruments and Regulators," 14 February 1941, OSRD GP, box 70, Collected Diaries, vol. 1. See also Bennett, "Emergence of a Discipline," 115, which mostly covers the period after 1945; and idem, "Development of Process Control Instruments."

4. WW diary, 8 October 1943; WW to Prof. P. Cromwell, New York University, 7 October 1943; and Hazen to Edward Moreland, 11 October 1943, OSRD, E-151, Applied Mathematics Panel General Records, box 5, Servomechanisms folder.

5. "Distribution List: 'Behavior and Design of Servomechanisms,'" OSRD7 GP, project file 1, box 1.

6. Gordon Brown, "Proposal," 15 July 1940, ibid.

7. Gordon Brown, "A Preamble to a List of Projects for an M.I.T. Group Working on Servomechanisms," ca. August–September 1940, ibid.

8. Warren Weaver, "Recommendation for Appropriation," 1 November 1940, ibid. The actual contract was not signed until 10 June 1941, but this was typical procedure for the NDRC at the time. Most project specifications were laid out in these informal "Recommendation for Appropriation" memoranda.

9. EJP diary, 9 November 1940, ibid.

10. Brown interview, 27 January 1970.

11. Gordon S. Brown, interview by Alex Pang, 24 July 1985, 50–52, 65, Computers at MIT Oral History Collection. In a meeting with Karl Compton in October 1941, Weaver observed that Brown's original proposal was made for Sperry, although it cannot be taken as authoritative since Weaver and Brown were at the time at each other's throats. Still, Weaver's observation matches the correlation of Brown's proposed projects with Sperry's interests and Brown's own recollections (see WW diary, 30 October 1941, OSRD7 GP, box 70, Collected Diaries, vol. 2).

12. Brown interview, 26 August 1994. Brown also tells this story in his Brown interview, 24 July 1985, 66.

13. Bureau of Ordnance, "Summary of Progress to July 1, 1942," RG 74, General Correspondence, 24–25.

14. Gillmor to Weaver, 19 November 1940, OSRD7 GP, project file 1, box 1. This letter thanks Weaver for receipt of the paper, although Gillmor doesn't appear on the distribution list.

15. Manuscript version of Wildes and Lindgren, *Century of Electrical Engineering and Computer Science,* 5-10 to 5-25. For an overview of the Servo Lab and Draper's lab, see Burchard and Killian, *Q.E.D,* chap. 9.

16. Diary of visit to MIT by TCF, SHC, and EJP, 5 December 1940, OSRD7 GP, project file 1, box 1.

17. Manuscript version of Wildes and Lindgren, *Century of Electrical Engineering and Computer Science,* 5-16. See also interviews by Richard Merz of Stephen Dodd Jr., 9 December 1969, Robert Everett, 26 January 1970, and Robert Wieser, 20 March 1970, Computer Oral History Collection, National Museum of American History.

18. Sperry's development program in the 1930s had produced directors that were accurate enough that the primary source of error was the "dead time" between the completion of the firing solution and the loading of the shell. Not only was this a source of unnecessary delay but with manual shell loading it varied from shell to shell. Shortening and standardizing, that is, mechanizing, the setting and loading (or "ramming") of the fuze was thus the surest way to reduce this uncertainty and improve accuracy (Weaver to Karl Compton, and Weaver to Bush, both 12 December 1940, OSRD7 GP, project file 1, box 1).

19. Sperry Company report, "Power Controls.". Sperry's devices were designated M-2 (for the 4.7-inch gun), M-3 and M-4 (for the 90 mm gun) and T-9 (for the 105 mm gun). The controls the army eventually asked Brown for help with were M-1 and M-5 for the 37 mm and 40 mm guns, respectively.

20. Gordon S. Brown and Jay W. Forrester, "Remote Control System," U.S. Patent 2,409,190 (15 October 1946). The patent lawyer signing the document was Herbert Thompson, Sperry's patent lawyer of many years.

21. The manuscript version of Wildes and Lindgren, *Century of Electrical Engineering and Computer Science,* 5-15 to 5-18, quotes Brown's accounts of these tests.

22. KTC to GSB, CSD, SHC, WW, TCF, and PRB, 1 March 1941, OSRD7 GP, project file 1, box 1.

23. SHC diary, 8 March 1941, OSRD 7, project file 1, box 1.

24. SHC diary, "Conference with G. S. Brown" and "Conference of Dr. Compton, G. S. Brown, and SHC," both 5 March 1941, and "Conference with H. L. Hazen and G. S. Brown," 6 March 1941, all in OSRD7 GP, box 70, Collected Diaries, vol. 1.

25. NDRC Index Card File, OSRD7, Index to Contracts. For a brief summary of this episode from the MIT point of view, see the manuscript version of Wildes and Lindgren, *Century of Electrical Engineering and Computer Science,* 5-15 to 5-16.

26. Bassett explained his anxieties to Caldwell during a visit to the company (SHC diary, 5–6 September 1941, OSRD7 GP, box 70, Collected Diaries, vol. 2). WW to KTC, 1 May 1942, reports a meeting when "the younger Sperry men talked rather frankly concerning the relationship between D-2 and the Sperry company with respect to [antiaircraft] developments." They felt the NDRC was "more or less ducking the Sperry Company" (OSRD7, Office Files of Warren Weaver, box 4, Sperry Gyroscope folder).

27. The company and the committee did, however, reconcile. Weaver noted in May 1942, "Mr. Bassett of the Sperry Company appeared at the door with flowers in his hand, love in his heart, and kisses on his lips." At that point Sperry officially acknowledged and supported the NDRC's efforts in fire control (WW diary, 21 May 1942, OSRD7 GP, box 70, Collected Diaries, vol. 3). This

memo records Sperry and Earl Chafee's experience licensing production of its mechanical gun director to other manufacturers. When Hazen took over as head of Division 7, he immediately appointed Bassett as a "part-time" member, formalizing the peace (HLH to PRB, 1 and 8 December 1942, OSRD7, Office Files of Preston Bassett, box 53).

28. WW to KTC, 3 March 1941, OSRD7 GP; WW to KTC, 1 May 1942, OSRD7, Office Files of Warren Weaver, box 4, Sperry Gyroscope folder; EJP diary, "Visit to Sperry," 14 May 1941, OSRD7 GP, box 70, Collected Diaries, vol. 1.

29. EJP diary, "Visit to Sperry," 14 May 1941.

30. Draper seems not to have experienced the friction Brown encountered, perhaps because Draper's field, aeronautics, was still controlled by NACA, a more established organization (recall that the NDRC's charter excluded all NACA terrain). On Draper's relationship to the NDRC, with whom he had no contracts, see Dennis, "Change of State," chap. 4.

31. When Sperry was having production troubles in 1917, Wilbur Van Auken offered to detail Blandy to the company to help smooth out production (Van Auken to Sperry, 10 November 1917, W. H. P. Blandy Papers, box 1, Personal Correspondence folder, Library of Congress). For biographical information on Blandy, see "The Navy's Gun Man," *Washington Sunday Star,* Magazine section, 19 April 1942, Blandy Papers, box 1, Biographical and Genealogical folder, and a host of other clippings in that folder. See also Blandy, "Possible Improvements in Our Gunnery Training"; and Graybar, "Buck Rogers of the Navy."

32. Antiaircraft Defense Board, December 1940, quoted in Rowland and Boyd, *U.S. Navy Bureau of Ordnance in World War II,* 220.

33. For a good summary of the state of BuOrd's work, as well as its problems, at the start of the war, see Poitras's notes after meeting with Commander France, of BuOrd, 27 February 1941, OSRD7 GP, box 70, Collected Diaries, vol. 1. Although Poitras may have favored his former employer, Ford Instrument, his assessment was still bleak.

34. Gannon, *Hellions of the Deep,* 73–93.

35. Blandy to J. R. Palmer, Commander, USS *Utah,* 10 November 1940, Blandy Papers, box 1, Personal Correspondence folder.

36. The reorganization had actually begun before Blandy was named chief of BuOrd, but it was his plan that his predecessor Furlong had begun to execute (Furer, *Administration of the Navy Department in World War II,* 319). Blandy describes the reorganization in detail, as well as much of the bureau's work during his tenure as chief, in "Final Report of present Chief of Bureau of Ordnance," 9 December 1943, Blandy Papers, box 1, Official Correspondence folder. For the official history, see U.S. Navy, *Administrative History,* 73:6–10. For a list of BuOrd officers at the start of the war, see Directory, Officers on Duty and Civilian Personnel, Bureau of Ordnance, Furlong Papers, box 4, General Correspondence, Military File.

37. U.S. Navy, *Administrative History,* 73:10–41, contains a remarkably frank discussion of the organization of research and its problems; the section titled "Administration of Research Activities and Contracts," 153–57, has contracting procedures and numbers.

38. M. E. Murphy, "Memorandum: Report of Fire Control Section (Re4), Summary of Activi-

ties and Accomplishments, and Recommendations for the Future." Reprinted as appendix A in ibid., 319, 330.

39. Ibid., 330, 312.

40. See Rowland and Boyd, *U.S. Navy Bureau of Ordnance in World War II,* chap. 11, for a detailed discussion of the production of these guns.

41. Murphy, "Memorandum: Report of Fire Control Section (Re4)," 311.

42. Blandy to Palmer, 6 January 1941, Blandy Papers, box 1, Personal Correspondence folder.

43. Rivero interview, 20 May 1975, 113.

44. Friedman, *U.S. Naval Weapons,* 86.

45. A number of accounts of this project survive. The most contemporary is Murphy, "Memorandum: Report of Fire Control Section (Re4)," 312–14. For the view from Sperry Gyroscope, see Morgan, "Navy's Mark 14 Gyro Gun Sight; see also Robert Ward, "Gunsight Mark 14 and Gun Director Mark 51," House Report, 20 January 1944, SGC Papers, box 40. Michael Dennis narrates MIT's role in the transition from instrument to production in "Change of State," chap. 4. For production figures, see "Bureau of Ordnance: Summary of Progress," 1 August 1945, RG 74 Ordnance Status Reports, box 7, 32. For the Mark 14 in the context of BuOrd fire control, see U.S. Navy, *Administrative History,* 73:160–68. See also Wildes and Lindgren, *Century of Electrical Engineering and Computer Science,* 214–15. For Webb, see Lambright, *Powering Apollo.*

46. Morgan, "Navy's Mark 14 Gyro Gun Sight," 17.

47. Brown called Caldwell in September 1941 to "do a little unofficial weeping." Caldwell noted, "He is not at all happy about the way things are going with Sperry and himself, but SHC did not press for details." Despite the unofficial nature of Brown's weeping, Caldwell wrote it up as a memo and distributed it to the committee (SHC diary, 8 September 1941, OSRD7 GP, project file 1, box 1).

48. WW diary, 30 October 1941.

49. HLH to WW, 15 August, and WW to HLH, 5 September 1941, OSRD7 GP, project file 1, box 1. Meanwhile, the original project, to build an automatic fuze setter (whose continuation was agreed on in Brown's absence), did near completion under the leadership of Donald P. Campbell and the oversight of Harold Hazen (Donald P. Campbell, "Report on a Relay Controller to Provide Proper Fuze Time on the Fuze Setter, M8, Corresponding to Director's Fuze Range," OSRD7 GP, project file 1, box 1).

50. For a summary of the Servo Lab's work in late 1941, see EJP diary, 8 December 1941, OSRD7 GP, box 70, Collected Diaries, vol. 2; and Gordon Brown, quoted in manuscript version of Wildes and Lindgren, *Century of Electrical Engineering and Computer Science,* 5-16.

51. KTC to WW, 20 December 1941, OSRD7 GP, project file 1, box 1. Two months previously, Weaver had written to Joseph Boyce, at MIT, asking him to report on Brown's work (WW to Boyce, 14 October 1941, OSRD7 GP, project file 1, box 1; WW to KTC, 18 December 1941, ibid., project file 35).

52. Sperry Company memorandum, "M-5 and M-6 Directors." For the Kerrison predictor in England, see Douch, "Use of Servos in the Army during the Past War"; Pile, *Ack-Ack,* 246–47, and

240 for a picture of the predictor in action. For a diagram of the internals of the Kerrison predictor, see Bromley, "Analog Computing Devices," 188–89; and for the machine's role in the Battle of Britain, see Routledge, *Anti-Aircraft Artillery*, 52–55.

53. HLH to EJP, 24 January 1942, OSRD7 GP, project file 35, box 30. See also Brown interview, 24 July 1985, 59–61; Memorandum of Agreement, 1 February 1942, OSRD7 GP, project file 35, box 30; Hazen, *Gunfire Control*, 40; and GSB to EJP, 3 July 1942, OSRD7 GP, project file 35, box 30. For the official NDRC history of the project, see Lawton M. McKenzie and Edward Poitras, "History, Section 7.3," 22 March 1946, 4–7, OSRD7, Office Files of Harold Hazen, box 6. For the Servo Lab's history of this project, as well as the final report on the servo, see "Report of Studies on Remote Control Systems M-1 and M-5," November 1942, MIT Servo Lab, Division Of Industrial Cooperation, Project 6047, in Servomechanisms Laboratory Papers, MIT Archives, AC-151, box 2, folder 8; hereafter referred to as Servo Lab Papers (these papers were the office files of Robert Everett and are fairly incomplete as documentation of the Servo Lab). The Servo Lab devised separate controls for driving the 40 mm gun by a computer from a data transmission system and for driving it by a human operator with a "handlebar input" (see reports in Servo Lab Papers, folders 6 and 7, respectively. See also "History of the Fire Control Sub-office," Frankford Arsenal, September 1943, 115–19, Army Ordnance Records, NARA, RG 156, box 36).

54. WW to Alan Waterman (vice chairman, Division D), 28 August 1942, OSRD7 GP, project file 35, box 30.

55. WW diary, 20 August 1942, OSRD7 GP; GSB to EJP, 22 August, and EJP to GSB, 26 August 1942, OSRD7 GP, project file 35, box 30.

56. Brown interview, 24 July 1985, 63.

57. Brown interview, 27 January 1970.

58. Brown interview, 24 July 1985, 75–76. For one description of the Servo Lab's operating environment, see Redmond and Smith, *Project Whirlwind*, 10–12.

59. Brown interview, 24 July 1985, 50, 65.

60. Ibid, 65–67, 71.

61. Ibid., 97.

62. Wieser, interview, 20 March 1970.

63. See, e.g., "Description and Operating Instructions of the Sperry MIT Automatic Remote Control System for the T-36 Gun Carriage," Division of Industrial Cooperation Project 6041, 11 August 1942, and several similar reports in Servo Lab Papers, box 2, folder 4. This report was not declassified until 1962.

64. Hall, "Early History of the Frequency Response Field," 1153. See also the manuscript version of Wildes and Lindgren, *Century of Electrical Engineering and Computer Science*, 5-19.

65. Herbert Harris Jr., "The Analysis and Design of Servomechanisms," OSRD 454, Report to the Services 23, MIT. This paper was revised and published as Harris, "Frequency Response of Automatic Control Systems."

66. Hall, *Analysis and Synthesis of Linear Servomechanisms*. Hall's thesis was published as a restricted report in 1943 and then reprinted in 1947 after it was declassified. For a technical discussion of Hall's paper, see Bennett, *History of Control Engineering, 1930–1955*, 140–43.

67. In a later version of his thesis published as a paper, Hall overcame his hesitation to retain the distinction between servos and feedback amplifiers and acknowledged his "transfer locus" to be identical to the Nyquist diagram. He even changed his title to "Application of Circuit Theory to the Design of Servomechanisms," further eroding the boundary between electronics and servomechanisms. A close comparison of this paper with Hall's thesis, though beyond the scope of this chapter, would detail the further merging of electronics and servo theory during 1943–46.

68. See, e.g., Stephen H. Dodd Jr., "Design and Test of a Hydraulic Transmission," 1945, Servo Lab Papers, box 1, folder 2; Servo Lab technical report, "Automatic Control Characteristics of a 0.682 cubic-inch per revolution Oilgear Hydraulic Transmission," June 1943, Division of Industrial Cooperation, Project 6152, ibid., box 2, folder 2. Both of these reports present frequency domain analysis. The only Servo Lab report in the archives written before 1943 that uses frequency response is a project under contract to Raytheon stabilizing a radar antenna on a ship. The project report lists no author, but it was likely done by Hall or Harris, since this was the type of project that led them to their frequency response work (ibid., box 1, folder 15).

*Chapter 9 Analog's Finest Hour*

1. When the ramps were destroyed or captured, the Germans began launching the missiles from aircraft. Now a pilot could drop his bomb from a safe distance over the ocean and let the control system do the piloting while he and his crew returned home.

2. Warren Weaver, foreword to "Final Report: D-2 Project #2, Study of Errors in T-10 Gun Director," OSRD7, Office Files Of Warren Weaver, 3.

3. D. B. Parkinson, from an account written on 5 January 1975 of his dream, which is in ATT 60 04 01 and partially reprinted in Fagen, *Early Years,* 135–36. For Parkinson biographical info, see *Wisconsin Bell Magazine,* n.d., ATT 60 04 01.

4. D. B. Parkinson, Notebook 16413, project file 23140, ATT. For other notebooks on this project, see B. T. Weber, 16042 and 18009; K. D. Swartzel Jr., 17512 and 16312; and C. A. Lovell, 17665 and 15627; all ATT notebooks cited below are from project file 23140. The control system also required an instrument of perception, and here Parkinson repeated a mistake made by both Arthur Pollen and Elmer Sperry in their first forays into fire control. He had an idea for a long-baseline rangefinder in which two widely separated operators, connected by a data transmission system, tracked a target. With more than one target present, however, it proved nearly impossible for both trackers to stay on the same one. Parkinson soon dropped the idea.

5. For a detailed chronology of this project, see "Check list for use in connection with record of laboratories work on N.D.R.C. and O.S.R.D. contracts, no. NDCrc-127," project file 23140, ATT. For specific meetings, see E. C. Wente diary, 27 June, 3 July, and 5 August 1940, ibid.; R. B. Colton to Kelley, 6 October 1944, OSRD7, Office Files of Warren Weaver, Project 2 folder; Colton to Chief Signal Officer, 5 September 1940, OSRD7 GP, box 67, Antiaircraft Artillery Board, Project 1214, Test of Electric Antiaircraft Director T-10 folder; Clarence Lovell, interview by Richard R. Merz, 19 March 1970, Computer Oral History Collection, National Museum of American History.

6. Lovell interview, 19 March 1970.

7. C. A. Lovell, 18 June 1940, Notebook 15627.

8. C. A. Lovell, 14 April 1941 and 17 July 1940, ibid.

9. D. B. Parkinson, 23 July 1940, Notebook 16413.

10. Lovell interview, 19 March 1970.

11. WW diary, 24 October 1940, OSRD7 GP, project file 2.

12. See WW to General Somers, 24 October 1940, OSRD7 GP, box 67, Antiaircraft Artillery Board, Project 1214, Test of Electric Antiaircraft Director T-10 folder.

13. WW diary of phone call to Somers, 6 November 1940, OSRD7 GP, project file 2; WW to Kelley, 9 November 1940, and Memorandum of Agreement between NDRC and Bell Labs, 19 May 1941, ibid.

14. C. A. Lovell to Ordnance Department, 15 January 1941, Project file 23140, ATT; Lovell and Parkinson, "An Electrical Director," 30 August 1940, OSRD7 GP, box 67, Antiaircraft Artillery Board, Project 1214, Test of Electric Antiaircraft Director T-10 folder; Lovell diary, 21 December 1940, Project file 23140, ATT.

15. EJP to Parkinson, 27 November 1940, Project file 23140, ATT.

16. K. D. Swartzel Jr., 9 April 1941 and April–June 1941, Notebook 17512, ibid.

17. Weaver, foreword to "Final Report: D-2 Project #2," 3.

18. "Final Report: D-2 Project #2," 129. Constructing this electrical computer proved no easy task. Among the many difficulties, none proved as challenging as the shaped potentiometers. The wire that wrapped them needed uniform resistance all down its length and had to maintain consistency despite temperature changes. The oddly shaped potentiometers that stored the firing tables required new equipment to smoothly wind their wire (Harvey Fletcher to R. R. Williams, 21 May 1941, project file 23140, ATT. For a more detailed discussion of the wire-winding machine, see Fagen, *Early Years,* 144–45).

19. "Final Report: D-2 Project #2," 37; Bruce T. Weber, 7 August–25 October 1940, Notebook 16042.

20. "Final Report: D-2 Project #2," app. 2, "Stability Considerations."

21. Ragazzini, Randall, and Russel, "Analysis of Problems in Dynamics by Electronic Circuits," 444. See also Ragazzini's introduction to the paper's reprint in McLeod, *Simulation,* 15; and Lovell, "Continuous Electrical Computation."

22. Caldwell, " History of Section D-2, NDRC," 10.

23. WW diary, 5 December 1940; D-2 diary of visit to Barber Coleman, 3 January 1941, OSRD7 GP, box 70, Collected Diaries, vol. 1. TCF to Somers, 10 March 1941, OSRD7 GP, project file 25, gives a brief history of the initiation of this project.

24. Caldwell, "History of Section D-2," 23. For Hazen's comments on Stewart, see Hazen, "Memoirs," 3–38.

25. Radiation Lab engineers used the machine to evaluate the Ford Mark 1 computer and decided that the rangekeeper was inadequate for high-speed targets and had basic flaws in its mechanisms, concluding that "the navy's standard testing procedure is inadequate." BuOrd treated these results "with considerable skepticism" and conducted its own tests, which purported to

show that the dynamic tester, designed to test army directors, had no validity when applied to navy equipment (OSRD7 GP, Office Files of Harold Hazen, box 82, Testing of Mark I Computer by NDRC folder; Getting, *All in a Lifetime*, 175–76. For the raw test data, see NARL, box 1149).

26. DJS to HLH, WW, EJP, and GRS, 31 December 1942, OSRD7 GP, project file 2. This memo summarizes Stewart's numerous objections to the Bell Labs project. "It is important to bear in mind that the Bell Telephone Laboratories, with unselfish and patriotic motives, has undertaken the development and construction of this instrument in accordance with a program which not only would be foolish under normal circumstances but is entirely at variance with the Bell Telephone Laboratories ordinary development procedure," he wrote. See also DJS diary, 20 September 1941, and Kelley to WW, 17 December 1941, ibid.

27. GRS diary, 25 December 1941, ibid.

28. WW to Fletcher, 31 October 1941, ibid.

29. WW to KTC, 11 November 1941, ibid.; Office of the Chief of Ordnance, "Catalog of Standard Ordnance Items," 1 May 1943, 331.

30. WW diary, 25 February 1942, OSRD7 GP, project file 2.

31. Stewart had proposed a director of his own, so his animus toward the Bell Labs machine may contain residual resentment, DJS to HLH, WW, EJP, and GRS, 31 December 1942.

32. WW diary, 23 January 1942, ibid.; see also DJS diary, 5 October 1942, OSRD7 GP, box 71, Collected Diaries, vol. 4.

33. Derivative models were designated M-8, M-10, M-12, M-13, and M-14. For Bell Labs accounts, see "Development of the Electrical Director"; "Electrical Gun Director Demonstrated"; and "Blow Hot—Blow Cold." For production numbers, see William J. Wuest, "History of Heavy AA Fire Control and Materiel" (Fort Bliss, Tex.: U.S. Army, The Artillery School, Antiaircraft and Guided Missiles Branch), 1951), ATT 84 05 02 03.

34. NDRC, "Antiaircraft Director T-15," Report to the Services 62 (contractor's report on OEMsr-353), August 1943, OSRD7. See also Fagen, *Early Years*, 151–55, based on this NDRC report, on Lovell, memo to Fagen, 3 January 1974, ATT 84 05 02 03, and on Fagen, "The War Years," manuscript in the same folder.

35. Colton, "Radar in the United States Army."

36. Guerlac, *Radar in World War II*, 103–10. A similar device developed by the army, the SCR-270, with a wavelength of 2.5 meters, was designed as an early-warning-and-search system. It was deployed in Hawaii in August 1940 and detected the attack on Pearl Harbor at a distance of more than 100 miles.

37. Ibid., 243–50.

38. Zimmerman, *Top Secret Exchange*, chap. 6.

39. Getting, *All in a Lifetime*, 37.

40. Ibid., 107.

41. Getting, "SCR-584 Radar and the Mark 56 Naval Gun Fire Control System," 924.

42. "Destroyer PCO/PXO Lectures," April 1944, Enclosure B, RG 38, Commander in Chief series, box 1162, NARA.

43. Getting, "SCR-584 Radar and the Mark 56 Naval Gun Fire Control System." On the details of the servo design, see reports by Godet, NARL, box 1165, Servomechanisms folder. For a technical summary of the '584 servos, see James, Nichols, and Phillips, *Theory of Servomechanisms*, 212–24; and Bennett, *History of Control Engineering, 1930–1955,* 143–46, which includes Godet's servo.

44. Getting, interview by Frederik Nebeker, 11 June 1991, Boston, Radiation Laboratory Oral History Interviews, IEEE Center for the History of Electrical Engineering, Rutgers University, New Brunswick, N.J., available at htttp://www.ieee.org:80/history_center/oral_histories/oh_rad_lab_menu.html.

45. Because of resource constraints on vehicle production, the prototype truck became a towed trailer in the production model (Ivan Getting, "Report of Anti-Aircraft Artillery Board on XT-1," Exhibit L, "Memorandum on the advisability of the use of trucks for the American GL system," 2 March 1942, NARA, RG 111, Office of the Chief Signal Officer, Classified Central Decimal File, 1940–48, RB 413.44, SCR-585/XT-1, box 1676; Getting interview, 11 June 1991).

46. "Report of A.A.B. Test on XT-1 at Fort Monroe, Virginia, February, 1942," Radiation Laboratory Report No. 359, MIT Archives; W. S. Bowen, "Report of Anti Aircraft Artillery Board on Project Number 1218, Radio Position Finder (MIT XT-1)," 20 March 1942, NARA, RG 111, Office of the Chief Signal Officer, Classified Central Decimal File, 1940–48. For firsthand accounts of the XT-1/SCR-584 development and its field deployment, see Henry Abajian, interview by Frederik Nebeker, 11 June 1991, Boston; Lee Davenport, interview by John Bryant, 12 June 1991, New Brunswick, N.J.; and Leo Sullivan, interview by Frederik Nebeker, 14 June 1991, Boston, all in Radiation Laboratory Oral History Interviews, IEEE Center for the History of Electrical Engineering, Radiation Laboratory Oral Histories.

47. WW diary re meeting with Loomis, 5 December 1940; WW to Loomis, 10 December 1940; and WW diary, 13 December 1940, OSRD7 GP, box 70, Collected Diaries, vol. 1. Weaver's diplomacy may have resulted from D-2's failed radar project, the range-only SCR-547 "Mickey," built by Western Electric to replace an optical rangefinder.

48. See TCF diary of meeting with Colonel Bowen, 3 July 1941, OSRD7 GP, box 70, Collected Diaries, vol. 2; Earl W. Chafee, "Memorandum of Conference in Fire Control Department," 24 September 1942, OSRD7, E-83, Office Files of Warren Weaver, box 4, Sperry Gyroscope folder.

49. WW diary, 12 November 1942, OSRD7 GP, box 72, Collected Diaries, vol. 5, emphasis in the original; WW to Lovell, 23 November 1942, and WW to Chafee, 1 December 1942, OSRD7, Office Files of Harold Hazen, box 9, Radiation Laboratory folder.

50. Chafee, "Study of the Requirements for a Satisfactory Antiaircraft Fire Control System." The report can also be found in OSRD7, Office Files of Harold Hazen, box 9, Radiation Laboratory folder.

51. WW to Fletcher, 28 February 1941, and Louis Ridenour to Lovell, 24 September 1941, project file 23140, ATT. GRS diary, 21 May 1941; Ridenour to Lovell, 6 August 1941; and Lovell to Ridenour, 23 September 1941, OSRD7 GP, project file 2.

52. Stibitz to Getting, 17 June 1942, and similar correspondence in NARL, box 1149, T-10 Gen-

eral folder; "Final Report: D-2 Project #2"; Getting, "SCR-584 Radar and the Mark 56 Naval Gun Fire Control System."

53. "Study of Errors in T-10 Gun Director," 72. For a Radiation Lab study of jitter in a tracking servo from radar data, see "Data Smoothing," Radiation Laboratory Report No. 673.

54. Weaver, foreword to "Final Report: D-2 Project #2," 3.

55. The competitors were a similar Bell Labs radar, the SCR-545, which was produced in limited numbers, and the Canadian GL-III-C, which had been designed in response to Tizard's initial assignment for gun-laying radar. The SCR-545 was the closest rival to the Radiation Lab set and included a long-wave search radar along with its microwave tracker (SHC diary, 26 March 1942, and J. B. Ridenour diary, 4 April 1942, OSRD7 GP, project file 2).

56. Davenport interview, 12 June 1991.

57. For the difficulties of producing the SCR-584 see Thompson et al., *Signal Corps: The Test,* 265–74; Getting, *All in a Lifetime,* 121–27; Guerlac, *Radar in World War II,* 481–83; Getting, Harris, Abajian, and Davenport oral histories, IEEE Center for the History of Electrical Engineering, Radiation Laboratory Oral Histories; and OSRD, boxes 1148–65.

58. Davenport interview, 12 June 1991.

59. *Radio Sets SCR-584-A and SCR-584-B Service Manual: Theory, Troubleshooting, and Repair,* War Department Technical Manual 11–1524 (Washington, D.C., July 1946).

60. "Operations of SCR-584 at Anzio," 26 May 1944, NARL, box 1164, Field Reports folder. Leo Sullivan, from the Radiation Lab, accompanied the SCR-584 to Anzio (Sullivan interview, 14 June 1991). For general information on the V-1 attacks, see Jones, *Wizard War,* chaps. 39 and 44; and Ordway and Sharpe, *Rocket Team.*

61. "Tactical Employment of Antiaircraft Artillery Units, Including Defenses Against Pilotless Aircraft (V-1)," General Board, United States Forces, European Theater Antiaircraft Artillery Section, Study Nos. 38, 39, NARA, RG 407.

62. "Operators Report: HQ Antwerp X Report #2E," 50th Anti-Aircraft Artillery Battalion, 6 December 1944, NARA, RG 407, Antiaircraft Action Reports, 3.

63. Pile, *Ack-Ack,* 314–15; Frederick Pile to George C. Marshall, quoted in Bush to Hazen, 31 August 1944, OSRD7, Office Files of Harold Hazen.

64. "Performance of SCR-584B Serial No. 2 in Great Britain, November 26, 1943," NARA, RG 111, Office of the Chief Signal Officer, Classified Central Decimal File, 1940–48; Abajian to Perry Lewis (president of AAA Board), 20 July 1944, NARL, box 1164.

65. "Operators Report: HQ Antwerp X Report #2E"; Getting, Davenport, and Abajian oral histories, IEEE Center for the History of Electrical Engineering, Radiation Laboratory Oral Histories.

66. Major Harris T. Richards, "Technical and Tactical Operation of the SCR-584 in the Anzio Nettuno Bridgehead," 2 April 1944, NARA, RG 11, RB 413.44, "SCR Series Radio Equipment Research and Development Case Files, 1923–54"; J. T. G. Milne, "The Performance of Radar SCR '584 in the Diver Belt," September 1944, Army Operational Research Group Memorandum No. 370, NARL, box 1164.

67. "Tactical Employment of Antiaircraft Artillery Units, Including Defenses Against Pilotless Aircraft (V-1)," Study Nos. 38, 42.

68. Baldwin, *Deadly Fuze.*

69. Guerlac, *Radar in World War II*, 859. For a personal account of the automatic system vs. the V-1, see Abajian interview, 11 June 1991. For the Antwerp statistics, see "Operators Report: HQ Antwerp X Report #2," 3. For a daily breakdown of the numbers, see "Tactical Employment of Antiaircraft Artillery Units, Including Defenses Against Pilotless Aircraft (V-1)," Study No. 38, p. 41.

70. Pile to Marshall, 12 August 1944, NARL, box 1164, Field Reports folder.

71. A number of SCR-584s were given to the Soviet Union, and for years after the war Soviet radars incorporated the American radar's design features. For a summary of SCR-584 projects, including a number of modifications, see NDRC, *NDRC Division 14 Final Project Report,* MIT Archives, 2-41 to 2-68. For the operational history of the SCR-584, see Thompson and Harris, *Signal Corps: The Outcome,* 474–77; Guerlac, *Radar in World War II,* 480–96, 853–62, 882–97, 1018–25; Getting, *All in a Lifetime,* 130–35; and Abajian, Davenport, Harris, and Getting oral histories, IEEE Center for the History of Electrical Engineering, Radiation Laboratory Oral Histories.

72. NARL, box 1164; OSRD7, E-84, Office Files of Section Heads, box 6, V-2 folder; Getting, *All in a Lifetime,* chap. 9.

73. Ferrell, "Electrical and Mechanical Analogies."

74. In July 1944 a group of four army officers and two Bell Labs engineers, including Clarence A. Lovell (who headed the T-10/M-9 design team), traveled to Europe for an observation tour of antiaircraft batteries and produced the report "Antiaircraft Artillery Fire Control," prepared by Bell Labs for the Army Ordnance Department in fulfillment of contract W-30-069-Ord-1448, 1 May 1945, ATT.

75. Conference on Navy Fire Control, 1 December 1943, NARL, box 1157.


*Chapter 10 Radar and System Integration at the Radiation Laboratory*

*Epigraph:* Bassett, "Review of Ground Anti-Aircraft Defense."

1. Battle Experience Secret Information Bulletin 24, World War II Command File, Entry #514, Naval Operational Archives, 81-33, 81-39.

2. Allison, *New Eye for the Navy,* 114–17; Rowland and Boyd, *U.S. Navy Bureau of Ordnance in World War II,* 415–16.

3. Bureau of Ordnance, "Source Book on the History of Fire Control Radar and Related Projects," RG 74, box 1, 3-4.

4. The Bell Labs microwave radars became the CXAS, or FA when applied to fire control. For a detailed, official history, see Bureau of Ordnance, "Source Book on the History of Fire Control Radar and Related Projects," 3-4. For prewar development of fire control radar, see Howeth, *History of Communications,* 463–67; and Tinus and Higgins, "Early Fire-Control Radars for Naval Vessels," 18. For operational experience and differentiation between model and mod numbers, see "Resume of Shipboard Fire Control Radar," *CIC Magazine,* August 1944, World War II Command

File, CNO, Naval Operational Archives. See also Fagen, *National Service in War and Peace,* chap. 2, "Radar"; the table of radar development programs on pages 68–69 shows the first antiaircraft fire control radar, a initiated in October 1940, a Rivero project. For a comprehensive list of naval radars and mark numbers, see Friedman, *Naval Radar,* 145–82; and Rowland and Boyd, *U.S. Navy Bureau of Ordnance in World War II,* chap. 17, "Fire Control Radar."

5. Rivero interview, 20 May 1975; Bureau of Ordnance, "Source Book on the History of Fire Control Radar and Related Projects," 3-9, 3-10.

6. Jurens, " Evolution of Battleship Gunnery," 255.

7. Hooper interview, 23–26 June 1970.

8. Ibid., 94–95; Scotty Campbell, quoted in Musicant, *Battleship at War.*

9. Hooper interview, 23–26 June 1970, 81. After the battle, the historian Samuel Eliot Morison boarded the *Washington,* and Hooper gave him the rangekeepers' plots of the battle. See also Musicant, *Battleship at War,* chap. 5, for an account of the battle from the *Washington*'s perspective and Hooper's role. Erling Hustvedt gives a personal account of the gunnery room aboard the *South Dakota* during the same battle in "Battleship Gunfire Control," courtesy John Tetsuro Sumida.

10. SHC diary, 30 September 1941, OSRD7 GP, box 70, Collected Diaries, vol. 2; EJP diary, 30 September 1941, OSRD7 GP, project file 1, box 1; Bureau of Ordnance, "Source Book on the History of Fire Control Radar and Related Projects," 6-8.

11. Bureau of Ordnance, "Source Book on the History of Fire Control Radar and Related Projects," chap. 6; Rowland and Boyd, *U.S. Navy Bureau of Ordnance in World War II,* 377–78; Fagen, *Early Years,* 67–72; Friedman, *U.S. Naval Weapons,* 83–84, 243.

12. "Report of a board convened by the Commander in Chief, U.S. Pacific Fleet, to make recommendations on ordnance matters based on experience in the war which resulted in the defeat of Japan," November 1942, World War II Command File, Captain George Kraker folder, Naval Operational Archives.

13. Rowland and Boyd, *U.S. Navy Bureau of Ordnance in World War II,* 421, 429. The three that became widely used were Marks 3, 4, and 8. Bureau of Ordnance, "Source Book on the History of Fire Control Radar and Related Projects," chaps. 7–9, describes in detail the troubles encountered with a number of these systems.

14. The low-angle addition was Mark 22, with a characteristic "orange peel" antenna aboard a Mark 37 director. For a detailed evaluation of naval radar in operation, see "Report of a board convened by the Commander in Chief U.S. Pacific Fleet, to make recommendations on radar and countermeasures based on experience in the war which resulted in the defeat of Japan," October 1945, Naval Operational Archives, World War II Command File, Charles F. Horne folder.

15. The failed projects were Marks 49, 46, and 50, respectively. For detailed accounts of these projects, see U.S. Navy, *Administrative History,* vol. 79, chap. 4, "Antiaircraft Fire Control"; and Murphy, "Memorandum: Report of Fire Control Section (Re4)." See also Bureau of Ordnance, "Source Book on the History of Fire Control Radar and Related Projects," chap. 11. Ivan Getting gives the most pessimistic assessment of the situation in, "Draft History, Section 7.6," 1946,

OSRD7, Office Files of Harold Hazen, box 6. Getting provides the perspective of a BuOrd outsider but also of an interested party frustrated with the bureau. The discussion in Getting's memoir, *All in a Lifetime,* 165–67, is based on this account. For the Mark 9 radar, see Guerlac, *Radar in World War II,* 279–81; "Mark 151 Director," March 1946, Radiation Laboratory Report No. S-75; and *NDRC Division 14 Final Project Report,* 4–54. For G.E. efforts, see Brittain, *Alexanderson,* 256–63. The British were having similar problems with blind firing (Kingsley, *Applications of Radar and Other Electronic Systems,* 71–77, 86).

16. IAG to KTC, "U.S.N. AA Director Mk. 56," 29 December 1943, OSRD E-39, Office Files of Karl Taylor Compton, box 51, Division 7 folder.

17. Division 7 meeting minutes, 3 February 1943; Getting, "SCR-584 Radar and the Mark 56 Naval Gun Fire Control System," 932.

18. HLH diary, 20, 21 April 1943, OSRD7 GP, Office Files of Harold Hazen, box 70; Division 7 meeting minutes, 28 April 1943. Guerlac mistakenly recounts these events as occurring in the summer of 1942 in *Radar in World War II,* 490, based on a misunderstanding of Getting's letter to Compton of 29 December 1943.

19. Getting, "Draft History, Section 7.6," 7. See also Fagen, *Early Years,* 158–62. Bell Labs built a prototype of this computer, designated Mark 8, which directly replaced the Ford Instrument Mark 1, but it was never put into production.

20. Getting, "Draft History, Section 7.6," 7.

21. Getting interview, 11 June 1991.

22. The complete 7.6 membership comprised the following: George Agins, vice president of Arma; R. F. Cooke, vice president of Ford Instrument; C. S. Draper, MIT; A. W. Horton, Bell Labs; R. M. Page, NRL; E. J. Poitras, Division 7 (Ford Instrument); R. B. Roberts, Section T, OSRD; A. L. Ruiz, Division 7 (G.E.).

23. Getting, *All in a Lifetime,* 201.

24. Getting, "Draft History, Section 7.6," 10.

25. IAG to KTC, "U.S.N. AA Director Mk. 56."

26. Dennis, "Change of State," 341. Getting had actually collaborated with Section T on these fire control projects and did not share Division 7's animus toward Tuve. Still, Getting needed Division 7 for his own project, so he did nothing to avert its fight with Tuve.

27. Division 7 meeting minutes, 9–10 April, 7–8 July 1943. See also Dennis, "Change of State," 340–46, on how this dispute played out in Section T.

28. Getting, "Draft History, Section 7.6," 8.

29. Division 7 meeting minutes, 28 April 1943.

30. For the design history of the Mark 56, see IAG diary, "Conference on Mark 56 Director," 10 June 1943, "Mk 56 Radar Discussions at Bureau of Ordnance," 15 July 1943, "Mk 56," 2 July 1943, and "Mk 56," 26 July 1943, OSRD7 GP, box 72, IAG Diary folder; Division 7 meeting minutes, "Minutes of Rochester Meeting," 5 January 1944; and Getting, *All in a Lifetime,* 177–81. For an operating description of the system, see U.S. Navy, Bureau of Navy Personnel, *Naval Ordnance and Gunnery,* vol. 2, *Fire Control,* 318–40. For the project history, see *Division 14 Final Report,* 4-55 to 4-63. For Svoboda's relay computers, see "Eloge: Antonin Svoboda, 1907–1980."

31. WW to IAG, 16 January 1945, OSRD7, Office Files of Ivan Getting, box 62.

32. Loomis to IAG, 9 March 1945, ibid. Getting recalls that he was never assigned to BuOrd (personal correspondence with author, 30 October 1996, 3 February 1997).

33. "Statement of Relationships between the Bureau of ordnance, U.S. Navy and the National Defense Research Committee, OSRD, on the Development and Production of the Gunfire Control System Mark 56," reprinted in Getting, *All in a Lifetime*, 186.

34. Ibid.

35. Getting to Furer, 26 April 1945, reprinted in Getting, *All in a Lifetime*, 182–85.

36. Ibid., 184–85.

37. In addition to the works cited in the text, see Brown and Campbell, *Principles of Servomechanisms;* and MacColl, *Fundamental Theory of Servomechanisms.* For an account of the postwar publishing effort and a comparative discussion of control textbooks, see Bissel, "Textbooks and Subtexts." Comparing the degrees of importance of these books is, of course, splitting hairs, although Bissel calls the Radiation Lab volume, James, Nichols, and Phillips's *Theory of Servomechanisms,* "perhaps the most influential of all the American publications of the 1940s."

38. Harold Hazen, "Fire Control Activities of Division 7, NDRC," in Hazen, *Gunfire Control,* 4. Stuart Bennett noted the "systems approach" in his comparison of British and American fire control work during the war in *History of Control Engineering, 1930–1955,* 125.


*Chapter 11 Cybernetics and Ideas of the Digital*

1. HLH to WW, "The Human Being as a Fundamental Link in Automatic Control Systems," 13 May 1941, OSRD7 GP.

2. EJP diary, 9 November 1940, OSRD7 GP, box 70, Collected Diaries, vol. 1.

3. Wiener, *I Am a Mathematician,* 190; Masani, *Norbert Wiener,* 168–69.

4. According to Gordon Brown, Wiener had been invited into fire control work after a meeting at MIT between Fry and Brown that November. Brown wished to discriminate Wiener's "fundamental" work, which belonged to the NDRC, from his own practical work, which belonged to Sperry, and he wanted the ability to freely apply Wiener's conclusions in the Servo Lab. Brown wrote to Nathaniel Sage, head of MIT's Division of Industrial Cooperation, on 18 December 1940, "Dr. Wiener is an authority on many aspects of the branch of mathematics that is related to this work. However, he is but meagerly informed on the techniques necessary to reduce to practice the matters which he can express mathematically. He is also but meagerly informed on the specific limits which must be met when the results of a mathematical investigation are reduced to practice" (OSRD7 GP, box 4, project file 6).

5. Julian Bigelow, interview by Steve J. Heims, 12 November 1968, Princeton, N.J., Steve J. Heims Papers, MC-361, box 1, folder 5, MIT Archives. At Sperry, Bigelow worked on railroad instrumentation, not on fire control (Julian Bigelow, interview by Richard R. Merz, 20 January 1971, Computer Oral History Collection, National Museum of American History).

6. Several published accounts narrate Wiener's work in prediction, including Wiener, *I Am a Mathematician,* 242–56; Bennett, *History of Control Engineering, 1930–1955,* 170–79; idem, "Nor-

bert Wiener and Control of Anti-Aircraft Guns"; Galison, "Ontology of the Enemy"; and P. Masani and R. S. Phillips, "Antiaircraft Fire Control and the Emergence of Cybernetics," in Wiener, *Norbert Wiener,* 141–79. Masani and Phillips's article has a mathematical analysis of an antiaircraft director system based on Wiener's theory and on input from Ivan Getting. Its conclusion, however, that "all told, the results of the air war fought in the years 1942 to 1944 with AA directors designed and operated along the lines of the principles described in this section were impressive," is misleading: Wiener's ideas on prediction did not make it into operational settings during the war. For an evaluation of the legacy of cybernetics, see Heims, *Constructing a Social Science for Postwar America;* and Arbib, "Cybernetics after Twenty-five Years."

7. Norbert Wiener, Final Report on Section D-2, project file 6, 1 December 1942, quoted in Masani and Phillips, "Antiaircraft Fire Control and the Emergence of Cybernetics," 152.

8. Wiener, *I Am a Mathematician,* 244.

9. For a technical explanation of this approach, see Bennett, *History of Control Engineering, 1930–1955,* 174; and idem, "Norbert Wiener and Control of Anti-Aircraft Guns," See also Kailath, "Norbert Wiener and the Development of Mathematical Engineering."

10. "Meeting at BTL of Wiener and Bigelow (and SHC) and BTL Group," 4 June 1941, OSRD7 GP, box 70, Collected Diaries, vol. 2.

11. "Summary of Project #6: Section D-2, NDRC," 1 October 1941, OSRD, E-151, Applied Mathematics Panel General Records, box 24.

12. Wiener, *Extrapolation, Interpolation, and Smoothing,* 3. This is the version published for unrestricted circulation.

13. Ibid., 116. For a discussion of the technical significance of this paper, see Masani, *Norbert Wiener,* 182–87.

14. For one example, see Gelb, *Applied Optimal Estimation.* Gelb cites Wiener's paper third in the introduction to the volume and also identifies the intimate similarity between prediction, filtering, and smoothing, all of which estimate the state of some system, differing only in that they work before, during, or after the data are available, respectively. See also Kailath, "Norbert Wiener and the Development of Mathematical Engineering"; and Zadeh and Ragazzini, "Extension of Wiener's Theory of Prediction," reprinted in Bellman, *Selected Papers on Mathematical Trends in Control Theory,* 150–62. For NDRC researchers' involvement with Wiener's work, see Bennett, *History of Control Engineering, 1930–1955,* 180–81. For Wiener's paper's significance in signal processing, see Nebeker, *Signal Processing,* chap. 2.

15. "Summary of Project #6: Section D-2, NDRC."

16. Stibitz, "Zeroth Generation," 204; WW diary, 1 July 1942, OSRD7 GP, box 71, Collected Diaries, vol. 4. See also Bennett, *History of Control Engineering, 1930–1955,* 178–79.

17. WW diary, 1 September 1942, OSRD7 GP, box 71, Collected Diaries, vol. 4.

18. NW to WW, 6 January, 18 June 1942, NW Papers, box 2, folder 64.

19. WW diary, 10 November 1942, OSRD7 GP, box 72, Collected Diaries, vol. 5.

20. Division 7 meeting minutes, 7–8 January, 3 February 1943. See also Galison, "Ontology of the Enemy," 244–45; and Bigelow interview, 12 November 1968, 8. Wiener's last words on the project to the NDRC are in NW to WW, 15 January and 28 January 1943. Wiener recognized that his

predictor barely exceeded the performance of competing smoothers, but he believed that there were too few data (only two courses for comparison) and that further work should continue in order to compare ten or a hundred courses.

21. Division 7 meeting minutes, 1 March 1944.

22. Cannon, *Wisdom of the Body.*

23. See, e.g., Wiener to J. B. S. Haldane, 22 June 1942, NW Papers, box 2, folder 64. This letter is marked "Not Sent." That May, Rosenblueth mentioned his conversations with Wiener and Bigelow in a presentation at a meeting on the physiology of the conditioned reflex, sponsored by the Macy Foundation (Heims, *Constructing a Social Science for Postwar America*, 14–15).

24. Heims, *John von Neumann and Norbert Wiener*, 184; Edwards, *Closed World*, 180–81; Pickering, "Cyborg History and the World War II Regime."

25. Coast Artillery Board report, 20 April 1940, quoted in Chafee, "Study of the Requirements for a Satisfactory Antiaircraft Fire Control System."

26. HLH diary, 23 September 1943, WW diary, 22–23 July 1941, and TCF diary, 3 April and 19 June 1941, all in OSRD7 GP; John V. Atanasoff, "Elements of Anti-Aircraft Fire Control," OSRD, E-26, Contractors Reports, box 777.

27. HLH to WW, "The Human Being as a Fundamental Link in Automatic Control Systems."

28. TCF diary of conversation with PRB, 19 June 1941, OSRD7 GP.

29. HLH to WW, "The Human Being as a Fundamental Link in Automatic Control Systems"; Fernberger to HLH, 27 May 1941, OSRD7, Office Files of Warren Weaver, box 3, MIT General folder. For the NDRC program, see Hazen, *Airborne Fire Control.*

30. Capshew, *Psychologists on the March*, chaps. 2 and 7; Peter S. Buck, "Adjusting to Military Life: The Social Sciences Go to War, 1941–1950," in Smith, *Military Enterprise and Technological Change*, 203–52. For aviation, see Miles, "Psychological Aspects of Military Aviation."

31. Report from Project 10 (Tufts College) to the NDRC, "Experiments with British Seamen," OSRD7 GP; Hazen, *Airborne Fire Control*, 126–27.

32. E. B. Ferrell, "Automatic Tracking as a Feedback Problem," 20 May 1942, OSRD7 GP, box 2.

33. On page 7 of *Cybernetics* Wiener cites MacColl, *Fundamental Theory of Servomechanisms.* In Wiener, *Norbert Wiener*, the following are the only references to feedback control other than Maxwell in about 75 papers: "Time Communication and the Nervous System" (220–42), a transcript of a speech Wiener gave to the New York Academy of Sciences meeting, "Teleological Mechanisms," in 1946, cites MacColl, *Fundamental Theory of Servomechanisms*; James, Nichols, and Phillips, *Theory of Servomechanisms*; Ahrendt and Taplin, *Automatic Feedback Control*; and Brown and Campbell, *Principles of Servomechanisms* as bibliography entries, not footnotes. "Automatization," from the *St. Louis Dispatch*, 5 December 1954, was co-authored with Donald Campbell, MIT engineering professor, Servo Lab engineer, and student of Gordon Brown. This article is clearly divided between Campbell's predictions for industry and Wiener's warnings. "Muscular Clonus: Cybernetics and Physiology," by Rosenblueth, Wiener, and J. Garcia Ramos (466–510), includes Nyquist diagrams (citing MacColl), but this paper seems to have been written by Garcia Ramos in 1985 from research notes.

34. Mayr, "Maxwell and the Origins of Cybernetics."

35. Wiener, *I Am a Mathematician*, 265, emphasis added.

36. Wiener, *Human Use of Human Beings*, chap. 1.

37. Warren Weaver oral history, RAC, 627; Lieutenant Colonel C. Thomas Sthole to NW, 23 July 1943, NW Papers, box 1, folder 57; NW to Bush, 21 September 1940, ibid., box 2, folder 58.

38. Norbert Wiener, "A Scientist Rebels," reprinted in Wiener, *Norbert Wiener*, 748. Note that in Masani, *Norbert Wiener*, the bibliography of Wiener's military work (p. 391) lists no contributions after 15 January 1943.

39. Masani, *Norbert Wiener*, 190.

40. George Stibitz, "Summary Report on Division 7.5, Relay Computers," OSRD7, Office Files of Harold Hazen, box 6.

41. Galison, "Ontology of the Enemy," 253.

42. Heims, *Constructing a Social Science for Postwar America*; Edwards, *Closed World*, chap. 6; Kay, "Cybernetics, Information, Life," 47.

43. Moreland, OSRD, to Richard Taylor, MIT, 18 August 1942, and Taylor to WW, 29 August 1942, OSRD7, Office Files of Harold Hazen, Differential Analyzer folder. Taylor's letter contains a survey of computing facilities available to the NDRC in 1942.

44. Kuehni and Peterson, "New Differential Analyzer," describes the G.E. machine, with commentary from C. N. Wygandt, of Penn. See also HLH diary, 8 December 1942, OSRD7 GP, box 72, HLH Diaries folder.

45. "Recommendation for Contract: Improvement of Differential Analyzers," 23 December 1942, OSRD7 GP, box 46, project file 62; Moreland to Taylor, 18 August 1942; Proposal from BRL to OSRD, 27 July 1942, OSRD7, Office Files of Harold Hazen, Differential Analyzer folder; WW diary, 28 October 1942, meeting with J. G. Brainerd of Moore School, OSRD7 GP, box 71, Collected Diaries, vol. 4.

46. Claude Shannon, "Theory and Design of Linear Differential Equation Machines," Report to the Services 20, January 1942, reprinted in Shannon, *Claude Elwood Shannon*, 514–59. See also Hazen, *Gunfire Control*, 59–60. Shannon's other four papers were "A Height Data Smoothing Mechanism," "A Study of the Deflection Mechanism and some Results on Rate Finders," Some Experimental Results on the Deflection Mechanism," and "Backlash in Overdamped Systems," all in Claude E. Shannon Papers, MIT Archives. See also Frankford Arsenal, Fire Control Design Section, "Description of Antiaircraft Director," 12 November 1940, OSRD7, Misc. Project Files, box 68; D-2 meeting minutes, 28 January 1941, OSRD7 GP, Collected Diaries, box 70.

47. For a detailed history of these projects, see Williams, "Computing with Electricity," 263–66; and the manuscript version of Wildes and Lindgren, *Century of Electrical Engineering and Computer Science*, 5-101 to 5-112. Bush's original memoranda have been lost, but they are summarized in his "Arithmetical Machine," 7 March 1940, reprinted in Randell, *Origins of Digital Computers*, 337–43; and see Claude Shannon, "A Height Data Smoothing Mechanism," 26 May 1941, Claude E. Shannon Papers, MIT Archives.

48. Crawford, "Automatic Control by Arithmetic Computation"; Perry Crawford, interview by author, 10 January 1995, Cambridge, Mass., notes in author's possession. While Crawford's ideas

were never implemented in a machine, he did play a significant role in digital computing at MIT. Late in the war he joined the navy's Special Devices Center (SDC), which built simulation and training devices. The SDC sponsored an analog flight simulator at MIT's Servo Lab, and Crawford was instrumental in convincing Jay Forrester to build a digital computer for the project, which eventually became the Whirlwind computer. For a detailed technical analysis, see Williams, "Computing with Electricity," 271–90.

49. Vladimir Zworykin memo describing RCA tasks, 6 June 1941, OSRD7 GP, project file 48, box 40; "Recommendation for Appropriation," 1 July 1942, ibid.; J. A. Rajchman, G. A. Morton, and A. W. Vance, "Report on Electronic Predictors for Anti-Aircraft Control," April 1942, reprinted in Randell, *Origins of Digital Computers,* 345–48. See also Williams, "Computing with Electricity," 266–71. Jan Rajchman later designed a magnetic core memory, licensed it to IBM, and challenged Jay Forrester's patent for the technology (see Pugh, *Memories That Shaped an Industry,* 81–87).

50. Blandy to Secretary of the Navy, [ca. spring 1941], OSRD7, Office Files of Warren Weaver, Electronic Computers folder.

51. TCF to WW, 27 June 1941, ibid., Thornton Fry folder.

52. WW to Zworykin, 20 January 1942, ibid., Electronic Computers folder.

53. Rajchman, Morton, and Vance, "Report on Electronic Predictors for Anti-Aircraft Control"; "Electronic Fire Control Computers," 16 December 1941, OSRD7 GP, project file 48, box 40.

54. "Electronic Fire Control Computers," 16 December 1941, OSRD7 GP, project file 48, box 40.

55. For a similar take on this problem, see Owens, "Where Are We Going, Phil Morse?"

56. "Conference on Electronic Fire Control Computing," 16 April 1942, OSRD7, Office Files of Warren Weaver, box 2. For another description of this meeting, see Williams, "Computing with Electricity," 303–9.

57. George Stibitz, critique of conference on electronic fire control computers, OSRD7, box 73, quoted in Williams, "Computing with Electricity," 309–15. John Atanasoff may have been the first to specifically apply the term *analog* to a computing machine, in 1940, although Bush and Hazen had frequently used the term when speaking about their circuit models (Burks and Burks, *First Electronic Computer,* 124). The *Oxford English Dictionary* is no help here, as its first listing of *analog* used in reference to a computer is Douglas Hartree in 1946.

58. "Recommendation for Appropriation," 1 July 1942. See also "Automatic Computer for Anti-Aircraft Fire Control," 6 July 1942, and DJS diary, 19 June 1942, both in OSRD7 GP, project file 48, box 40.

59. WW to Zworykin, 20 July 1942, OSRD7 GP, project file 48, box 40.

60. WW to Zworykin, 11 November 1942, ibid.

61. WW to HLH, 20 November 1942, ibid.

62. HLH to Paul E. Klopsteg, 6 February 1943, and George Beggs Jr. to HLH, 22 March 1943, OSRD7 GP.

63. Brainerd, "Genesis of the ENIAC." The proposal was dated 2 April 1943, just two days after the RCA project terminated (Stern, *From ENIAC to UNIVAC,* 18–19).

64. The issue of ENIAC's similarity to a differential analyzer is tied up with the contentious historical and legal arguments surrounding the ENIAC's origins, which are well covered in Williams, "Computing with Electricity," 354–72; and Burks and Burks, *First Electronic Computer,* 109–11.

65. SCH to WW, 15 May 1943, OSRD GP, box 80, Ballistics—General Correspondence folder; HLH diary, 14 April 1943, OSRD GP, project file 62, box 46. For participant accounts of ENIAC development, see Goldstine, *Computer;* and Brainerd, "Genesis of the ENIAC."

66. Stern implies that the MIT engineers were envious of Penn's "more powerful" differential analyzer; in fact, they felt that the university had not adequately acknowledged its intellectual debt to MIT (Stern, *From ENIAC to UNIVAC,* 21; Goldstine, *Computer,* 150). For a balanced, detailed discussion that explains Caldwell's skepticism, see Williams, "Computing with Electricity," 376–86.

67. HLH diaries, 5 March, 14 April 1943, OSRD7 GP, box 72, HLH Diaries folder.

68. Division 7 meeting minutes, 6 October, 3 November 1943; HLH diary, 14 April 1943. For a clear, concise statement of Division 7's position on the ENIAC proposal, including Caldwell's opinion and references to the RCA project, see HLH to Moreland, 10 May 1943, OSRD7 GP, project file 62, box 46. Weygandt, in his discussion of G.E.'s Kuehne and Peterson's "New Differential Analyser," wrote that "those of us who have spent a great deal of time and effort in trying to design a servo system for this purpose [a differential analyzer] realize the difficulty of the job. . . . [It] is a difficult one because of the wide speed range which must be covered and also because in the setup of a problem a number of the servo mechanisms may be cascaded. . . . A servo system which is stable in itself may not remain stable when interconnected with other similar systems."

69. Rajchman, quoted in Stern, *From ENIAC to UNIVAC,* 44.

70. "ENIAC Progress Report," 31 December 1943, quoted in Stern, *From ENIAC to UNIVAC,* 75. For ENIAC programming, see Marcus and Akera, "Exploring the Architecture of an Early Machine."

71. Manuscript version of Wildes and Lindgren, *Century of Electrical Engineering and Computer Science,* 5-127. This project and the Center for Analysis itself were eventually terminated as a result of the increasing prominence of the Whirlwind project, run by the Servo Lab.

72. It would take Vannevar Bush's 1945 report to the president, *Science: The Endless Frontier,* to add the crucial ingredient to the postwar research paradigm: government support of basic research.

73. See OSRD7 GP, April–November 1942, project file 25, for problems with dynamic testers.

74. George Stibitz, "Computer," 1940, reprinted in Randell, *Origins of Digital Computers,* 247–52.

75. "Project Recommended for Appropriation, No. 60, Section D-2—Fire Control, Simplified Dynamic Tester," OSRD7 GP, project file 60, box 45. For Stibitz's computers in general, see Ceruzzi, *Reckoners.*

76. George Stibitz, "Proposed Dynamic Tester," 19 October 1942, OSRD7 GP, project file 60, box 44; idem, "Zeroth Generation," 167–68.

77. George Stibitz, "Relay Servo Circuit," 28 October 1942, OSRD7 GP, project file 60, box 45. Stibitz copied this memo to Poitras, Stewart, Weaver, Mooney (Division 7's liaison with the Antiaircraft Artillery Board) and MacNair of Bell Labs.

78. George Stibitz, "Nyquist Loop for Tape Servo," 19 February 1943, and idem, "Equivalent Feedback Amplifier for Tape Servo," 14 February 1943, ibid. In 1944 Claude Shannon performed a similar analysis for a different application, with results equivalent to Stibitz's (see Shannon, "Feedback Systems with Periodic Loop Closure, Memorandum for File," 16 March 1944, ibid.).

79. Daniel Silverman to GRS, 29 April 1943, ibid. See also Noble, *Forces of Production,* 88. Noble states, incorrectly, that Bell Labs never used the tape dynamic tester for its original purpose.

80. GRS diary, 10 February 1943, OSRD7 GP, project file 60, box 45.

81. Caldwell to Stewart, 28 June 1945, ibid., project file 70, box 50.

82. GRS to SHC (copies to WW and DJS), 9 June 1941, and George Stibitz, "An Application of the Relay Interpolator," 10 June 1943, ibid.

83. Applied Mathematics Panel, NDRC, "A Statement Concerning the Future Availability of a New Computing Device," AMP Note No. 7, November 1943, ibid.; Contract Proposal, "Relay Interpolator," 10 August 1943, ibid.

84. For all these memoranda, see OSRD7 GP, project file 70, box 50, Relay Interpolator BTL folder. For a technical description of the RI see Cesareo, "Relay Interpolator."

85. George Stibitz, "Unified Theory of the Relay Interpolator," OSRD7 GP, project file 70, box 50, Relay Interpolator BTL folder; idem, "Zeroth Generation," 181.

86. George Stibitz, "Outline of Relay Ballistic Computer," 7 July 1943, OSRD7 GP, project file 74, box 54; Juley, "Ballistic Computer." The "Ballistic Computer," or Model III, could calculate errors in one of two ways. First, the machine could replicate the calculations in the director and compute the correct gun orders based on the three inputs, then subtract them from the director's gun orders to find the errors. This method had the advantage of isolating errors in each of the three variables for separate analysis; Stibitz called these errors Class 1. The second method, in contrast, took as input the gun orders produced by the director and essentially performed the ballistics calculation to determine the point and time at which the shell would explode. The machine then interpolated the position of the target plane at that particular time and produced a distance by which the shell missed (or hit) the target when it exploded. This technique lumped the errors in all three output variables together as Class 2 errors but had the advantage of producing a "miss distance," which quantitatively compared the performance of different directors.

87. George Stibitz, "Progress on AAB Computer," 13 May 1944, OSRD7 GP, project file 74, box 54; Franz L. Alt, "A Bell Telephone Laboratories Computing Machine," *Mathematical Tables and Other Aids to Computation* 3, no. 21 (1948): 1–13, 69–84, reprinted in Randell, *Origins of Digital Computers,* 257–86; Stibitz, "Zeroth Generation," chap. 9; Ceruzzi, *Reckoners,* 105.

88. Stibitz, "Zeroth Generation," 106.

89. George Stibitz, "Computation," [early 1943?], OSRD7 GP, project file 74, box 54.

90. Bernard Holbrook, interview by Uta C. Merzbach, 10 May 1969, Computer Oral History Collection, National Museum of American History.

91. George Stibitz, "Relay Interpolator as a Differential Analyzer," OSRD7 GP, project file 70, box 50, Relay Interpolator BTL folder, emphasis added.

92. Stibitz, "Zeroth Generation," 109.

93. Stibitz, "Relay Interpolator as a Differential Analyzer," emphasis added. Howard Aiken made a similar distinction between architecture and components (see Cohen, *Howard Aiken*, 38).

94. Von Neumann, "First Draft of a Report on the EDVAC."

*Chapter 12 Conclusion*

1. George A. Philbrick, "Aiming Controls in Aerial Ordnance," in Hazen, *Airborne Fire Control*, 4, 19, 23–24. The MIT-educated Philbrick worked for the Foxboro Company, a supplier of process instruments and controls. In the late 1930s Philbrick began building a flexible, electronic analog computer for studying industrial systems. He called this machine an "automatic analyzer" and named it Polyphemus because of its one-eyed oscilloscope output (Holst, "George A. Philbrick and Polyphemus," 144–45; Philbrick, "Designing Industrial Controllers by Analog").

2. Philbrick, "Aiming Controls in Aerial Ordnance," 24, 48.

3. Ibid., 63–65. As he notes in ibid., 51, Philbrick was elaborating ideas he had had before the war. See also Redmond and Smith, *Project Whirlwind*.

4. Philbrick, "Designing Industrial Controllers by Analog." *Electronic Design*, 16 December 1995, 8, reprints the original 1955 announcement of the Philbrick K2-X operational amplifier. The magazine's retrospective calls the device "an industry classic" and notes that a number of the leading analog electronics designers had gotten their start at Philbrick Researches. For a rich explication of Philbrick's analog computers and their applications in the 1950s, see Paynter, *Palimpsest on the Analog Art*.

5. Philbrick, "Aiming Controls in Aerial Ordnance," 65.

6. Sapolsky, *Science and the Navy*; Rees, "Computing Programs of the Office of Naval Research"; Norberg and O'Neill, *Transforming Computer Technology*.

7. Quoted in Weizenbaum, *Computer Power and Human Reason*, 30; Redmond and Smith, *From Whirlwind to MITRE*.

8. See "Special Issue: SAGE" for a number of personal accounts and oral histories of SAGE and related projects; Valley, "How the Sage Development Began"; Edwards, *Closed World*, chap. 3, for a discussion of SAGE as a Cold War icon of technology and its connection to other large-scale command and control systems; and Hughes, *Rescuing Prometheus*, chap. 2.

9. Getting, *All in a Lifetime*; Aerospace Corporation, *Aerospace Corporation*.

10. Van Valkenburg, "In Memoriam: Hendrik W. Bode"; Hendrik Bode Papers.

11. Gordon Brown, quoted in Wildes and Lindgren, *Century of Electrical Engineering and Computer Science*, 342.

12. Hooper and Rivero biographies, Admiral Biographies Collection, Naval Operational Archives; Blandy Papers.

13. Bissel, "Spreading the Word"; James, Nichols, and Phillips, *Theory of Servomechanisms*; Brown and Campbell, *Principles of Servomechanisms*; MacColl, *Fundamental Theory of Servomechanisms*; Ahrendt and Taplin, *Automatic Feedback Control*; Bissel, "Textbooks and Subtexts." Before the war, the American Society of Mechanical Engineers established a Committee on In-

dustrial Instruments and Regulators; in 1946 the American Institute of Electrical Engineers, historically the home of power engineers, established a subcommittee on servomechanisms; and in a parallel move in 1952, the Institute of Radio Engineers founded a Feedback Control Systems Committee (Bennett, "Emergence of a Discipline").

14. I am indebted to Stuart Bennett for this observation (personal communication to author, 20 September 2001).

15. Gerovitch, *From Newspeak to Cyberspeak;* Bowker, "How to Be Uiversal"; Hughes and Hughes, *Systems, Experts, and Computers.*

16. See, e.g., the Scientific American book *Automatic Control,* with articles by Brown, Campbell, William Pease, Warren Weaver, and Louis Ridenour, in addition to Arnold Tustin and Wassily Leontief; Noble, *Forces of Production;* and Reintjes, *Numerical Control.*

17. McSwain, "Social Reconstruction of a Reverse Salient in Electric Guitar Technology," 206.

18. Stibitz, "Zeroth Generation," 109.

19. Small, "General Purpose Electronic Analog Computing."

20. The popular mathematical software MATLAB, for example, includes a simulation package with integrators, multipliers, nonlinear components, and a host of other elements identical to those illustrated for an analog computer in Paynter, *Palimpsest on the Analog Art,* 41.

21. Blackman, Bode, and Shannon, "Data Smoothing and Prediction in Fire-Control Systems." See also Bode and Shannon, " Simplified Derivation of Linear Least Square Smoothing and Prediction Theory," 425, which addresses Wiener's prediction in more detail; and Blackman, *Linear Data-Smoothing and Prediction,* an extension of the 1948 work.

22. Shannon, "Mathematical Theory of Communication"; Shannon and Weaver, *Mathematical Theory of Communication.*

23. Shannon and Weaver, *Mathematical Theory of Communication.* For Shannon's early work see Shannon to Bush, 16 February 1939, in Shannon, *Claude Elwood Shannon,* 455–56.

24. Shannon, "Mathematical Theory of Communication," 53n. The relationship between Shannon's and Wiener's work is more complex than described here. In a 1987 interview Shannon said, "I don't think Wiener had much to do with information theory. He wasn't a big influence on my ideas there [at MIT], though I once took a course from him" (Shannon, *Claude Elwood Shannon,* xix). Semantic confusion sometimes exists over the "Weaver-Shannon" or the "Wiener-Shannon" theory of communication. The former, which derives from Shannon and Weaver, *Mathematical Theory of Communication,* is inaccurate because Weaver only translated Shannon's work to make it more accessible (Weaver claimed no more).

25. Wiener, *Cybernetics,* 6.

# Bibliography

*Major Archival Sources*

AT&T Archives, Warren, New Jersey
American Telephone and Telegraph Archives
Bell Telephone Laboratories Archives

Hagley Museum and Library, Wilmington, Delaware
Elmer Ambrose Sperry Papers
Sperry Gyroscope Company Papers

Harvard University Archives, Cambridge, Massachusetts
Hendrik Bode Papers

IEEE Center for the History of Electrical Engineering, Rutgers University, New Brunswick, New
    Jersey
Radiation Laboratory Oral History Interviews

Library of Congress, Washington, D.C.
W. H. P. Blandy Papers
Vannevar Bush Papers
Charles Stark Draper Papers
William R. Furlong Papers
Edwin Hooper Papers

MIT Archives, Cambridge, Massachusetts
Gordon Brown Papers
Vannevar Bush Papers
Samuel Caldwell Papers
Computers at MIT Oral History Collection
Harold Hazen Papers
Steve J. Heims Papers
Servomechanisms Laboratory Papers
Claude E. Shannon Papers
Norbert Wiener Papers
Karl Wildes Papers

National Archives and Records Administration, College Park, Maryland
RG 38. Records of the Office of the Chief of Naval Operations
RG 74. U.S. Navy, Bureau of Ordnance Records
RG 111. U.S. Army, Office of the Chief Signal Officer
RG 156. U.S. Army, Office of the Chief of Ordnance
RG 227. Office of Scientific Research and Development
RG 407. U.S. Army, Antiaircraft Action Reports

National Archives, Waltham, Massachusetts
RG 181. MIT Radiation Laboratory Records

National Museum of American History, Smithsonian Institution, Washington, D.C.
Computer Oral History Collection
Lloyd Espenschied Papers
Technical Manual and Catalog Collection

Naval Operational Archives, Washington, D.C.
Admirals Biographies Collection
World War II Command File

Naval War College Library, Newport, Rhode Island

Rockefeller Archive Center, Sleepy Hollow, New York
Rockefeller Foundation Archives
Warren Weaver Papers

*Unpublished Manuscripts, Theses, and Dissertations*

Barry, Constantine. "The Parallel Operation of Alternators through Long Transmission Lines." S.B. thesis, Massachusetts Institute of Technology, 1927.

Brown, Gordon S. "The Cinema Integraph: A Machine for Evaluating a Parametric Product Integral." Sc.D. diss., Massachusetts Institute of Technology, 1938.

———. "A Photocell Receiver and a Direct Current Vacuum-tube Amplifier for the Cinema Integraph." S.M. thesis, Massachusetts Institute of Technology, 1934.

———. "Transient Behavior and Design of Servomechanisms." Research report, Servomechanisms Laboratory, Massachusetts Institute of Technology, 1943.

Brown, Gordon S., and Kenneth Germeshausen. "The Effect of Controlled Field Switching on the Pulling-into-Step of a Synchronous Induction Motor." S.B. thesis, Massachusetts Institute of Technology, 1931.

Clymer, Ben A. "Mechanical Integrators." M.S. thesis, Ohio State University, 1946.

Crawford, Perry. "Automatic Control by Arithmetic Computation." S.M. thesis, Massachusetts Institute of Technology, 1942.

Davy, S. J. "A Case Study: The American Bosch Arma Corporation." Term paper, Sloan School of Management, Massachusetts Institute of Technology, 1958.

Dennis, Michael. "A Change of State: The Political Cultures of Technical Practice at the MIT Instrumentation Laboratory and the Johns Hopkins Applied Physics Laboratory, 1930–1945." Ph.D. diss., Johns Hopkins University, 1991.

Edgerton, Harold. "Abrupt Change in Load on a Synchronous Machine." S.M. thesis, Massachusetts Institute of Technology 1927.

———. "Benefits of Angular-Controlled Field Switching on the Pulling-into-Step Ability of Salient-Pole Synchronous Motors." Sc.D. diss., Massachusetts Institute of Technology, 1931.

Gerovitch, Vyacheslav. "Speaking Cybernetically: The Soviet Remaking of an American Science." Ph.D. diss., Massachusetts Institute of Technology, Program in Science, Technology, and Society, 1999.

Green, Cecil. "A Static Study of the No Load Flux Distribution in a Salient Pole Alternator." S.B. and S.M. thesis, Massachusetts Institute of Technology, 1924.

Hazen, Harold L. "The Extension of Electrical Engineering Analysis through Reduction of Computational Limits by Mechanical Means." Sc.D. diss., Massachusetts Institute of Technology, 1931.

———. "Memoirs: An Informal Story of My Life and Work." MS, Massachusetts Institute of Technology Archives, Cambridge, 1976.

Hedeman, Walter R. "Numerical Solutions of Integral Equations on the Cinema Integraph." Sc.D. thesis, Massachusetts Institute of Technology, 1939.

Hooper, E. B., and A. G. Ward. "Control of an Electro-Hydraulic Servo Unit." S.M. thesis, MIT, 1940.

Howard, John H. "Measurement and Analysis of Errors in the Cinema Integraph." S.M. thesis, Massachusetts Institute of Technology, 1939.

Hustveldt, Erling. "Battleship Gunfire Control." MS, University of Maryland, 15 March 1990.

Kailath, Thomas. "Norbert Wiener and the Development of Mathematical Engineering." MS, Stanford University, Palo Alto, 1996.

Owens, Larry. "Straight-Thinking: Vannevar Bush and the Culture of American Engineering." Ph.D. diss., Princeton University, 1987.

Poitras, Edward J., and James A. St. Louis. "Photoelectric Efficiency for Visible and Ultra-Violet Radiation." M.S. thesis, MIT, 1929.

Rivero, H., and L. M. Mustin. "A Servo Mechanism for a Rate Follow-up System." S.M. thesis, MIT, 1940.

Segal, Jérôme. "Théorie de l'information: Sciences, techniques, et société de la seconde guerre mondiale á l'aube du XXie siècle." Ph.D. diss., Université Lyon II, 1998.

Stibitz, George. "The Zeroth Generation: A Scientist's Recollections (1937–1955) from the Early Binary Relay Digital Computers at Bell Telephone Laboratory and OSRD to a Fledgling Minicomputer at the Barber Coleman Company." MS, National Museum of American History, Smithsonian Institution, Washington, D.C., 1993.

Terman, Frederick E. "Characteristics and Stability of Transmission Systems." Sc.D. diss., Massachusetts Institute of Technology, 1924.

Wang, Sherman. "A Study of Synchronous Machines Not Running at Synchronous Speed." S.B. thesis, Massachusetts Institute of Technology, 1929.

Wildes, Karl, and Nilo Lindgren. "A Century of Electrical Engineering and Computer Science at MIT, 1882–1982." Wildes Papers, Massachusetts Institute of Technology Archives, Cambridge, 1985.

Williams, Bernard. "Computing with Electricity, 1935–1945." Ph.D. diss., University of Kansas, 1984.


Published Sources

Aerospace Corporation. *Aerospace Corporation, Its Work: 1960–1980*. El Segundo, Calif., 1980.

Affel, H. A., C. S. Demarest, and C. W. Green. "Carrier Systems on Long Distance Telephone Lines." *Bell System Technical Journal* 7 (July 1928): 564–627.

Ahrendt, William, and John Taplin. *Automatic Feedback Control*. New York: McGraw-Hill, 1951.

AIEE Committee on Communication. "Annual Report: Electrical Communication." *Transactions of the AIEE* 46 (1927): 713–17.

———. "Annual Report: Recent Advances in the Communication Art." *Transactions of the AIEE* 44 (1925): 707–13.

AIEE Committee on Power Transmission and Distribution. "Annual Report." *Transactions of the AIEE* 46 (1927).

Aitken, Hugh. *The Continuous Wave: Technology and American Radio, 1900–1932*. Baltimore: Johns Hopkins University Press, 1985.

Alder, Ken. *Engineering the Revolution: Arms and Enlightenment in France, 1763–1815*. Princeton: Princeton University Press, 1997.

Allison, David Kite. *New Eye for the Navy: The Origin of Radar at the Naval Research Laboratory*. Washington, D.C.: Naval Research Laboratory, Technical Information Division, 1981.

Arbib, Michael. "Cybernetics after Twenty-five Years: A Personal View of System Theory and Brain Theory." *IEEE Transactions on Systems, Man, and Cybernetics* 5 (May 1975): 359–65.

Arnold, Harold D. "Systematized Research." *Bell Laboratories Record*, June 1928, 316–17.

Aspray, William. *Computing before Computers*. Ames: Iowa State University Press, 1990.

———. "Edwin L. Harder and the Anacom: Analog Computing at Westinghouse." *IEEE Annals of the History of Computing* 15, no. 2 (1993): 35–52.

———. "The Scientific Conceptualization of Information: A Survey." *Annals of the History of Computing* 7 (April 1985): 117–40.

Baitsell, George. *Science in Progress*. New Haven: Yale University Press, 1945.

Baldwin, Ralph. *The Deadly Fuze: The Secret Weapons of World War II*. San Rafael: Presidio, 1980.

Bassett, Preston. "Review of Ground Anti-Aircraft Defense." *Sperryscope* 11, no. 7 (1948): 16–20.

———. "Servomechanisms: Controlling Vehicles in the Air." *Sperryscope* 13, no. 1 (1953): 20–24.

Baxter, James Phinney. *Scientists against Time*. Boston: Little, Brown, 1946.

Bellman, Richard Ernest, and Robert E. Kalaba, eds. *Selected Papers on Mathematical Trends in Control Theory.* New York: Dover, 1964.

"The Bell System Exhibit at the Century of Progress Exposition." *Bell Laboratories Record,* July 1933.

Beniger, James R. *The Control Revolution: Technological and Economic Origins of the Information Society.* Cambridge: Harvard University Press, 1986.

Benjamin, Walter, and Hannah Arendt. *Illuminations.* New York: Harcourt Brace & World, 1968.

Bennett, Stuart. "The Development of Process Control Instruments, 1900–1940." *Transactions of the Newcomen Society* 63 (January 1992): 133–64.

———. "Development of the PID Controller." *IEEE Control Systems,* December 1993, 58–65.

———. "The Emergence of a Discipline: Automatic Control, 1940–1960." *Automatica* 12 (1976): 113–21.

———. "Harold Hazen and the Theory and Design of Servomechanisms." *International Journal of Control* 5, no. 42 (1985): 989–1012.

———. *A History of Control Engineering, 1800–1930.* Stevenage, England: Peter Peregrinus for the IEE, 1979.

———. *A History of Control Engineering, 1930–1955.* Stevenage, England: Peter Peregrinus for the IEE, 1993.

———. "The Industrial Instrument—Master of Industry, Servant of Management: Automatic Control in the Process Industries, 1900–1940." *Technology and Culture* 32, no. 1 (1991): 69–81.

———. "Nicholas Minorsky and the Automatic Steering of Ships." *IEEE Control Systems,* November 1984, 10–15.

———. "Norbert Wiener and Control of Anti-Aircraft Guns." *IEEE Control Systems,* December 1994, 58–62.

Bijker, Wiebe E., Thomas Parke Hughes, and T. J. Pinch, eds. *The Social Construction of Technological Systems: New Directions in the Sociology and History of Technology.* Cambridge: MIT Press, 1987.

Bissel, Chris C. "Modeling Sampled-Data Systems: A Historical Outline." *Transactions of the Institute of Measurement and Control* 7 (April–June 1985): 159–64.

———. "Spreading the Word: Aspects of the Evolution of the Language of the Measurement of Control." *Measurement and Control* 27 (June 1994): 154.

———. "Textbooks and Subtexts: A Sideways Look at the Post-War Control Engineering Textbooks, Which Appeared Half a Century Ago." *IEEE Control Systems,* April 1996, 71–78.

Black, Harold S. "Feedback Amplifiers." *Bell Laboratories Record,* June 1934, 294.

———. "Inventing the Negative Feedback Amplifier." *IEEE Spectrum* 14 (December 1977): 54–60.

———. "Stabilized Feedback Amplifiers." *Bell System Technical Journal* 13 (January 1934): 1–18.

Blackman, R. B. *Linear Data-Smoothing and Prediction in Theory and Practice.* Reading, Mass.: Addison-Wesley, 1965.

Blackman, R. B., H. W. Bode, and C. E. Shannon. "Data Smoothing and Prediction in Fire-Control Systems." In Hazen, *Gunfire Control,* 71–155.

Blandy, W. H. P. "Possible Improvements in Our Gunnery Training." *U.S. Naval Institute Proceedings* 51, no. 271 (1925): 1696–1702.

"Blow Hot—Blow Cold: The M-9 Never Failed." *Bell Laboratories Record,* December 1946, 454–56.

Bode, Hendrik W. "Feedback: The History of an Idea." In Bellman and Kalaba, *Selected Papers on Mathematical Trends in Control Theory,* 106–23.

———. "General Theory of Electric Wave Filters." *Journal of Mathematics and Physics* 13 (November 1934): 275–362.

———. *Network Analysis and Feedback Amplifier Design.* New York: Van Nostrand, 1945.

———. "Relations between Attenuation and Phase in Feedback Amplifier Design." *Bell System Technical Journal* 19 (July 1940): 421–54.

———. *Synergy: Technical Integration and Technological Innovation in the Bell System.* Murray Hill, N.J.: Bell Laboratories, 1971.

———. "Variable Equalizers." *Bell System Technical Journal* 17 (April 1938): 229–44.

Bode, Hendrik W., and C. E. Shannon. "A Simplified Derivation of Linear Least Square Smoothing and Prediction Theory." *Proceedings of the IRE* 38 (1950): 417–25.

"Bomber Defense from a Little Black Box." *Sperryscope* 9, no. 12 (1943): 3–12.

Borst, John M. "The Decibel, Its Definition, Derivation and Application." *Radio News* 13 (November 1931): 384–85.

Bowker, Geof. "How to Be Universal: Some Cybernetic Strategies, 1943–70." *Social Studies of Science* 23 (February 1993): 107–27.

Boyce, Joseph C. *New Weapons for Air Warfare: Fire-Control Equipment, Proximity Fuzes, and Guided Missiles.* Boston: Little, Brown, 1947.

Brainerd, John G. "Genesis of the ENIAC." *Technology and Culture* 17, no. 3 (1976): 482–88.

Brittain, James E. *Alexanderson: Pioneer in American Electrical Engineering.* Baltimore: Johns Hopkins University Press, 1992.

———. "The Introduction of the Loading Coil: George A. Campbell and Michael I. Pupin." *Technology and Culture* 11, no. 1 (1970): 36–57.

Bromley, Allan G. "Analog Computing Devices." In Aspray, *Computing before Computers,* 156–99.

———. "Difference and Analytical Engines." In Aspray, *Computing before Computers,* 59–98.

Brooks, John. *Telephone: The First Hundred Years.* New York: Harper & Row, 1976.

Brown, Gordon S. "Eloge: Harold Locke Hazen, 1901–1980." *Annals of the History of Computing* 3 (January 1981): 4–12.

Brown, Gordon S., and Donald P. Campbell. *Principles of Servomechanisms: Dynamics and Synthesis of Closed-Loop Control Systems.* New York: J. Wiley & Sons, 1948.

Brown, Gordon S., Donald P. Campbell, William Pease, Warren Weaver, Louis B. Ridenour, Arnold Tustin, and Wassily Leontief. *Automatic Control: The Scientific American Book.* New York: Simon & Schuster, 1948.

Brown, John K. *The Baldwin Locomotive Works, 1831–1915: A Study in American Industrial Practice.* Baltimore: Johns Hopkins University Press, 1995.

Brown, Louis. *A Radar History of World War II: Technical and Military Imperatives*. Philadelphia: Institute of Physics, 1999.

Burchard, John E., and James Rhyne Killian. *Q.E.D.: MIT in World War II*. New York: J. Wiley & Sons, 1948.

Burks, Alice R., and Arthur W. Burks. *The First Electronic Computer: The Atanasoff Story*. Ann Arbor: University of Michigan Press, 1988.

Bush, Vannevar. "The Differential Analyzer: A New Machine for Solving Differential Equations." *Journal of the Franklin Institute* 212 (1931): 447–88.

———. "Gimbal Stabilization." *Journal of the Franklin Institute* 203 (1919): 199–215.

———. "A New Type of Differential Analyzer." *Journal of the Franklin Institute* 240 (October): 255–326.

———. *Operational Circuit Analysis*. New York: J. Wiley & Sons, 1929.

———. *Pieces of the Action*. New York: Morrow, 1970.

———. *Science: The Endless Frontier*. Cambridge: MIT Press, 1945.

Bush, Vannevar, and R. D. Booth. "Power System Transients." *Transactions of the AIEE* 44 (1925): 229–40.

Bush, Vannevar, F. D. Gage, and H. R. Stewart. "A Continuous Integraph." *Journal of the Franklin Institute* 211 (1927): 63–84.

Bush, Vannevar, and Harold Hazen. "Integraph Solution of Differential Equations." *Journal of the Franklin Institute* 211 (1927): 575–615.

Campbell-Kelly, Martin, and William Aspray. *Computer: A History of the Information Machine*. New York: Basic Books, 1996.

Cannon, Walter B. *The Wisdom of the Body*. New York: Norton, 1932.

Capshew, James. *Psychologists on the March: Science, Practice, and Professional Identity in America, 1929–1969*. Cambridge: Cambridge University Press, 1999.

Carlson, W. Bernard. "Academic Entrepreneurship and Engineering Education: Dugald C. Jackson and the MIT-GE Cooperative Engineering Course, 1907–1932." *Technology and Culture* 29, no. 3 (1988): 536–67.

Carson, John Renshaw. *Electric Circuit Theory and the Operational Calculus*. New York: McGraw-Hill, 1926.

———. "Theory of Transient Oscillations in Electric Networks." *Transactions of the AIEE* 38 (1919).

Ceruzzi, Paul E. *A History of Modern Computing*. Cambridge: MIT Press, 1998.

———. *Reckoners: The Prehistory of the Digital Computer*. Westport, Conn.: Greenwood Press, 1983.

———. "When Computers Were Human." *Annals of the History of Computing* 13 (1991): 237–44.

Cesareo, O. "The Relay Interpolator." *Bell Laboratories Record*, December 1946, 5–9.

Channell, David. *The Vital Machine: A Study of Technology and Organic Life*. Oxford: Oxford University Press, 1991.

Chapuis, Robert J. *One Hundred Years of Telephone Switching, 1878–1978*. Amsterdam: North Holland, 1982.

Charlesworth, H. P. "General Engineering Problems of the Bell System." *Bell System Technical Journal* 4 (October 1925): 515–41.

Chase, J. V. "Accuracy of Fire at Long Ranges." *U.S. Naval Institute Proceedings* 46, no. 8 (1920): 1175–95.

Cherry, E. Colin. "A History of the Theory of Information." *Proceedings of the IEEE* 98 (September 1951): 383–93.

Clark, A. B., and B. W. Kenall. "Carrier in Cable." *Bell System Technical Journal* 12 (July 1933): 251–62.

Clymer, Ben A. "The Mechanical Analog Computers of Hannibal Ford and William Newell." *IEEE Annals of the History of Computing* 15, no. 2 (1993): 19–22.

Cohen, Bernard. *Howard Aiken: Portrait of a Computer Pioneer.* Cambridge: MIT Press, 1999.

Cohen, Nathan. "Recollections of the Evolution of Realtime Control Applications to Electric Power Systems." *Automatica* 20 (1984): 145–62.

Colpitts, E. H. "Dr. H. D. Arnold." *Bell Laboratories Record,* August 1928, 411–13.

Colpitts, E. H., and O. B. Blackwell. "Carrier Current Telephony and Telegraphy." *Journal of the AIEE* 40 (1921): 301–15.

Colton, Roger B. "Radar in the United States Army: History and Early Development at the Signal Corps Laboratories, Fort Monmouth, N.J." *Proceedings of the IRE* 33 (1945): 740–53.

*Communications Milestone: Negative Feedback.* Murray Hill, N.J.: Bell Laboratories, 1971. Film.

Constant, Edward W. *The Origins of the Turbojet Revolution.* Baltimore: Johns Hopkins University Press, 1980.

Coon, Horace Campbell. *American Tel & Tel: The Story of the Great Monopoly.* New York: Longmans, Green, 1939.

Cortada, James W. *Before the Computer: IBM, NCR, Burroughs, and Remington Rand and the Industry They Created, 1865–1956.* Princeton: Princeton University Press, 1993.

Crary, Jonathan. *Suspensions of Perception: Attention, Spectacle, and Modern Culture.* Cambridge: MIT Press, 1999.

———. *Techniques of the Observer: On Vision and Modernity in the Nineteenth Century.* Cambridge: MIT Press, 1990.

Crimi, Alfred. *Crimi: A Look Back, a Step Forward.* New York: Center for Migration Studies, 1988.

Crowell, Benedict. *America's Munitions, 1917–1918.* Washington, D.C.: U.S. Government Printing Office, 1919.

Dagognet, François. *Etienne-Jules Marey: A Passion for the Trace.* Cambridge, Mass.: Zone Books, 1992.

Dahl, J. F. "Improved Equipment for Information Service." *Bell Laboratories Record,* March 1930, 328–32.

Daston, Lorraine, and Peter Galison. "The Image of Objectivity." *Representations* 40 (fall 1992): 81–128.

Deleuze, Gilles, and Seán Hand. *Foucault.* Minneapolis: University of Minnesota Press, 1988.

"Development of the Electrical Director." *Bell Laboratories Record,* January 1944, 225–40.

Dorf, Richard C. *Modern Control Systems.* 5th ed. Reading, Mass.: Addison-Wesley, 1995.

Douch, E. J. "The Use of Servos in the Army during the Past War." *Journal of the IEE* 94 (March 1947): 177–89.

Draper, C. S., G. P. Bentley, and H. H. Willis. "The M.I.T.-Sperry Apparatus for Measuring Vibration." *Journal of the Aeronautical Sciences* 4 (May 1937): 281–85.

Dreyer, Frederic C. *The Sea Heritage: A Study of Maritime Warfare.* London: Museum Press, 1955.

Dudley, Homer. "The Carrier Nature of Speech." *Bell System Technical Journal* 19 (October 1940): 495–515.

Dupree, A. Hunter. "The *Great Instauration* of 1940." In Holton, *Twentieth-Century Sciences,* 443–67.

———. *Science in the Federal Government: A History of Policies and Activities to 1940.* Cambridge, Mass.: Harvard University Press, 1957.

Edwards, Paul N. *The Closed World: Computers and the Politics of Discourse in Cold War America, Inside Technology.* Cambridge: MIT Press, 1996.

Eksteins, Modris. *Rites of Spring: The Great War and the Birth of the Modern Age.* New York: Anchor Books, 1990.

"Electrical Gun Director Demonstrated." *Bell Laboratories Record,* December 1943, 157–67.

"Eloge: Antonin Svoboda, 1907–1980." *Annals of the History of Computing* 2 (October 1980): 284–92.

Espenschied, Lloyd. "Application of Radio to Wire Transmission Engineering." *Bell System Technical Journal* 1 (October 1922): 117–41.

Espenschied, Lloyd, Joseph Gray Jackson, and John G. Brainerd. "Discussion of James G. Brittain, 'The Introduction of the Loading Coil.'" *Technology and Culture* 11, no. 4 (1970): 596–603.

Espenschied, Lloyd, and M. E. Strieby. "Systems for Wide-Band Transmission over Coaxial Lines." *Bell System Technical Journal* 13 (October 1934): 654–79.

Fagen, M. D. *The Early Years.* Vol. 1 of *A History of Engineering and Science in the Bell System,* ed. E. F. O'Neill. Murray Hill, N.J.: Bell Laboratories, 1975.

———. *National Service in War and Peace.* Vol. 2 of *A History of Engineering and Science in the Bell System,* ed. E. F. O'Neill. Murray Hill, N.J.: Bell Laboratories, 1975.

Ferguson, Eugene S. *Engineering and the Mind's Eye.* Cambridge: MIT Press, 1992.

Ferrell, E. B. "Electrical and Mechanical Analogies." *Bell Laboratories Record,* October 1946, 372–73.

———. "The Servo Problem as a Transmission Problem." *Proceedings of the IRE* 33 (1945): 763–67.

Findley, Paul B. "The Apparatus Development Department." *Bell Laboratories Record,* May 1926, 113–20.

———. "Mathematical Research." *Bell Laboratories Record,* September 1925, 15–18.

———. "The Research Department." *Bell Laboratories Record,* June 1926, 164–70.

———. "The Systems Development Department." *Bell Laboratories Record,* April 1926, 69–73.

Fletcher, Harvey. "The Nature of Speech and Its Interpretation." *Bell System Technical Journal* 1 (July 1922): 129.

———. "Physical Measurements of Audition and Their Bearing on the Theory of Hearing." *Bell System Technical Journal* 2 (October 1923): 145.

———. "The Theory of the Operation of the Howling Telephone with Experimental Confirmation." *Bell System Technical Journal* 5 (January (1926): 27–49.

———. "Useful Numerical Constants of Speech and Hearing." *Bell System Technical Journal* 4 (1925): 375–86.

Florence, Ronald. *The Perfect Machine.* New York: Harper Collins, 1994.

Ford Instrument Company, Division of the Sperry Corporation. "News Release: For Release in the Event of Mr. Ford's Death." New York, n.d.

Forman, Paul. "Behind Quantum Electronics: National Security as a Basis for Physical Research in the United States, 1940–1960." *Historical Studies in the Physical and Biological Sciences* 18, no. 1 (1987): 149–229.

Fortescue, C. L. "Discussion of Bush and Booth, 'Power System Transients.'" *Transactions of the AIEE* 44 (1925): 97–103.

———. "Transmission Stability: Analytical Discussion of Some Factors Entering into the Problem." *Transactions of the AIEE* 26 (1927): 984–1003.

Foster, Ronald M. "A Reactance Theorem." *Bell System Technical Journal* 3 (April 1924): 266.

Friedman, Norman. *Naval Radar.* Annapolis, Md.: Naval Institute Press, 1981.

———. *U.S. Battleships: An Illustrated Design History.* London: Arms and Armour Press, 1986.

———. *U.S. Naval Weapons: Every Gun, Missile, Mine, and Torpedo Used by the U.S. Navy from Eighteen Eighty-Three to the Present Day.* Annapolis, Md.: Naval Institute Press, 1982.

Friis, H. T., and A. G. Jensen. "High Frequency Amplifiers." *Bell System Technical Journal* 3 (April 1924): 181–205.

Fry, Thornton. "Industrial Mathematics." *Bell System Technical Journal* 20 (July 1941): 258.

Fuller, A. T. "The Early Development of Control Theory." *Transactions of the American Society of Mechanical Engineers, Journal of Dynamic Systems, Measurement and Control,* September 1976, 224–35.

Furer, Julius Augustus. *Administration of the Navy Department in World War II.* Washington, D.C.: U.S. Government Printing Office, 1959.

Galison, Peter. *Image and Logic: A Material Culture of Microphysics.* Chicago: University of Chicago Press, 1997.

———. "The Ontology of the Enemy: Norbert Wiener and the Cybernetic Vision." *Critical Inquiry* 21 (winter 1994): 228–66.

Gannon, Robert. *Hellions of the Deep: The Development of American Torpedoes during World War II.* University Park: Pennsylvania State University Press, 1996.

Gelb, Arthur, ed. *Applied Optimal Estimation.* Cambridge: MIT Press, 1974.

Gerovitch, Vyacheslav. *From Newspeak to Cyberspeak: A History of Soviet Cybernetics.* Cambridge: MIT Press, 2002.

Getting, Ivan. *All in a Lifetime: Science in the Defense of Democracy.* New York: Vantage Press, 1989.

———. "SCR-584 Radar and the Mark 56 Naval Gun Fire Control System." *IEEE Transactions on Aerospace and Electronic Systems* 11, no. 5 (September 1975): 922–36.

Gherardi, Bancroft. "The Dean of Telephone Engineers." *Bell Laboratories Record,* September 1930.

Gibson, William. *Neuromancer.* New York: Ace Books, 1984.

Gitelman, Lisa. *Scripts, Grooves, and Writing Machines: Representing Technology in the Edison Era.* Stanford: Stanford University Press, 1999.

Goldstine, Herman H. *The Computer: From Pascal to Von Neumann.* Princeton: Princeton University Press, 1972.

Graybar, Lloyd J. "The Buck Rogers of the Navy: Admiral William H. P. Blandy." In *New Interpretations in Naval History: Selected Papers from the Ninth Naval History Symposium,* ed. William R. Roberts and Jack Sweetman, 335–50. Annapolis, Md.: Naval Institute Press, 1991.

Green, Constance M., Harry Thompson, and Peter C. Roots. *The Ordnance Department: Planning Munitions for War.* The U.S. Army in World War II: The Technical Services. Washington, D.C.: Department of the Army, Office of the Chief of Military History, 1955.

Green, C. W., and E. I. Green. "A Carrier Telephone System for Toll Cables." *Bell System Technical Journal* 17 (January 1938).

Guerlac, Henry. *Radar in World War II.* New York: Tomash Publishers, American Institute of Physics, 1987.

Hall, Albert C. *The Analysis and Synthesis of Linear Servomechanisms.* Cambridge: Technology Press MIT, 1943.

———. "Application of Circuit Theory to the Design of Servomechanisms." *Journal of the Franklin Institute* 242 (1946): 279–307.

———. "Early History of the Frequency Response Field." In *Frequency Response,* ed. Ralph Oldenberger, 1153–54. New York: Macmillan, 1956.

Hallenbeck, Francis. "The Inspection Engineering Department." *Bell Laboratories Record,* August 1926, 243–47.

Hamilton, B. P., H. Nyquist, M. B. Long, and W. A. Phelps. "Voice-Frequency Carrier Telegraph System for Cables." *Transactions of the AIEE* 44 (1925): 327–39.

Hammond, John Winthrop, and Arthur Pound. *Men and Volts: The Story of General Electric.* New York: J. B. Lippincott, 1941.

"Hannibal Ford." *New York Times,* 19 March 1955.

Haraway, Donna J. *Simians, Cyborgs, and Women: The Reinvention of Nature.* New York: Routledge, 1991.

Harriss, Herbert. "The Frequency Response of Automatic Control Systems." *Transactions of the AIEE* 65 (1946): 539–46.

Hart, David M. *Forged Consensus: Science, Technology, and Economic Policy in the United States, 1921–1953.* Princeton: Princeton University Press, 1998.

Hartley, Ralph V. L. "Transmission of Information." *Bell System Technical Journal* 7 (July 1928): 535–63.

———. "TU Becomes 'Decibel.'" *Bell Laboratories Record,* December 1928, 137–39.

Hartley, Ralph V. L., and Thornton Fry. "The Binaural Location of Complex Sounds." *Bell System Technical Journal* 1 (November 1922): 33–42.

Hartree, Douglas R. *Calculating Instruments and Machines.* Urbana: University of Illinois Press, 1949.

Hartree, Douglas R., Arthur Porter, A. Clallender, and A. B. Stevenson. "Time-Lag in a Control System." *Proceedings of the Royal Society of London* 161 (1937): 460–76.

Hazen, Harold L. "Design and Test of a High-Performance Servo-Mechanism." *Journal of the Franklin Institute* 218 (1934): 543–80.

———. "Discussion of H. A. Thompson, 'A Stabilized Amplifier for Measurement Purposes.'" *Electrical Engineering* 57 (July 1938): 383.

———. "Electrical Water-Level Control and Recording Equipment for Model of Cape Cod Canal." *Electrical Engineering* 56 (1937): 237–44.

———. "Theory of Servo-Mechanisms." *Journal of the Franklin Institute* 218 (1934): 279–331.

———. "Working Mathematics by Machinery." *Technology Review* 34 (1932): 323–46.

———, ed. *Airborne Fire Control.* Vol. 3 of *Summary Technical Report of Division 7, NDRC.* Washington, D.C.: Office of Scientific Research and Development, National Defense Research Committee, 1946.

———. *Gunfire Control.* Vol. 1 of *Summary Technical Report of Division 7, NDRC.* Washington, D.C.: Office of Scientific Research and Development, National Defense Research Committee, 1946.

———. *Rangefinders and Tracking.* Vol. 2 of *Summary Technical Report of Division 7, NDRC.* Washington, D.C.: Office of Scientific Research and Development, National Defense Research Committee, 1946.

Hazen, Harold L., and Gordon S. Brown. "The Cinema Integraph: A Machine for Evaluating a Parametric Product Integral." *Journal of the Franklin Institute* 230 (1940): 19–44, 183–205.

Hazen, Harold L., J. J. Jaeger, and Gordon S. Brown. "An Automatic Curve Follower." *Review of Scientific Instruments* 7 (1936): 353–57.

Hazen, Harold L., and Hugh Spencer. "Artificial Representation of Power Systems." *Journal of the AIEE* 44 (1925).

Hecht, Gabrielle. *The Radiance of France: Nuclear Power and National Identity after World War II.* Cambridge: MIT Press, 1998.

Heims, Steve J. *Constructing a Social Science for Postwar America: The Cybernetics Group, 1946–1953.* Cambridge: MIT Press, 1993.

———. *John von Neumann and Norbert Wiener: From Mathematics to the Technologies of Life and Death.* Cambridge: MIT Press, 1980.

Hewlett, E. M. "The Selsyn System of Position Indication." *General Electric Review* 24 (March 1921): 210–18.

Hochheiser, Sheldon. "What Makes the Picture Talk: AT&T and the Development of Sound Motion Picture Technology." *IEEE Transactions on Education* 35 (November 1992): 278–85.

Hoddeson, Lillian. "The Emergence of Basic Research in the Bell Telephone System, 1875–1915." *Technology and Culture* 22, no. 4 (1981): 512–44.

Hodges, Andrew. *Alan Turing: The Enigma.* New York: Simon & Schuster, 1983.

Holst, Per A. "George A. Philbrick and Polyphemus: The First Electronic Training Simulator." *Annals of the History of Computing* 4 (April 1982): 144–45.

Holton, Gerald, ed. *The Twentieth-Century Sciences: Studies in the Biography of Ideas.* New York: Norton, 1970.

Horsburgh, E. M., ed. *Handbook of the Napier Tercentenary Celebration, or Modern Instruments and Methods of Calculation.* Vol. 3. Los Angeles: Tomash, 1982.

Hounshell, David A. *From the American System to Mass Production, 1800–1932: The Development of Manufacturing Technology in the United States.* Baltimore: Johns Hopkins University Press, 1984.

Howeth, L. S. *History of Communications-Electronics in the United States Navy.* Washington, D.C.: U.S. Navy Bureau of Ships and Office of Naval History, 1963.

Hughes, Agatha C., and Thomas Parke Hughes, eds. *Systems, Experts, and Computers: The Systems Approach in Management and Engineering, World War II and After.* Cambridge: MIT Press, 2000.

Hughes, Thomas Parke. *Elmer Sperry: Inventor and Engineer.* Baltimore: Johns Hopkins Press, 1971.

———. *Networks of Power: Electrification in Western Society, 1880–1930.* Baltimore: Johns Hopkins University Press, 1983.

———. *Rescuing Prometheus.* New York: Norton, 1998.

Hunter, Louis, and Lynwood Bryant. *A History of Industrial Power in the United States.* Vol. 3, *The Transmission of Power.* Cambridge: MIT Press, 1991.

Hutchins, Edwin. *Cognition in the Wild.* Cambridge: MIT Press, 1995.

Jahn, R. F. "Employees Honor Hannibal C. Ford." *Sperryscope* 12, no. 2 (1950): 11–12.

James, Hubert M., Nathaniel B. Nichols, and Ralph S. Phillips. *Theory of Servomechanisms.* Radiation Laboratory Series, 25. New York: McGraw-Hill, 1947.

Jewett, Frank Baldwin. "Electrical Communication, Past, Present, and Future: Speech to National Academy of Sciences." *Bell Telephone Quarterly* 14 (1935): 167–99.

Joel, Amos E., Jr., ed. *Switching Technology (1925–1975).* Vol. 3 of *A History of Engineering and Science in the Bell System,* ed. E. F. O'Neill. Whippany, N.J.: Bell Laboratories, 1982.

Johnson, E. D. "Transmission Regulating System for Toll Cables." *Bell Laboratories Record* 7, no. 5 (1929): 183–87.

Johnson, J. B. "Thermal Agitation of Electricity in Conductors." *Physical Review* 7 (1928): 97–113.

Jones, R. V. *The Wizard War.* New York: Coward, McCann, & Geoghegan, 1978.

Jordan, John M. *Machine-Age Ideology: Social Engineering and American Liberalism, 1911–1939.* Chapel Hill: University of North Carolina Press, 1994.

Juley, Joseph. "The Ballistic Computer." *Bell Laboratories Record,* December 1946, 5–9.

Jurens, William J. "The Evolution of Battleship Gunnery in the U.S. Navy, 1920–1945." *Warship International,* no. 3 (1991): 240–71.

Jury, Eliahu I. "On the History and Progress of Sampled-Data Systems." *IEEE Control Systems,* February 1987, 16–21.

Kay, Lily E. "Cybernetics, Information, Life: The Emergence of Scriptural Representations of Heredity." *Configurations* 5 (1997): 23–91.

———. *Who Wrote the Book of Life? A History of the Genetic Code, Writing Science.* Stanford: Stanford University Press, 2000.

Keller, Evelyn Fox. *Refiguring Life: Metaphors of Twentieth-Century Biology.* New York: Columbia University Press, 1995.

Kelley, Merton J. "Career of the 1957 Lamme Medalist Harold S. Black." *Electrical Engineering* 77 (1958): 720–22.

Kellogg, E. W. "Design of Non-Distorting Power Amplifiers." *Electrical Engineering* 44 (1925): 490.

Kern, Stephen. *The Culture of Time and Space, 1880–1918.* Cambridge: Harvard University Press, 1983.

Kevles, Daniel J. *The Physicists: The History of a Scientific Community in Modern America.* New York: Knopf, 1978.

Kingsley, F. A., ed. *The Applications of Radar and Other Electronic Systems in the Royal Navy in World War II.* London: Macmillan, 1995.

Kline, Ronald R. "Harold Black and the Negative-Feedback Amplifier." *IEEE Control Systems,* August 1993, 82–85.

———. *Steinmetz: Engineer and Socialist.* Baltimore: Johns Hopkins University Press, 1992.

Kohler, Robert E. *Partners in Science: Foundations and Natural Scientists, 1900–1945.* Chicago: University of Chicago Press, 1991.

Kuehni, H. P., and H. A. Peterson. "A New Differential Analyzer." *Transactions of the IRE* 63 (May 1944): 221–28, discussion 429–31.

Lambright, W. Henry. *Powering Apollo: James Webb of NASA.* Baltimore: Johns Hopkins University Press, 1995.

Latour, Bruno. "Drawing Things Together." In *Representation in Scientific Practice,* ed. Michael Lynch and Stevel Woolgar. Cambridge: MIT Press, 1990.

———. *Science in Action: How to Follow Scientists and Engineers through Society.* Cambridge: Harvard University Press, 1987.

Layton, Edwin T. *The Revolt of the Engineers: Social Responsibility and the American Engineering Profession.* Cleveland: Case Western Reserve University Press, 1971.

Lecuyer, Christophe. "The Making of a Science Based Technological University: Karl Compton,

James Killian, and the Reform of MIT, 1930–1957." *Historical Studies in the Physical Sciences* 23, no. 1 (1992): 153–80.

Leslie, Stuart W. *The Cold War and American Science: The Military-Industrial-Academic Complex at MIT and Stanford.* New York: Columbia University Press, 1993.

Leslie, Stuart W., and Bruce Hevly. "Steeple Building at Stanford: Physics, Electrical Engineering, and Microwave Research." *IEEE Proceedings* 73 ( July 1989): 1169–80.

Levin, Miriam. "Contexts of Control." In *Cultures of Control,* ed. Levin, 13–40. Amsterdam: Harwood, 2000.

Licklider, J. C. R. "Man-Computer Symbiosis." *IRE Trans. on Human Factors in Electronics* 1 (March 1960): 4–11.

Lindbergh, Charles. *The Spirit of St. Louis.* New York: Scribner's, 1954.

———. *We.* New York: G. P. Putnam's Sons, 1927.

Lipartito, Kenneth. "When Women Were Switches: Technology, Work, and Gender in the Telephone Industry, 1890–1920." *American Historical Review* 99, no. 4 (October 1994): 1075–1111.

Lovell, Clarence A. "Continuous Electrical Computation." *Bell Laboratories Record,* March 1947.

Lozier, John C. "The Oldenberger Award Response: An Appreciation of Harry Nyquist." *Transactions of the American Society of Mechanical Engineers, Journal of Dynamic Systems, Measurement and Control,* June 1976, 127–28.

Lyng, J. J. "The Development of Apparatus." *Bell Laboratories Record,* May 1926, 113–20.

Mabon, Prescott C. *Mission Communications: The Story of Bell Laboratories.* Murray Hill, N.J.: Bell Laboratories, 1975.

MacColl, Leroy. *Fundamental Theory of Servomechanisms.* New York: Van Nostrand, 1945.

Mackenzie, Donald. *Inventing Accuracy: A Historical Sociology of Nuclear Missile Guidance.* Cambridge: MIT Press, 1990.

Mahoney, Michael. "The History of Computing in the History of Technology." *Annals of the History of Computing* 10 (1988): 113–25.

Marcus, Mitchell, and Atsushi Akera. "Exploring the Architecture of an Early Machine: The Historical Relevance of the ENIAC Machine Architecture." *IEEE Annals of the History of Computing* 18, no. 1 (1996): 17–24.

Martin, W. H. "The Transmission Unit and Telephone Transmission Reference Systems." *Bell System Technical Journal* 3 ( July 1924): 400–408.

———. "Transmitted Frequency Range for Telephone Message Circuits." *Bell System Technical Journal* 9 ( July 1930): 483–86.

Masani, Pesi. *Norbert Wiener, 1894–1964.* Basel: Burkhäuser Verlag, 1990.

Mason, W. P. "Electrical and Mechanical Analogies." *Bell System Technical Journal* 20 (October 1941): 405–14.

Massachusetts Institute of Technology. *Five Years at the Radiation Laboratory: Presented to Members of the Radiation Laboratory by the Massachusetts Institute of Technology, Cambridge, 1946.* Cambridge, 1947.

Mayr, Otto. *Feedback Mechanisms in the Historical Collections of the National Museum of History and Technology.* Washington, D.C.: Smithsonian Institution Press, 1971.

———. *Liberty, Authority, and Automatic Machinery in Early Modern Europe.* Baltimore: Johns Hopkins University Press, 1993.

———. "Maxwell and the Origins of Cybernetics." In *Philosophers and Machines,* ed. Otto Mayr, 168–88. New York: Science History Publications, 1976.

———. *The Origins of Feedback Control.* Cambridge: MIT Press, 1970.

McBride, William R. "The 'Greatest Patron of Science'? The Navy-Academia Alliance and U.S. Naval Research, 1896–1923." *Journal of Military History* 56 (January 1992): 7–33.

———. "Strategic Determinism in Technology Selection: The Electric Battleship and U.S. Naval-Industrial Relations." *Technology and Culture* 33, no. 2 (1992): 248–77.

———. *Technological Change in the United States Navy, 1865–1945.* Baltimore: Johns Hopkins University Press, 2000.

McConnel, Paul. "Some Early Computers for Aviators." *Annals of the History of Computing* 13 (1991): 155–74.

McDowell, C. S. "Naval Research." *U.S. Naval Institute Proceedings* 45 (1919): 895–908.

———. "The 200-Inch Telescope." *Scientific American* 159 (1938): 69–71.

McFarland, Stephen Lee. *America's Pursuit of Precision Bombing, 1910–1945.* Washington, D.C.: Smithsonian Institution Press, 1995.

McGinn, Robert E. "Stokowski and the Bell Telephone Laboratories: Collaboration in the Development of High-Fidelity Sound Reproduction." *Technology and Culture* 24, no. 1 (1983): 43.

McLeod, John, ed. *Simulation: The Dynamic Modeling of Ideas and Systems with Computers.* New York: McGraw-Hill, 1968.

McNeill, William H. *The Pursuit of Power: Technology Armed Force and Society since A.D. 1000.* Chicago: University of Chicago Press, 1982.

McSwain, Rebecca. "The Social Reconstruction of a Reverse Salient in Electric Guitar Technology: Noise, the Solid Body, and Jimi Hendrix." In *"I Sing the Body Electric": Music and Technology in the Twentieth Century,* ed. Hans-Joachim Braun, 198–210. Baltimore: Johns Hopkins University Press, 2000.

"A Mechanical Brain." *Bell Laboratories Record,* November 1926, 78–81.

"Mechanical Brains: Working in Metal Boxes, Computing Devices Aim Guns and Bombs with Inhuman Accuracy." *Life,* 24 January 1944, 66–72.

Miles, Walter. "Psychological Aspects of Military Aviation." In Baitsell, *Science in Progress,* 1–48.

Miller, John A. *Men and Volts at War: The Story of General Electric in World War II.* New York: McGraw-Hill, Whittlesey House, 1947.

Millikan, Robert Andrews. *Autobiography.* New York: Prentice-Hall, 1950.

Millman, S. *Communications Sciences.* Vol. 5 of *A History of Engineering and Science in the Bell System,* ed. E. F. O'Neill. Murray Hill, N.J.: Bell Laboratories, 1984.

Mills, John. "Communication with Electrical Brains." *Bell Telephone Quarterly* 13 (1934): 47–57.

Minorsky, Nicholas. "Directional Stability of Automatically Steered Bodies." *Journal of the American Society of Naval Engineers* 34, no. 2 (1922): 280–309.

———. "The Principles and Practice of Automatic Control." *Engineer* 163 (January–April 1937).

Mohler, Stanley R., and Bobby H. Johnson. *Wiley Post, His Winnie Mae, and the World's First Pressure Suit.* Annals of Flight. Washington, D.C.: Smithsonian Institution, 1978.

Morgan, Thomas. "The Navy's Mark 14 Gyro Gun Sight." *Sperryscope* 10, no. 8 (1945): 15–17.

———. "Sperry and Subcontracting." *Sperryscope* 9, no. 8 (1941): 1–8.

Morison, Elting Elmore. *Admiral Sims and the Modern American Navy.* Boston: Houghton Mifflin, 1942.

———. *Men, Machines, and Modern Times.* Cambridge: MIT Press, 1966.

Mumford, Lewis. *The Myth of the Machine: The Pentagon of Power.* New York: Harcourt Brace Jovanovich, 1970.

———. *Technics and Civilization.* New York: Harcourt Brace Jovanovich, 1934.

Murray, Williamson, and Allan R. Millett, eds. *Military Innovation in the Interwar Period.* Cambridge: Cambridge University Press, 1996.

Musicant, Ivan. *Battleship at War: The Epic Story of the USS Washington.* New York: Harcourt Brace Jovanovich, 1986.

Nahin, Paul J. *Oliver Heaviside, Sage in Solitude: The Life, Work, and Times of an Electrical Genius of the Victorian Age.* New York: IEEE Press, 1987.

Nance, H. H., and O. B. Jacobs. "Transmission Features of Transcontinental Telephony." *Journal of the AIEE* 45 (1926): 1061–69.

Nealey, J. B. "Integration of the Automatic Pilot System and the Norden Bombsight." *Aero Digest,* 1 June 1945, 98.

Nebeker, Frederik. *Signal Processing: The Emergence of a Discipline, 1948–1998.* New York: IEEE Press, 1998.

Nelson, Daniel. *Frederick W. Taylor and the Rise of Scientific Management.* Madison: University of Wisconsin Press, 1980.

Noble, David F. *America by Design: Science, Technology, and the Rise of Corporate Capitalism.* New York: Knopf, 1977.

———. *Forces of Production: A Social History of Industrial Automation.* New York: Knopf, 1984.

Norberg, Arthur L., and Judy E. O'Neill. *Transforming Computer Technology: Information Processing for the Pentagon, 1962–1986.* Baltimore: Johns Hopkins University Press, 1996.

Nyquist, Harry. "Certain Factors Affecting Telegraph Speed." *Bell System Technical Journal* 3 (April 1924): 324–46.

———. "Certain Topics in Telegraph Transmission Theory." *Transactions of the AIEE* 47 (1928): 617–44.

———. "Discussion of H. S. Black, 'Stabilized Feed-Back Amplifiers.'" *Electrical Engineering* 53 (1934): 1311–12.

———. "Regeneration Theory." *Bell System Technical Journal* 11 (1932): 126–47.

———. "Thermal Agitation of Electric Charge in Conductors." *Physical Review* 7 (1928): 97–113.

O'Connell, Robert. *Sacred Vessels: The Cult of the Battleship and the Rise of the U.S. Navy.* Boulder, Colo.: Westview Press, 1991.

O'Neill, E. F., ed. *Transmission Technology (1925–1975).* Vol. 7 of *A History of Science and Engineering in the Bell System,* ed. E. F. O'Neill. Murray Hill, N.J.: AT&T Bell Laboratories, 1985.

Oppelt, W. "On the Early Growth of Conceptual Thinking in Control System Theory: The German Role up to 1945." *IEEE Control Systems,* November 1984, 16–22.

Ordway, Frederick I., and Mitchell R. Sharpe. *The Rocket Team.* Cambridge: MIT Press, 1982.

Owens, Larry. "The Counterproductive Management of Science in the Second World War: Vannevar Bush and the Office of Scientific Research and Development." *Business History Review* 68 (winter 1994): 515–76.

———. "Mathematicians at War: Warren Weaver and the Applied Mathematics Panel, 1942–1945." In *The History of Modern Mathematics,* vol. 2, *Institutions and Applications,* ed. David E. Rowe and John McCleary, 287–305. Boston: Academic Press, 1989.

———. "MIT and the Federal 'Angel': Academic R&D and Federal-Private Cooperation before World War II." *Isis* 81 (1990): 188–213.

———. "Vannevar Bush and the Differential Analyzer: The Text and Context of an Early Computer." *Technology and Culture* 27, no. 1 (1986): 63–95.

———. "Where Are We Going, Phil Morse? Changing Agendas and the Rhetoric of Obviousness in the Transformation of Computing at MIT, 1939–1957." *IEEE Annals of the History of Computing* 18, no. 4 (1996): 34–41.

Padfield, Peter. *Aim Straight: A Biography of Admiral Sir Percy Scott.* London: Hodder & Stoughton, 1966.

———. *Guns at Sea.* New York: St. Martin's Press, 1974.

Pang, Alex Soojunk-Kim. "Edward Bowles and Radio Engineering at MIT, 1920–1940." *Historical Studies in the Physical and Biological Sciences* 20, no. 2 (1990): 313–37.

Paynter, Henry M. "The Differential Analyzer as an Active Mathematical Instrument: Keynote Speech to the 1989 American Control Conference." *IEEE Control Systems,* December 1989, 3–7.

———. *A Palimpsest on the Analog Art.* Boston: G. A. Philbrick Researchers, 1955.

Peterson, E., J. G. Kreer, and L. A. Ware. "Regeneration Theory and Experiment." *Bell System Technical Journal* 13 (October 1934): 680–700.

Philbrick, George. "Designing Industrial Controllers by Analog." *Electronics* 21 (June 1948): 108–11.

Pickering, Andrew. "Cyborg History and the World War II Regime." *Perspectives on Science* 3, no. 1 (1995): 1–49.

Pierce, J. R. "The Early Days of Information Theory." *IEEE Transactions on Information Theory* 19 (January 1973): 3.

Pile, Sir Frederick. *Ack-Ack, Britain's Defence against Air Attack during the Second World War.* London: Harrap, 1949.

Piore, Michael J., and Charles F. Sabel. *The Second Industrial Divide: Possibilities for Prosperity.* New York: Basic Books, 1984.

Porter, Arthur. "The Servo Panel—A Unique Contribution to Control-Systems Engineering." *Electronics and Power* 11 (1965): 330–33.

Post, Wiley. "Destination—New York." *Sperryscope* 7, no. 2 (1933).

Post, Wiley, and Harold Gatty. *Around the World in Eight Days: The Flight of the Winnie Mae.* London: John Hamilton, 1932.

Pugh, Emerson. *Memories That Shaped an Industry.* Cambridge: MIT Press, 1984.

Pursell, Carroll. "Science Agencies in World War II: The OSRD and Its Challengers." In *The Sciences in the American Context: New Perspectives,* ed. Nathan Reingold, 360–78. Washington, D.C.: Smithsonian Institution Press, 1979.

Rabinbach, Anson. *The Human Motor: Energy, Fatigue, and the Origins of Modernity.* New York: Basic Books, 1990.

Ragazzini, John R., Robert H. Randall, and Frederick A. Russel. "Analysis of Problems in Dynamics by Electronic Circuits." *Proceedings of the IRE* 35 (1947): 444.

Randell, Brian, ed. *The Origins of Digital Computers: Selected Papers.* Berlin: Springer-Verlag, 1982.

Redmond, Kent C., and Thomas M. Smith. *From Whirlwind to MITRE: The R&D Story of the Sage Air Defense Computer.* Cambridge: MIT Press, 2000.

———. *Project Whirlwind: The History of a Pioneer Computer.* Bedford, Mass.: DEC Press, 1980.

Rees, Mina. "The Computing Program of the Office of Naval Research, 1946–1953." *Annals of the History of Computing* 4 (April 1982): 102–20.

Reich, Leonard. "Industrial Research and the Pursuit of Corporate Security: The Early Years of Bell Labs." *Business History Review* 54 (winter 1980): 504–29.

———. *The Making of American Industrial Research: Science and Business at GE and Bell, 1876–1926.* Cambridge: Cambridge University Press, 1985.

———, ed. *The Sciences in the American Context: New Perspectives.* Washington, D.C.: Smithsonian Institution Press, 1979.

Reingold, Nathan. "Vannevar Bush's New Deal for Research: Or the Triumph of the Old Order." *Historical Studies in the Physical and Biological Sciences* 17, no. 2 (1987): 299–344.

Reintjes, Francis J. *Numerical Control: Making a New Technology.* New York: Oxford University Press, 1991.

Reynolds, Quentin James, and Wilfrid S. Rowe. *Operation Success.* New York: Duell Sloan & Pearce, 1957.

Rhodes, Richard. *The Making of the Atomic Bomb.* New York: Simon & Schuster, 1986.

Richardson, Kenneth I. T. *The Gyroscope Applied.* New York: Philosophical Library, 1955.

Ridenour, Louis B. *Radar System Engineering.* Radiation Laboratory Series, 1. New York: McGraw-Hill, 1948.

Rosenberg, David Alan. "Officer Development in the Interwar Navy: Arleigh Burke—The Making of a Naval Professional, 1919–1940." *Pacific Historical Review* 44 (1975): 503–26.

Rosenblueth, Arturo, Norbert Wiener, and Julian Bigelow. "Behavior, Purpose, and Teleology." *Philosophy of Science* 10 (1943): 18–24.

Routledge, N. W. *Anti-Aircraft Artillery, 1914–55.* New York: Brassey's, 1994.

Rowland, W. B., and B. Boyd. *U.S. Navy Bureau of Ordnance in World War II.* Washington, D.C.: Department of Navy, Bureau of Ordnance, 1953.

Rule, Bruce H. "Electrical Features of the 200-Inch Telescope." *Electrical Engineering* 61 (1942): 67–78.

Rydell, Robert W. *World of Fairs: The Century-of-Progress Expositions.* Chicago: University of Chicago Press, 1993.

Sanderson, John. "The Sperry Corporation: A Financial Biography." *Sperryscope* 12, no. 9 (1952): 2–7.

Sapolsky, Harvey M. *Science and the Navy: The History of the Office of Naval Research.* Princeton: Princeton University Press, 1990.

Sapolsky, Harvey M., and Eugene Gholz. "Restructuring the Defense Industry." *International Security* 24 (winter 1999): 5–51.

Schaffer, Simon. "Babbage's Intelligence: Calculating Engines and the Factory System." *Critical Inquiry* 21 (winter 1994): 203–27.

Schurig, H. L., O. R. Hazen, and M. F. Gardner. "The M.I.T. Network Analyzer: Design and Application to Power System Problems." *Transactions of the AIEE* 49 (1930): 872–75.

Schuyler, G. L. "The Distribution of Shots in Long Range Salvos." *U.S. Naval Institute Proceedings* 55, no. 311 (1929): 24–26.

———. "A Note on Salvo Dispersion." *U.S. Naval Institute Proceedings* 37, no. 139 (1911): 1005–10.

Scott, Lloyd N. *The Naval Consulting Board of the United States.* Washington, D.C.: U.S. Government Printing Office, 1920.

Scranton, Philip. *Endless Novelty: Specialty Production and American Industrialization, 1865–1925.* Princeton: Princeton University Press, 1997.

Serres, Michel. *The Parasite.* Baltimore: Johns Hopkins University Press, 1982.

Serres, Michel, and Bruno Latour. *Conversations on Science, Culture, and Time.* Studies in Literature and Science. Ann Arbor: University of Michigan Press, 1995.

Servos, John W. "The Industrial Relations of Science: Chemical Engineering at MIT, 1900–1939." *Isis* 71 (1980): 531–49.

Shannon, Claude Elwood. *Claude Elwood Shannon: Collected Papers.* Ed. N. J. A. Sloane and A. D. Wyner. New York: IEEE Press, 1993.

———. "A Mathematical Theory of Communication." *Bell System Technical Journal* 27 (July–October 1948): 379–423, 623–56.

———. "Mathematical Theory of the Differential Analyzer." *Journal of Mathematics and Physics* 20 (December 1941): 337–54.

———. "A Symbolic Analysis of Relay Switching Circuits." *Transactions of the AIEE* 57 (1938): 713–23.

Shannon, Claude Elwood, and Warren Weaver. *The Mathematical Theory of Communication.* Chicago: University of Illinois Press, 1949.

Shapin, Steven, and Simon Schaffer. *Leviathan and the Air-Pump: Hobbes, Boyle, and the Experimental Life.* Princeton: Princeton University Press, 1985.

Shaw, Thomas. "The Conquest of Distance by Wire Telephony." *Bell System Technical Journal* 23 (October 1944): 337–421.

Small, James. "General-Purpose Electronic Analog Computing: 1945–1965." *IEEE Annals of the History of Computing* 15, no. 2 (1993): 9–18.

Smith, Merritt Roe. "Army Ordnance and the 'American System' of Manufacturing, 1815–1861." In Smith, *Military Enterprise and Technological Change, 39–86.*

———. "The Military Roots of Mass Production: Firearms and American Industrialization, 1815–1913." Working paper, MIT Program in Science, Technology, and Society, Cambridge, 1995.

———, ed. *Military Enterprise and Technological Change: Perspectives on the American Experience.* Cambridge: MIT Press, 1985.

Smith, Michael. "Selling the Moon: The U.S. Manned Space Program and the Triumph of Commodity Scientism." In *The Culture of Consumption: Critical Essays in American History, 1880–1980,* ed. Richard Wightman and T. J. Jackson Lears Fox, 177–209. New York: Pantheon, 1983.

"Special Issue: SAGE." *Annals of the History of Computing* 5, no. 4 (October 1983).

Sperry, Elmer. "Automatic Steering." *Transactions of the Society of Naval Architects and Marine Engineers* 30 (1922): 53–61.

Sperry, Elmer, Jr. "Description of the Sperry Automatic Pilot." *Aviation Engineering,* January 1932, 16–18.

"Sperry: The Corporation." *Fortune,* May 1940.

"The Sperry Corporation." *Sperryscope* 9, no. 7 (1941): 1–8.

"The Spinal Cord of a Nation." *Bell Laboratories Record,* October 1926, 426–35.

Steinberg, J. C., H. C. Montgomery, and M. B. Gardner. "Results of the World's Fair Hearing Tests." *Bell System Technical Journal* 14 (July 1940): 533–40.

Stern, Nancy. *From ENIAC to UNIVAC: An Appraisal of the Eckert-Mauchly Computers.* Bedford, Mass.: DEC Press, 1981.

Stewart, Irvin. *Organizing Scientific Research for War: The Administrative History of the Office of Scientific Research and Development.* Boston: Little, Brown, 1948.

Stoller, Hugh M. "Speed Control for the Sound-Picture System." *Bell Laboratories Record,* November 1928, 101–5.

———. "Synchronization and Speed Control of Synchronized Sound Pictures." *Bell System Technical Journal* 8 (January (1929): 184–95.

Stoller, Hugh M., and E. R. Morton. "Synchronization of Television." *Bell System Technical Journal* 6 (October 1927): 604–15.

Stoller, Hugh M., and J. R. Power. "A Precision Regulator for Alternating Voltage." *Transactions of the AIEE* 48 (1929): 808–11.

Stone Stone, John. "The Practical Aspects of the Propagation of High Frequency Electric Waves along Wires." *Journal of the Franklin Institute* 174 (1912): 353–84.

Strieby, M. E. "A Million-Cycle Telephone System." *Bell System Technical Journal* 16 (January 1937): 1–9.

Sumida, John Tetsuro. *In Defence of Naval Supremacy: Finance, Technology, and British Naval Policy, 1889–1914.* London: Routledge, 1989.

———. "The Quest for Reach: Development of Long-Range Gunnery in the Royal Navy, 1901–1912." In *Tooling for War: Military Transformation in the Industrial Age,* ed. Stephen D. Chiabotti, 49–96. Chicago: Imprint, 1996.

Svoboda, Antonin. *Computing Mechanisms and Linkages.* Radiation Laboratory Series, 27. New York: McGraw-Hill, 1948.

Taylor, Frederick Winslow. *The Principles of Scientific Management.* New York: Harper & Brothers, 1911.

Tellegen, B. D. H. "Inverse Feedback." *Phillips Technical Review* 2 (October 1937): 289–94.

Terman, Frederick Emmons. *Radio Engineer's Handbook.* New York: McGraw-Hill, 1943.

Thompson, Emily. *The Soundscape of Modernity.* Cambridge: MIT Press, forthcoming.

Thompson, George Raynor, and Dixie R. Harriss. T*he Signal Corps: The Outcome (Mid-1943 through 1945).* The U.S. Army in World War II: The Technical Services. Washington, D.C.: Department of the Army, Office of the Chief of Military History, 1966.

Thompson, George Raynor, Dixie R. Harriss, Pauline M. Oakes, and Dulany Terrett. T*he Signal Corps: The Test (December, 1941 to July, 1943).* The U.S. Army in World War II: The Technical Services. Washington, D.C.: Department of the Army, Office of the Chief of Military History, 1957.

Thompson, William. "Mechanical Integration of the Linear Differential Equations of the Second Order with Variable Coefficients." *Proceedings of the Royal Society of London* 24 (1876): 269–71.

Tinus, W. C., and W. H. C. Higgins. "Early Fire-Control Radars for Naval Vessels." *Bell System Technical Journal* 25 (January 1946): 1–47.

Tomayko, James E. "Helmut Hoelzer's Fully Electronic Analog Computer." *Annals of the History of Computing* 7 (1985): 227–40.

Turing, Alan. "On Computable Numbers, with an Application to the *Entscheidungsproblem.*" *Proceedings of the London Mathematical Society* 2, no. 42 (1937): 230–65.

U.S. Naval Academy. *Fire Control Installations.* United States Naval Academy, Postgraduate School, Pub. no. 105. Annapolis, Md., [ca. 1940].

———. *Notes on Fire Control.* Washington, D.C.: U.S. Navy, Bureau of Naval Personnel, 1940.

U.S. Naval Academy, Department of Ordnance and Gunnery. *Naval Ordnance and Gunnery.* Vol. 1, *Naval Ordnance.* Washington, D.C.: U.S. Navy, Bureau of Naval Personnel, 1955.

U.S. Navy. *Administrative History of the U.S. Navy in World War II.* Vol. 73, *Research and Development.* Washington, D.C., 1947.

———. *Administrative History of the U.S. Navy in World War II.* Vol. 79, *Fire Control (Except Radar).* Washington, D.C., 1947.

U.S. Navy, Bureau of Naval Personnel. *Naval Ordnance and Gunnery.* NAVPERS 16116-A. Washington, D.C., 1946.

———. *Naval Ordnance and Gunnery.* Vol. 2, *Fire Control.* NAVPERS 10798. Washington, D.C., 1955.

U.S. Navy, Bureau of Ordnance. *Navy Ordnance Activities: World War, 1917–1918.* Washington, D.C., 1920.

Valley, George E. "How the Sage Development Began." *Annals of the History of Computing* 7, no. 3 (July 1985): 196–226.

Van Auken, Wilbur. *Notes on a Half Century of United States Naval Ordnance, 1880–1930.* Washington, D.C.: George Banta, 1939.

Van Valkenburg, M. E. "In Memoriam: Hendrik W. Bode, 1905–1982," *IEEE Transactions on Automatic Control* 29, no. 3 (March 1984): 193–94.

Varian, Dorothy. *The Inventor and the Pilot: Russel and Sigurd Varian.* Palo Alto, Calif.: Pacific Books, 1983.

Virilio, Paul. *War and Cinema: The Logistics of Perception.* London: Verso, 1989.

Vonnegut, Kurt. *Player Piano.* New York: Holt Rinehart & Winston, 1966.

Von Neumann, John. "First Draft of a Report on the EDVAC, 1945, Reprint." *IEEE Annals of the History of Computing* 15, no. 4 (1993): 28–43.

Ward, John W. "The Meaning of Lindbergh's Flight." *American Quarterly* 10 (1958): 3–16.

Ward, Roswell. "Hannibal Ford, Sperry Pioneer." *Sperryscope* 9, no. 11 (1943): 12–13.

Wasserman, Neil H. *From Invention to Innovation: Long-Distance Telephone Transmission at the Turn of the Century.* Baltimore: Johns Hopkins University Press, 1985.

Weaver, Warren. *Scene of Change: A Lifetime in American Science.* New York: Scribner's, 1970.

Wells, G. M. "New Fire Control for Divisional Weapons." *Army Ordnance* 11, no. 65 (1931).

Weizenbaum, Joseph. *Computer Power and Human Reason: From Judgment to Calculation.* San Francisco: W. H. Freeman, 1976.

Wiener, Norbert. *Cybernetics: Or Control and Communication in the Animal and the Machine.* Cambridge: Technology Press, 1948.

———. *The Extrapolation, Interpolation, and Smoothing of Stationary Time Series.* Cambridge: MIT Press, 1949.

———. *The Human Use of Human Beings: Cybernetics and Society.* Boston: Houghton Mifflin, 1950.

———. *I Am a Mathematician: The Later Life of a Prodigy.* Garden City, N.Y.: Doubleday, 1956.

———. *Norbert Wiener: Collected Works, with Commentaries.* Ed. Pesi Masani. Vol. 4. Cambridge: MIT Press, 1985.

———. "A Scientist Rebels." *Atlantic Monthly* 179 (January 1946): 46.

Wildes, Karl, and Nilo Lindgren. *A Century of Electrical Engineering and Computer Science at MIT, 1882–1982.* Cambridge: MIT Press, 1985.

Woodward, W. R. "Calculating Short-Circuit Currents in Networks—Testing with Miniature Networks." *Electric Journal,* August 1919, 344–45.

Yavetz, Ido. *From Obscurity to Enigma: The Work of Oliver Heaviside, 1872–1889*. Basel: Birkhauser Verlag, 1995.

Zachary, G. Pascal. *Endless Frontier: Vannevar Bush, Engineer of the American Century*. New York: Free Press, 1997.

Zadeh, Lotfi A., and John R. Ragazzini. "An Extension of Wiener's Theory of Prediction." *Journal of Applied Physics* 21, no. 7 ( July 1950): 645–55.

Zimmerman, David. *Top Secret Exchange: The Tizard Mission and the Scientific War*. Montreal: McGill-Queens University Press, 1996.

# Index

Abajian, Henry, 245, 254–55

Aberdeen Proving Ground (U.S. Army), 94, 193, 198, 281, 297, 303. *See also* Army Ordnance Department; Ballistics Research Laboratory

"Abrupt Change in Load on a Synchronous Machine" (Edgerton), 148

abstraction, 3, 11, 15–16, 143, 146, 163, 166; of signals, 106–7, 135–36; and systems, 67, 114

accuracy, 1, 15–16, 66, 74, 140, 150, 192, 194, 206, 220, 240, 244–45, 350n115; and differential analyzers, 157, 159, 171; and digital computing, 298, 319; and electronic computing, 293–96; and electronics, 236; in Ford systems, 35–39; of gun directors, 95, 98, 242–43, 371n18; and human operators, 75, 98, 100, 154, 194; of SCR-584 systems, 253, 255–57

acoustics, 132, 146, 235, 284, 357n44. *See also* sound

actor-network theory, 16

actuators, 75, 77, 80, 82

Advanced Research Projects Agency (ARPA), 311

Ahrendt, William, 315

AIEE. *See* American Institute of Electrical Engineers

Aiken, Howard, 390n93

"Aiming Controls in Aerial Ordnance" (Philbrick), 309

*Airborne Fire Control* (NDRC), 308

aircraft, 47, 61, 66, 81–82, 85, 187, 218; automatic pilots for, 71, 76–80; bombers, 70, 100; dive bombers, 68, 177, 217–18; and fire control, 68, 100; instruments for, 76, 99, 178–79; Sperry on, 71, 351n3

aircraft carriers, 21, 55

Aircraft Stability and Control Analyzer, 313

Alexanderson, Ernst F., 50–51, 356n23

algebra: Boolean, 173–74, 318; relay, 174, 300, 304

alternating current (AC), 49, 143, 149, 243

alternators, 356n23

altitude, 77, 90–91, 94, 237; assumption of constant, 84–85, 87, 254, 279, 312

American Institute of Electrical Engineers (AIEE), 208–9, 391n13

American Society of Mechanical Engineers (ASME), 208–9, 390n13

American Telephone and Telegraph (AT&T), 13, 107–13, 120, 125, 132–34, 136, 310, 320, 355n10, 358n50; and MIT, 144; transcontinental line of, 111–13, 116, 128. *See also* Bell Telephone Laboratories

amplidyne, 51, 246, 268

amplification, 94, 103, 115–16, 156–58, 239

amplifiers, 51, 54, 105, 167, 169; audions as, 354n8; and carrier multiplex system, 117, 123, 125; closed *vs.* open-loop, 168–69; distortion in, 116–18, 235, 356n30; and feedback, 118–20, 124, 174, 317; feed-forward, 118; in Ford integrator, 38–39; and frequency domain, 125–27, 228; gain in, 74, 116, 118–19, 121–27, 130, 158, 239, 356n30; human operators as, 98, 285; linear, 116–18, 168; operational, 239; Philbrick, 308, 390n4; regenerative, 121; repeater, 111–12, 122–23, 131, 154, 157, 357n50; and servomechanisms, 141–42, 168–70, 177, 246; stability of, 125–29, 239; telephone, 111–12, 131, 312; torque, 158–59, 288–89; and vacuum tubes, 50, 117–19, 123, 168

automatism, 2

aviation instruments, 100, 102–3

azimuth, 83, 205, 238, 240, 248, 335; and Sperry gun director, 88, 90–92

Babbage, Charles, 155

back coupling, 155, 235, 289

Bainbridge, Kenneth T., 245

ballistic cams, 89–90, 195, 289, 295; manufacture of, 94–98, 236

Ballistic Computer (Model III), 389n86

ballistics, 19, 27, 197; and arithmetical computers, 290–91; calculations of, 88–89, 238, 288–89, 296, 303. *See also* firing tables

Ballistics Research Laboratory (BRL; Aberdeen), 288, 297, 335

bandwidth, 134–36, 231, 275, 355n13

Barber Coleman Company, 199, 327, 329, 331, 334–35; dynamic testers of, 239–42, 249, 299, 301

Bassett, Preston, 78, 178, 215–16, 260, 284, 311; and NDRC, 194, 199, 201–2, 330, 371n27

battleships, 26, 32, 49, 60, 62, 67, 100, 218, 345n52; *vs.* aircraft carriers, 21; Ford Rangekeepers on, 19–21, 40, 56–57; Sperry systems on, 24–25, 44, 46–47; treaty limitations on, 47

Battle Tracer, 30–35, 309, 323, 344n42, 353n59; and Sperry's problems, 44–46

Bausch and Lomb, 60, 83, 203, 327, 331–32, 334

Baxter, James Phinney, 307

"Behavior, Purpose, and Teleology" (Wiener, Rosenblueth, and Bigelow), 286

behaviorism, 282, 286

Bell, Alexander Graham, 107, 112, 115, 359n75

Bell Telephone Laboratories, 7–8, 17–18, 66, 99, 105–37, 146, 232, 244–45, 250–54, 262, 267, 274, 278–79, 289, 293, 300, 302–3, 310–11, 377nn26,31, 379n55, 389n79; and antiaircraft systems, 98, 194, 215, 233–37, 258, 312, 315; and Brown's paper, 210; engineering cultures of, 13, 114, 120, 122–23, 136–37; establishment of, 113–14; and feedback amplifiers, 228, 235, 237,

239; and fire control, 266, 281, 292, 304; gun directors of, 98, 233–39, 241–43; and MIT, 169, 290; and NDRC, 187, 191, 195, 198–99, 203, 206, 277, 304–5, 328–29, 331–34; organization of, 113–14, 356n17; and Rad Lab, 250–51, 266; and Shannon, 290; and Stibitz, 193; Systems Development Department of, 113–14, 122–23; technical agenda of, 114–16

Beniger, James, 11

Benjamin, Walter, 15

Bennett, Stuart, 11, 74, 123, 140, 383n38

Bigelow, Julian, 277–80, 282, 284, 286, 320, 385n23

biology, 170, 283; homeostasis in, 13, 282; molecular, 189, 192

Black, Harold, 3, 11, 17, 157, 168–69, 179, 209, 228, 237, 286, 310, 312, 319, 357n35, 359n62; and feedback, 38, 123–25, 131, 135–37, 159, 239; and negative feedback amplifier, 105–6, 117–23, 132, 135, 164

Blackman, Richard B., 319–20

Blandy, William H. P. "Spike," 217–20, 268, 315, 372nn31,36; and fire control, 262, 266, 292

blind firing systems, 206, 264–68, 270

Bode, Hendrik, 17, 105, 135, 137, 169, 209, 274, 277, 279, 282, 286, 312, 319–20, 359n62; and feedback amplifiers, 106, 121, 123, 129–31, 239; and frequency domain, 228, 250; and gun directors, 235, 237, 243; and network theory, 127–29, 131; and Nike antiaircraft missiles, 315

Bode plot, 106, 131

bombs: atomic, 287, 315; guided, 83, 205; and NDRC, 188, 205; nuclear, 316–17; V-1 buzz, 231–33, 254–58

bombsights, 43, 68, 77, 83–84, 99, 205

Booth, R. D., 146–47, 149, 361n13

Bowen, William S., 194, 328–29

Bowles, Edward, 249, 252

Brainerd, John, 296, 331

Breuchaud, Jules, 33–34

Bristol Company, 203, 208, 327, 330, 332–34

Britain, 11, 55, 195, 212, 343n20; antiaircraft problem in, 61, 177, 217, 220; Battle of, 193, 224, 231–

carrier multiplex system, 114–17, 123, 125, 128

Carson, John, 109, 145, 153

Carty, John J., 110–11, 113, 131, 354n8

cathode-ray tubes (CRT), 232

cavity magnetron, 244–45, 262

Center for Analysis (MIT), 171, 191–92, 291, 388n71

Century of Progress Exposition (Chicago; 1933–34), 134

Chafee, Earl W., 194, 216, 249, 289, 372n27; and Sperry's fire control work, 85, 87, 89, 92, 94; term *computer* used by, 353n41

"The Characteristics and Stability of Transmission Systems" (Terman), 147

Chrysler, 251–52

Churchill, Winston, 254

circuits, 210–11, 253, 289–91, 295, 318; Bush on, 146, 153, 166, 173–74, 318; and differential analyzer, 160–61, 173–74; digital, 304; electroacoustic, 357n44; integrated, 317

closed cycle (closed loop) systems, 142, 165, 167–69. *See also* feedback

Coastal Artillery Board (CAB), 194

Cold War, 2, 9, 87, 311, 315

Colpitts, E. H., 111

Columbia University, 200, 239, 285, 329–30, 333, 335

Combat Information Centers, 264

commercial sector, 76–77, 83; business machines in, 163; and Sperry, 71, 75, 81

communications, 2, 26, 29, 131–37, 188, 197, 200, 207, 210–11, 283, 300, 312, 316; and control, 3–6, 18, 171, 287–88, 299, 302, 306, 319–21; and fire control, 9, 28, 67, 237, 319–20; history of, 7–8, 10, 14; and information, 206, 320–21; and noise, 16, 206, 319–21; as signals, 112, 132, 136–37, 206, 228, 319–21; unity of, 137

communications engineering, 4, 6, 195, 233, 251, 257–58, 284, 303, 319, 321

communications theory, 206, 279, 312; and feedback, 136–37; and frequency, 227–28; and radar, 6, 228; and servomechanisms, 169, 277, 286; and Weiner, 287, 391n24

compasses, repeater, 25, 29–30, 36, 38, 45, 56, 72. *See also* gyrocompasses

Compton, Karl Taylor, 145, 157, 187–88, 224, 245, 249, 267; and NDRC, 193, 214–15

*computer* as term, 8, 62, 87, 353n41

computers, 15, 22, 56, 183, 233, 237–39; analog, 304, 308–9; and antiaircraft systems, 94–98, 217; arithmetical, 290–91, 299; at Bell Labs, 8, 18, 293; complex number, 193, 300, 302, 317; digital, 4, 288, 304–5, 387n48; digital relay, 277, 334; EDVAC, 306; electrical, 236; electronic, 195, 236, 251, 288, 293–96; as electronic integrators, 39; fire control, 289–90, 293–96, 299; manufacturing of, 304–5, 319; Mark 1, 62–63, 65, 263, 343n18, 371n19, 382n19; Mark 8, 382n19; mechanical, 70, 182–83, 196, 268, 274, 289; mechanical *vs.* digital, 304; mechanical *vs.* electronic, 11, 94–98; network, 152; relay, 288, 299–302, 304–6, 317; Sperry, 70, 82, 87, 283; and telephony, 304–6, 317; Whirlwind digital, 76, 207, 265, 308, 312–13, 315, 317, 387n48, 388n71. *See also* ENIAC

computing, 193, 200, 291, 305, 311, 316, 335; and control, 319–22; and control systems, 313–17; and differential analyzer, 174; electrical, 96, 98, 243; electronic, 293–96, 298, 335; and feedback, 98, 294, 317, 319–22; and fire control, 10, 82, 233–36, 293–96, 298, 317; history of, 7–8, 10–11, 14, 161–62, 317; and human-machine interactions, 10, 306, 317, 321. *See also* analog computing; digital computing; numerical computing

computron, 292

Conant, James, 187–88

Confidential Instruments Laboratory (Draper), 227, 230, 267

Constant, Edward, 12

control, 2–11, 174, 196, 200, 207, 210, 275, 291, 312, 316, 319–22; and communications, 3–6, 18, 171, 287–88, 299, 302, 306, 319–21; and communications engineering, 258, 321; and engineering cultures, 13, 17, 175; and feedback, 283, control (*cont.*)

310, 319–22; of machines by human operators, 72, 74, 77; *vs.* regulation, 209; technologies of, 2–3, 8, 11, 22, 206; textbooks on, 315; trading of, 80, 103; values of, 11. *See also* engineering: control; remote control

control systems, 1–68, 87, 98, 139, 141–42, 167, 182, 184, 203, 207–12, 226, 228, 241, 249–51, 258, 270, 274–75, 298–99, 306–19, 327, 335, 361n13; and animated machinery, 71; and antiaircraft problem, 178, 312; and Brown's paper, 209–10; and feedback, 8, 137, 165, 233, 251, 307; human operators in, 23, 233, 276–77, 284, 308, 342n10; and NDRC, 186, 188, 191, 193–97, 200, 334; and representation, 14–17; and signals, 228; and Sperry, 24, 81, 99, 103–4; and symbols, 14–16

control theory, 6, 10, 96, 165, 167, 287; defined, 340n17; diffusion of, 209–10; and feedback, 105–6, 138–41; and governors, 139–41; of Hazen, 139; and Servo Lab, 211–12

coordinates: Cartesian, 36, 67, 88–89, 248; polar, 36, 67, 88–89, 243

Crawford, Perry O., 291, 312, 315, 386n48

Crimi, Alfred, 69–70, 101

Crooke, R. E., 201, 369n46

crossbar switch, 171–72

crosstalk, 109, 116–17

cruisers, 40, 44, 47, 56, 59, 62, 67

culture: of BuOrd, 67; gyro, 360n1; institutional, 201, 205; of NDRC, 198, 201; popular, 316–17, 391n16; technical, 305–6, 317. *See also* feedback culture

cybernetics, 4–6, 10; and digital computing, 276–306; and feedback, 283, 321; and helmsman, 76; and NDRC, 206, 283–86; origins of, 283, 286–88; popular images of, 316; and Wiener's predictor, 282–83

*Cybernetics* (Wiener), 4, 6–7, 286, 307, 320

cyberspace, 3–4, 10

cyborgs, 4

data entry, 163–64, 170, 172–73

data flow, 53–54, 66, 92, 206

"Data Smoothing and Prediction in Fire-Control Systems" (Blackman, Bode, and Shannon), 319

Davenport, Lee, 245–46, 252, 255

Davis, Arthur P., 54, 210

D-Day, 254

decibel, 112

Defense Advanced Research Projects Agency (DARPA), 311

de Forest, Lee, 111, 354n8

Delco, 96, 224

Deleuze, Gille, 342n10

Depression, the, 61, 99, 189, 358n50

"Design and Test of a High-Performance Servo-Mechanism" (Hazen), 141

destroyers, 40, 62

Dickieson, Alton C., 120–21, 359n62

Difference Engine (Babbage), 155

differential analyzers, 157–61, 170–74, 181, 258, 274–75, 278, 291, 297–98, 310; and analog computing, 172, 319; and Bell Labs gun director, 235; and Bush, 157–61, 170–71, 174, 288, 319; and calculating machines, 157–58, 288; and circuits, 160–61, 173–74, 289–90; and ENIAC, 296–99, 388nn64,68; and feedback, 158, 174, 289; and fire control, 175, 177, 288–90; flexibility of, 159–60; and graphical techniques, 161–62, 196; and Hazen, 163; improvements in, 288–89; and industry, 157–58; and NDRC, 191–92, 288–89, 331, 335; notation for, 160–61, 289–90; Relay Interpolator as, 302–4; and servomechanism theory, 168; as teaching tool, 161. *See also* Rockefeller Differential Analyzer

differential equations, 109, 153, 155, 161, 227, 235, 303

*digital* as term, 174, 295

Digital Computer Laboratory (MIT), 315

digital computing, 4, 7–10, 17–18, 97, 168, 291, 300, 316, 321; and accuracy, 298, 319; *vs.* analog, 10, 23, 139, 293–97, 304, 306, 318–19; and cybernetics, 276–306; and electronics, 306, 313; at MIT, 306, 315, 387n48; and NDRC, 206,

304–6, 318; and Relay Interpolator, 303; and representation, 8, 135, 168; and Rockefeller Differential Analyzer, 173, 298; and signals, 134–35; and Stibitz, 174, 297, 304–6, 319; and switching, 51–53, 304–5

direct-current (DC), 45, 51, 109, 149

"Directional Stability of Automatically Steered Bodies" (Minorsky), 140

director fire, 25–28, 31–32, 84–85, 100; defined, 342n18; and Navy, 28–30; and synchronous systems, 48–49

distortion, 111, 128, 231, 360n77; in amplifiers, 116–18, 235, 356n30; in representation, 14–15

*Dr. Strangelove* (film), 2, 316

Draper, Charles Stark, 13, 102, 200, 209, 230, 268, 366n12, 372n30, 382n22; and Brown, 212–14; and BuOrd, 220–21; and Four Horsemen, 177; gunsight of, 221–24, 227, 260, 264–65, 311; laboratory of, 227, 230, 267; and Sperry, 178–79, 202

Dreyer, Frederic, 28

Dreyer Table, 28

Dudley, Homer, 122, 132

Dupree, Hunter, 185

dynamics, 6, 275, 316

dynamic systems, 141, 239, 241, 263, 286, 289, 340n17, 358n54; stability in, 120, 123, 135, 139

dynamic testers, 301, 376n25; and Army, 240–42; Barber Coleman, 239–42, 249, 299, 301; and NDRC, 239–41, 300–302, 329, 331, 334; programming of, 299–300; tape, 300–302, 317, 331, 389n79; Texas Tester, 301

E. W. Bliss Company, 345n50

Earhart, Amelia, 80

Eastman Kodak, 60, 83, 203, 293, 327–28, 334

Eckert, J. Presper, 10

Edgerton, Harold, 148–49

Edison, Thomas, 5, 145, 149

Edwards, Paul, 9, 333

EG&G Inc., 149

"Electrical Mathematics" (Parkinson and Lovell), 235

electric power, 1, 8, 31, 35, 50–51, 56, 67, 94, 105, 112, 133, 139, 153, 196, 204, 206, 235, 238, 258, 300, 328; as circuits, 289–90; and communications, 136–37; in computing, 96, 98, 236, 243; growth of networks of, 143; in motors, 48, 72, 75, 77, 155, 182; and networks, 128, 143, 147, 173, 243; and operational calculus, 146; as pulses, 292, 298, 318, 320; and servomechanism theory, 168; as signals, 7, 17, 106–7, 134, 146, 197; transmission by, 62, 106–7, 132, 134, 143, 146, 197. *See also* engineering: electrical

"Electronic Fire Control Computers" (1942 conference), 293–95

electronics, 7, 50, 113, 191, 195, 280, 292–96; in amplifiers, 105, 164, 168–69; in antiaircraft systems, 215; in computing, 39, 195, 236, 251, 288, 293–96, 298, 306, 313, 335; digital, 306, 313; and fire control, 48, 67, 257, 313; in gun directors, 98, 194, 215, 236–37, 292; *vs.* mechanical methods, 11, 94–98, 124–25, 355n13; reliability of, 295; and representation, 294, 318; and servomechanisms, 164, 168, 229, 375n67

elevation, 205; in gun directors, 90–92, 237–38, 335; in radar, 248, 265

engineering: and analog computing, 310; control, 6, 10–11, 175, 316, 340n17, 361n13; electrical, 48, 105, 142–45, 162, 279; and feedback culture, 142; interwar, 310; at MIT, 315; postwar, 316; and Rockefeller Foundation, 189; and society, 144; systems, 10, 113–14, 122–23, 316

engineering cultures, 5–8, 11–14, 17, 23, 61–62, 106, 131, 135–36, 142, 168, 174–75, 200, 274, 297, 305–6, 317; assumptions of, 67–68; at Bell Labs, 13, 114, 120, 122–23, 136–37; at BuOrd, 13, 67, 138; and fire control, 12–13, 66, 309–10; at Ford Instrument, 21, 67, 138; local, 8; at MIT, 13, 139, 147–49, 161, 178, 208; at Sperry, 13, 60, 70–71, 82, 138; and wartime research, 185–86, 311

ENIAC (Electronic Numeric Integrator and Calculator), 193, 295, 300; and differential analyzers, 296–99, 388nn64,68; limitations of, 306; manufacture of, 305; and NDRC, 277, 296–99, 304–5, 388n68 ·

entrepreneurs, independent, 28–31, 36. *See also* industry

equilibrium, 50, 71, 140. *See also* stability

Everett, Robert, 213, 313

expertise, 311

"The Extension of Engineering Analysis through Reduction of Computational Limits by Mechanical Means" (Hazen), 162

*The Extrapolation, Interpolation, and Smoothing of Stationary Time Series with Engineering Applications* (Wiener), 279, 282, 287, 320

Farcot, Jean Joseph Léon, 165

feedback, 3–13, 25–26, 50, 67, 70, 74, 135–41, 156, 164, 177, 184, 189, 191, 197, 200, 209–11, 277–80, 282–86, 288, 294, 307–22; in amplifiers, 105–7, 114, 117–24, 132, 135, 164, 169, 174, 285, 310, 317; and antiaircraft problem, 312; in automatic pilots, 123, 138; and Bell Labs, 66, 235; biological, 282–83; and Black, 38, 123–25, 131, 135–37, 159, 239; and Bode, 131, 135, 137; closed *vs.* open, 168–69; and computing, 98, 294, 317, 319–22; and control, 8, 137, 165, 233, 251, 283, 307, 310, 319–22; and differential analyzers, 158, 174, 289; in gun directors, 92, 98, 235, 238–39, 289; Maxwell on, 286; in music, 317; and network theory, 127–29, 131, 168; and Nyquist, 106, 121, 123, 125, 128–29, 131, 135, 137, 174, 229, 239; positive, 121; in radar, 232, 246, 262–63; and radar-director connection, 250–51; in rangekeepers, 36–40, 56, 155, 289, 325–26; in regulators, 123–25, 137–38; and servomechanisms, 165–66, 168–70, 180, 230; Shannon on, 294, 389n78; and stability, 76, 106, 118–20, 125–26, 138, 147, 358n54; in telephony, 110, 120, 357n44; theory of, 8, 18, 110, 137, 168–70, 191, 229–30, 279–80, 358n54. *See also* amplifiers, feedback

feedback, negative: and linearity, 168; and stability, 125

feedback culture, 138–39; and servomechanisms, 164–65, 167; and theory, 141–42

feedback loops: and amplifier gain, 239; and control, 165, 307; in fire control systems, 255; in gun directors, 235, 335; in gyropilots, 74–75; in rangekeepers, 326; and smoothing, 290; in Sperry systems, 75, 80, 89–90, 103, 345n52; in V-1 bombs, 254

Fernberger, Samuel, 283–84

Ferrell, Enoch, 231, 258, 285

fire: continuous aim, 55, 66, 154; pointer, 28

fire control, 202; accuracy in, 350n115; on aircraft, 100; antiaircraft, 9, 60, 71, 100, 103, 177–78, 191, 194, 196–97, 199, 205, 215–16, 263, 276–77; arithmetical, 312; and Arma, 55; assumptions in, 66–68; automatic, 245; and Bell Labs, 136, 233–37, 304; British, 30, 42, 81, 244–45, 325; and Bush, 290; and communications systems, 319–20; and computing, 10, 290–91, 293–96, 298, 317; and differential analyzers, 171, 288–90; and digital *vs.* analog modes, 318; documentation of, 53–54, 63; electronic, 292–96, 298; elements of, 22–23; and engineering cultures, 13, 309–10; and Ford Instrument, 34, 265, 290; and interwar engineering, 55–60, 66–68, 310; naval, 19–68, 87–88, 92, 99, 102, 136, 171, 181–84, 193–95, 210–11; and NDRC, 188, 191–97, 199, 245, 249, 327–35, 367n11; pre–World War I, 28–29; and radar, 232, 260–75; seacoast artillery, 352n36; and secrecy, 156, 208–9; and Servo Lab, 207–30; and servomechanism theory, 176–80; simplification of, 220–23; and smoothing and prediction, 319; and Sperry, 30–33, 82, 87–88, 99–100, 102–3, 215–16; and switching, 51–54; synchronous systems for, 48–51; treaties on, 57; and wartime research, 311; and Weaver, 188–89; and Wiener, 277–78, 286. *See also particular systems*

firing tables: and arithmetical computers, 290–91, 299; and differential analyzers, 258, 288–89; and director fire, 27; and electronic computing, 295; errors in, 258; and gun directors, 88–89, 237; mechanical, 89; numerical, 94–96; and Relay Interpolator, 303

Fiske, Bradley, 28, 344n34

Fletcher, Harvey, 132, 236, 328–29, 331, 333, 357n44

flight simulators, 291, 308, 312–13, 387n48

Florez, Luis de, 76

follow-the-pointer mode, 32–33, 238, 245, 309, 346n58, 348n88. *See also* pointer matching

Ford, Hannibal Choate, 11, 17, 22, 25, 43–44, 60, 81, 155, 191; and Battle Tracer, 30; and Bush, 24, 175; and Elmer Sperry, 24, 34, 346n66; founding of firm by, 33–34; integrator of, 38, 156, 171; and NDRC, 201–2, 369n46; and Palomar telescope, 181–82; and rangekeeper, 20–21

Ford, Henry, 5

Ford Instrument Company, 17, 19–68, 138, 186, 191, 208, 210, 212, 242, 265, 267, 290–91, 310; and antiaircraft problem, 61, 218–19; and BuOrd, 19–68, 202, 216; Dead Reckoning Tracers of, 344n42; and dynamic tester, 241, 299; founding of, 33–34; and Four Horsemen, 179; and G.E., 47–48, 50; and manufacturing, 40–41, 317; and Mark 37 system, 61–62, 66; and Mark 45 system, 264–65; naval inspectors at, 41–42; and NDRC, 195, 199–202, 369n46; and Palomar telescope, 181–82; and Sperry, 44, 46, 60, 81–82, 99, 215–16, 345n52. *See also particular systems*

Ford Marine Appliance Corporation, 33–34

Ford Motor Company, 40, 317; and Sperry gun directors, 95–96, 194, 215

foreign trade: BuOrd limits on, 43; and Sperry, 89, 352n35

Forrester, Jay, 10, 213, 227, 291, 312, 387n48; and magnetic core memory, 387n49

Four Horsemen (Hooper, Mustin, Ward, Rivero), 175–84, 187, 192, 262

Fourier analysis, 110, 146, 277, 303

Foxboro Company, 208, 281, 284, 390n1; and NDRC, 199–200, 203, 330, 332

France, 212

Frankford Arsenal, 83–84, 237, 281, 289, 293

Franklin Institute, 198, 203, 330, 334–35

frequency domain, 146, 241, 276, 312, 320; and amplifiers, 125–27, 228; and Bode, 228, 250;

and feedback theory, 137; and networks, 128, 135; and radar, 228; and Servo Lab, 210; and servomechanisms, 168–69, 179, 230, 238; and transient analysis, 227–29

frequency response, 135, 238, 250, 284, 375n68; and servomechanisms, 227–29

Friis, H. T., 119, 121

Fry, Thornton C., 113, 169, 210, 280, 289–90, 292–93, 300, 311, 383n4; and Black, 120; and NDRC, 191, 194, 196–99, 249, 328–33; and Stibitz, 193

Furlong, William R., 29, 48–49, 182, 216, 372n36

fuze setting, 192, 205, 213, 335, 371n18, 373n49. *See also* proximity fuze


Gage, F. D., 138, 153, 162

Galison, Peter, 14, 287

Gatty, Harold, 78

General Electric Company (G.E.), 48, 50, 55, 59, 67, 96, 138, 142, 149, 153, 179, 206, 208, 210, 246, 265, 267, 310, 355n10; and antiaircraft problem, 61, 217–19; and BuOrd, 22, 47–50; and differential analyzers, 158, 288; and fire control, 60, 251, 265; and Mark 56 system, 268, 271, 273; and MIT, 144, 148, 150–51; and NDRC, 195, 198, 201, 327, 331–34; and Sperry Gyroscope, 47–48, 99–100, 348n87; and synchronization, 47–54, 216

General Motors, 81, 329

generators, 143, 148–49, 333

George Philbrick Researches, 308

Germany, 13–14, 20, 31, 49, 54, 220

Germeshausen, Kenneth, 149

Gerovitch, Vyacheslav, 316

Getting, Ivan, 11, 18, 250, 275, 306, 312–13, 315–16, 382n26, 384n6; and BuOrd, 270–73, 381n15, 383n32; and integrated systems, 265–68; on integration, 259; and M-9/SCR-584 system, 255, 258; and Mark 56 system, 260–61, 266–70; and NDRC, 199, 249, 266, 332; and radar systems, 246–48, 252; and radar textbooks, 274; and Rad Lab, 245, 267–68, 270–73

networks as, 131–33; numerically controlled, 301, 316; precision, 181; rapid arithmetical, 291–92; representation of world by, 56, 82, 98, 277, 288, 290–91, 318–19. *See also* human-machine interactions

MacKenzie, Donald, 9

McMath-Hulbert Observatory, 327, 334

MacNair, Walter, 200, 243, 315

Macy conferences, 287, 385n23

Mahood, David H., 54–55

main batteries, 51, 56, 59, 217; and antiaircraft systems, 21–22, 62; and fire control radar, 262–63

management, scientific, 217

"Manhattan Engineering District," 210

Manhattan Project, 145, 185–86, 188

manufacturing, 67, 95, 98, 309, 369n52; and antiaircraft problem, 219–20; of ballistic cams, 94–98, 236; and computers, 294–95, 304–5, 317, 319, 321; and control systems, 12, 103–4; and cybernetics, 283; of ENIAC, 305; of Ford Rangekeeper, 40–41; and labor, 236, 306; and machine representation, 82, 98, 235–36, 304–5, 319; of Mark 56 System, 266, 268–70, 273; mass, 95, 98, 219; and NDRC, 200, 202–3, 225, 334; precision *vs.* mass, 95; of SCR-584, 251–53; and Sperry, 70, 99–100, 102–4, 317, 372n27

Marey, Etienne-Jules, 5

Mark 37 Gun Fire Control System, 61–66, 175–76, 192, 219, 263–64, 381n14

Mark 56 Gun Fire Control System, 206, 260–61, 271, 274, 332–33; organizational problems with, 266–68; production of, 266, 268–70, 273

Masani, Pesi, 287

Mason, Max, 180–82, 184

Massachusetts Institute of Technology (MIT), 66, 98, 148, 150–51, 157, 239, 246, 254, 310–11, 315, 388n66; and analog computing, 138–74, 288; and arithmetical computing, 290–91; and Bell Labs, 144, 169, 290; and BuOrd, 18, 219–21, 223; Bush laboratory at, 7, 17, 145–50; Center for Analysis at, 171, 191–92, 291, 388n71; and digital computing, 306, 315, 387n48; and

electronic computing, 293, 298; engineering cultures at, 13, 139, 147–49, 161, 178, 208; Four Horsemen at, 175–84, 187, 192, 262; and NDRC, 8, 187, 191, 199, 203, 206, 304, 328–30, 333, 335; and Palomar telescope, 180–84; and radar, 232, 263; and Rockefeller Differential Analyzer, 170–71; and Servo Lab, 211, 227; and servomechanism theory, 168, 208–9; and Sperry, 213, 221; and stability problem, 144–45, 147–49. *See also* Radiation Laboratory; Servomechanisms Laboratory

"A Mathematical Theory of Communication" (Shannon), 7, 320

mathematics: at Bell Labs, 113; Bush on, 145–46; and dynamic tester, 241; and feedback, 141; and fire control, 196, 235; and NDRC, 202, 328, 334; of operational calculus, 110, 145–46, 227, 279

Mauchly, John W., 10, 193, 296

Maxwell, James Clerk, 5, 139, 165, 286

Mayr, Otto, 11, 139

measurement, 83; and accuracy, 350n115; in Ford Rangekeeper, 37

message, 4–5, 302, 312, 320–21; and transmission theory, 134–35

Metal Mike (Sperry gyropilot), 72–76

Metropolitan-Vickers, 158

Michelson, Albert A., 111

microwaves, 194–95, 202, 232, 244–45, 334

Midvale Steel Company, 217

military: and automatic pilots, 77; and electronics, 292–93; and fire control systems, 221, 252; and industry, 71, 311; and NDRC, 195–96, 198, 202; secrecy in, 13, 202, 208–9; and Sperry, 70–71, 81, 100, 103; and universities, 257, 311; and wartime research, 311. *See also* Bureau of Ordnance, U.S. Navy; Navy, U.S.

military-industrial complex, 9

Millikan, Robert, 111, 132, 180–81, 189, 355n10

Minorsky, Nicholas, 96, 140–41, 166, 177, 274, 286

missiles, 316; ballistic, 87, 258, 313; guided, 312, 315; Nike antiaircraft, 315; Patriot, 313

Mitchell, Billy, 60

modeling, 149–51, 178, 321; and analog computing, 163; artificial, 174; hydraulic, 153; of networks, 146, 149–50; of power systems, 150–51

modulation, 210

Moore School of Electrical Engineering (University of Pennsylvania), 158, 288, 331, 335, 388n66; and ENIAC, 296–99

Moore's Law, 317

Morgan, Thomas, 25, 30, 81, 221, 350n2

Morison, Samuel Eliot, 381n9

Morristown trial, 125–27, 132, 151, 358n50

Mumford, Lewis, 1–4, 11, 16, 106, 163, 318; on dissociation, 15, 112, 321

Murphy, M. Emerson, 219–21, 266, 293

music, 132, 136, 317

Mustin, Lloyd, 176–79, 208, 210, 212, 220, 366n12

National Advisory Committee on Aeronautics (NACA), 186–87, 303, 372n30

National Bureau of Standards, 45–46, 55, 202, 209, 240

National Cash Register (NCR), 83, 291, 298, 302

National Defense Research Committee (NDRC), 8–9, 18, 180, 185–206, 260, 291, 298–99, 311, 367n12; and antiaircraft problem, 312; and antiaircraft systems, 195–97, 199, 331; Applied Psychology Panel of, 284; and Army, 187, 192, 200, 219, 297; and Brown, 213, 223–26; and BuOrd, 207, 219, 221; contracts from, 196, 198, 200–201, 203–5, 207, 224, 237, 240, 242–43, 245, 278–79, 288–89, 292–93, 300, 303, 327–35; and cybernetics, 283–86, 288; D-2 (fire control) division of, 191–97, 207–30, 249; and differential analyzers, 288–89; and digital computing, 304–6; and digital *vs.* analog modes, 318; Division 7 of, 226, 268; Division 7.6 of, 266–68, 270, 382n22; and Division 7–Section T dispute, 267–68; divisions of, 188, 197–200, 266; and dynamic testers, 239–41, 300–302, 329, 331, 334; and electronic computing, 293–96; and electronic fire control, 292–

93; and ENIAC, 296–99, 388n68; and fire control systems, 257–58, 262, 265–68, 273; and fundamental research, 197–98, 202–3, 281–82; and Getting, 199, 249, 266, 272, 332; goals of, 201–3; and gun directors, 98, 236–37, 241–43; and human-machine interactions, 276–77; management style of, 200–201; and microwave research, 244–45; organization of, 197–200, 226, 367n11; and radar, 262, 265–68, 270; and Rad Lab, 265–66, 270; and Relay Interpolator, 302–4; and secrecy, 202, 208–9, 214–16; Section T of, 267–68; and Servo Lab, 207, 210–12, 229–30; and Sperry, 214–16, 371nn26,27; *Summary Technical Report* of, 307–8; and Tuve, 382n26; and Wiener, 277–83, 287, 383n4. *See also under particular topics and organizations*

National Research Council, 181

Naval Gun Factory (Washington, D.C.), 61–62, 179, 195

Naval Research Laboratory (NRL), 261–62, 267, 303

navigation, 32, 176, 245, 261, 315

Navy, U.S., 175, 205, 209, 224, 270, 273; *vs.* Army, 193, 202, 244, 377n25; Bureau of Engineering of, 261–62; Bureau of Navigation of, 176; and director fire, 28–30; fire control in, 21, 28–29, 40–44, 66, 197; and fire control systems, 235, 255; inspectors from, 41–42; and integrated systems, 261–65; and NDRC, 187, 192–94, 200, 219; and Palomar telescope, 180–84; and proximity fuze, 257; and secrecy, 216; and Sperry, 30, 81–82. *See also* Bureau of Ordnance, U.S. Navy

NDRC. *See* National Defense Research Committee

neotechnic era, 1, 4

network analysis, 127

network analyzer, 151–53, 156, 164, 169, 310; and differential analyzer, 157, 159–60, 174

networks, 18, 114, 136, 304, 310; and analog computing, 318; and articulation, 131–32; and

tronics, 164, 168, 229, 375n67; and feedback amplifiers, 168–70, 179, 228–29, 239, 301, 375n67; and fire control, 176–80, 233–36, 263, 309; and frequency domain, 168–69, 179, 230, 238; and frequency response, 227–29; in gun directors, 237, 289, 312; for gunsights, 220–21; Hazen on, 38, 141–42, 146, 158, 164–71, 174, 179, 189; human, 91–94, 98, 154, 166, 191, 235, 246, 283, 309, 321; and human operators, 165, 194, 284–85; hydraulic, 335; and integraph, 154–55; and loading problem, 156–57; manual, 32, 38, 91–94, 98, 112, 276; and NDRC, 191–92, 196, 199, 328–30, 335; Oilgear, 224, 226; and Palomar telescope, 181–82; and prediction, 279; and radar, 232, 246, 263; and secrecy, 209; and Servo Lab, 226; Sperry, 255; and stability, 145, 165, 167–68, 231; tape-controlled, 300; and telephony, 168, 211, 228, 230; textbooks on, 274; theory of, 17, 141–42, 145–46, 168–70, 176–80, 191, 230; and transient analysis, 168–69, 179, 230

Servomechanisms Laboratory (MIT), 8, 18, 180, 207–30, 251, 281, 286, 383n4; and antiaircraft problem, 312; and Brown, 223–26, 265; and flight simulators, 291, 308, 387n48; and frequency response, 238; funding of, 210–12, 225; identity of, 226–27; and NDRC, 210–12, 214, 224, 229–30, 335; and numerically controlled machine tools, 316; and postwar research, 315; textbooks from, 274, 315; and Whirlwind computer, 388n71

Shannon, Claude Elwood, 7, 10, 291, 300, 318–20; on binary logic, 52; and differential analyzers, 173–74, 289–90; on feedback, 294, 389n78; and information theory, 135; and NDRC, 196, 328; and Nyquist, 134, 360n78; and Wiener, 391n24

signals, 240, 275; abstract, 106–7, 135–36; and antiaircraft problem, 312; and communications, 112, 132, 136–37, 206, 228, 319–21; continuous and discrete, 134, 174; electrical, 7, 17, 106–7, 134, 146, 197; equivalency of, 135–36; and fire

control, 258–59; information as, 134–36, 320; and linear amplifier, 118; measurement of, 134–35; and network theory, 131; and radar, 246, 251; and radar-director connection, 250–51; and Relay Interpolator, 303; and servomechanisms, 230–31; sound as, 132–33; transmission of, 105–37, 258

Silicon Valley, 147

simulation, 17, 150, 321, 391n20; and analog computing, 319, 387n48; and NDRC, 334; power-system, 362n34. *See also* modeling

simulators, 307–9, 321; flight, 291, 308, 312–13, 387n48

Singer Sewing Machine, 96, 224

singing, 119–20, 357n43

Smith, Edward S., 208–9

Smith, Merritt Roe, 201, 369n52

Smith, Sinclair, 181–82

smoothing, 250–51, 282, 294; and communications and control, 319–20; and fire control, 309, 313; and NDRC, 329, 332, 334; and prediction, 319, 384n14, 385n20; and Relay Interpolator, 303; Shannon on, 289–90

society: and engineering, 144; and feedback, 282–83; and technology, 16, 341n27; and telephone networks, 131

sound, 106; in films, 131–32; high fidelity, 132; and networks, 132–33; reproduction of, 133, 359n71, 360n77. *See also* acoustics

sound location equipment, 83–84, 99

Special Devices Center (SDC; U.S. Navy), 387n48

speech, human, 132–34, 136

Sperry, Elmer, Jr., 76–77, 208

Sperry, Elmer Ambrose, 9, 11, 33, 80, 286, 345n47, 375n4; on airplanes, 71, 351n3; and Anschütz gyroscopic compass, 54; and antiaircraft systems, 83–85; and control systems, 103; and director fire, 29–30; and gyropilot, 71–73; and Hannibal Ford, 24, 34, 346n66; on human operators, 140, 166; and Pollen's system, 42, 44; and Sperry Gyroscope, 45–46, 99

Sperry, Lawrence, 76

Sperry Corporation, 80–82, 310; and aircraft, 68, 78, 81; engineering cultures of, 60, 70–71, 82; and Ford Instrument, 60, 81–82; sales of, 102

Sperry Gyroscope Company, 7–9, 17, 34–35, 53–55, 69–105, 136, 177, 200, 208, 229, 268, 278, 309; advertisements of, 72–75; and antiaircraft systems, 71, 82–99, 157, 178, 205, 215, 221; and Army, 77, 87, 216, 249; automatic controls of, 68, 99–100, 103; autopilots of, 71–80, 351n21; and British fire control, 30; and Brown, 202, 212–14, 223, 370n11, 373n47, 383n4; and BuOrd, 22, 24–25, 30, 44–47, 50, 60, 81, 102, 216, 219, 221; and computers, 94–98, 291; and control systems, 24, 81, 99, 103–4, 142; and Draper, 178–79, 202; and Draper gunsight, 221–24, 227, 260, 265, 311; and dynamic tester, 241, 299; and electrical computers, 236; engineering cultures of, 13, 60, 70–71, 82, 138; and fire control, 30–33, 68, 82, 87–88, 99–100, 102–3, 215–16; and Ford Instrument, 44, 46, 215–16, 345n52; foreign trade of, 46, 89, 352n35; founding of, 24–25; and G.E., 47–48, 99–100, 348n87; management problems of, 46; and manufacturing, 70, 99–100, 102–4, 317, 372n27; and Minorsky, 140; naval inspectors at, 41–42; and NDRC, 194–95, 198–99, 201–2, 206, 214–16, 327, 371nn26,27; patents of, 50; products of, 45, 47, 99; and radar, 232, 249; and research on human operators, 283; and seacoast artillery fire control, 352n36; and Servo Lab, 207, 210, 227, 230; and Sperry Corporation, 60, 80–81; in World War II, 68, 71, 87, 99–103, 186, 311; and XT-1/SCR-584, 252. *See also particular systems and products*

Sperry Gyroscope Ltd., 30, 81

stability, 13, 164, 192, 197, 246, 289, 303, 312, 316; of amplifiers, 125–29, 135–36, 239; Bush on, 147–49; as consistent performance, 121–23; and control, 140, 178; and differential analyzers, 158, 171, 174, 298; dynamic, 120, 123, 135, 139; as engineering problem, 142–45, 153, 158, 171; and feedback, 76, 106, 118–20, 125–26, 138, 147,

358n54; and fire control systems, 65–66, 175, 216, 219, 237; and frequency response, 228; Hazen on, 142; and human operators, 276, 285; and hydraulics, 225–26; interpretations of, 121–23; mathematics of, 123; of motion, 123; and negative feedback amplifier, 118–20, 125; of networks, 128–29, 143–45, 147–49, 151, 250; Nyquist criterion for, 127, 285, 358n54; *vs.* oscillation, 50, 121–24; of power systems, 143–45, 149, 361n13; and prediction, 277–80; psychological, 285; Routh-Hurwitz criteria for, 139–40, 358n54; and servomechanisms, 145, 165, 167–68, 231; in telephony, 120, 357n44; and transient phenomena, 146–47, 228; of transmission, 126–27; in V-1 bombs, 254; and Weaver, 189

"Stabilized Feedback Amplifiers" (Black), 121, 125

stable elements, 55, 59–62, 67, 343n18

standardization, 112, 202, 239–41

Standard Oil, 331

Stanford University, 147, 187

Star Wars, 313

state-space methods, 316

statistics, 279, 287, 312, 334–35

steady-state analyses, 110, 144, 146, 150–51, 178; *vs.* dynamic systems, 141

steady-state error, 213

Steinmetz, Charles, 96, 110, 140, 143–44, 354n7

step-by-step systems, 45, 51, 85

step function, 109

Stewart, Duncan J., 199, 241–42, 293, 377nn26,31; and dynamic tester, 240, 301; and NDRC, 328–33

Stewart, Herbert R., 138, 153, 162

Stibitz, George, 8, 10–11, 13, 18, 250, 297, 389n78; and Ballistic Computer (Model III), 389n86; and computing, 193, 293, 295, 305; digital approach of, 304–6, 318–19; on *digital* as term, 174, 295; and dynamic tester, 299–302; and NDRC, 193, 199–200, 277, 280, 287, 299, 334; and relay computers, 300–306, 317; and Relay Interpolator, 302–4; and T-10 director, 241–

Van Auken, Wilbur R., 29, 372n31

van der Bijl, H. J., 111

Varian, Russel & Sigurd, 100, 249

velocity lag, 213–14, 224, 229

Vickers, Harry, 82, 210

Vickers Inc., 81–82, 94, 100

Vocoder (speech synthesizer), 132

Voder (speech synthesizer), 132–33

Vonnegut, Kurt, 316

Ward, Alfred, 176–77, 179, 210–11

Washington Naval Treaty (1922), 47

Waterbury Tool Company, 81–82, 94, 177, 210

Watt, James, 11, 121

watt-hour meter, 154, 156

Weaver, Warren, 8, 18, 184, 209, 270, 310–11, 316, 378n47, 391n16, 391n24; on Applied Mathematics Panel, 197–98, 226, 240–42, 292, 297; and Bassett, 371n27; on Bell Labs, 233; and Brown, 180, 210–16, 223–26, 370n11, 373n51; and Bush, 180, 189–90, 368n15, 369n40; and computing, 293–96, 305; and differential analyzer, 170; and electronic fire control computers, 293–98; and ENIAC, 297–98; and fire control, 180, 191–97, 244; and gun directors, 236–37, 242; and human-machine interactions, 276; and human operators, 284; and information theory, 320; and long-term research, 305; and molecular biology, 189; and NDRC, 188–97, 199, 201–2, 249, 259, 284, 328–33, 368n40; and Palomar telescope, 180–81; and postwar research, 315; and Rad Lab, 249, 266; and RCA, 292–93; and Rockefeller Differential Analyzer, 170–71; and Rockefeller Foundation, 188–90, 367n15; and Servo Lab, 210–11; on smoothing networks, 250–51, 282; and Sperry, 214–16; and Wiener, 279–82, 287

Webb, James E., 221

Western Electric, 111, 113, 117, 267, 300; and antiaircraft problem, 204; and audion, 354n8; and cavity magnetron, 245; and digital computers, 304–5; and gun directors, 236, 242; and NDRC, 203, 327–29, 331–33; and radar, 244,

262–63, 378n47; Standard Operating Procedures manuals of, 304

Westinghouse, 24, 151, 181, 251; and NDRC, 224–25, 330, 332–34

Whirlwind digital computer, 76, 265, 308, 315, 317, 387n48, 388n71; and antiaircraft problem, 312–13; and Servo Lab, 207

Wiener, Norbert, 76, 145, 193, 274, 383n4; and antiaircraft problem, 82, 280–83, 312; and Bush, 146, 286; and cinema integraph, 164; and computing, 291, 306; on control and communication, 320–21; and cybernetics, 4–6, 8, 10, 14, 18, 286–87, 316; and fire control, 280–83, 292, 319–20; on homeostasis, 13, 282; and human-machine interactions, 276–77, 307; and human operators, 284–86; on the message, 4, 302; and NDRC, 203, 277, 280–82, 296, 298, 328–29, 334; and origins of cybernetics, 283, 286–88; and prediction, 244, 277–80, 320, 384nn6,20; and Rosenblueth, 282, 385n23; and Shannon, 391n24; on social systems, 316

Wieser, Robert, 227

Wildes, Karl, 199–200

Willis, Hugo, 178, 194, 212–13, 216

Wilson, Carroll, 210

Wilson, Thomas, 83–85

Wilson, Woodrow, 112

*Winnie Mae* (airplane), 78

*The Wisdom of the Body* (Cannon), 282

Wittkuhns, Bruno, 96, 98

world's fairs, 132–33, 163

World War I, 5; aircraft instruments in, 76; battle of Jutland in, 20–21; fire control during, 47, 83, 92; research during, 42; and technology, 28–29

World War II, 2, 5; "famous secret weapons" of, 43; fire control systems in, 56, 61–62; and Four Horsemen, 177; research during, 7–10, 82, 185–206, 231–59, 298–99, 310–11; and servomechanism theory, 179–80; and Sperry, 99–103; Sperry in, 68, 71, 87, 99–103, 186, 311

Zworykin, Vladimir, 292–93, 295–96, 331